Power Analysis for Experimental Research

A Practical Guide for the Biological, Medical and Social Sciences

Power analysis is an essential tool for determining whether a statistically significant result can be expected in a scientific experiment prior to the experiment being performed. Many funding agencies and Institutional Review Boards now require power analyses to be carried out before they will approve experiments, particularly where they involve the use of human subjects. This comprehensive, yet accessible, book provides practicing researchers with step-by-step instructions for conducting power/ sample size analyses, assuming only basic prior knowledge of summary statistics and the normal distribution. It contains a unified approach to statistical power analysis, with numerous easy-to-use tables to guide the reader without the need for further calculations or statistical expertise. This will be an indispensable text for researchers and graduates in the medical and biological sciences needing to apply power analysis in the design of their experiments.

R. BARKER BAUSELL is Professor at the University of Maryland School of Medicine and Director of Research of the Complementary Medicine Program. He has served as the editor-in-chief of the peer-reviewed methodology journal *Evaluation and the Health Professions* for the past 25 years and has conducted statistical research in a number of disciplines related to the medical sciences.

YU-FANG LI is a Data Analyst at the Department of Veteran's Affairs, Puget Sound Health Care System in Seattle, Washington. She previously worked as a statistical consultant with a particular interest in meta-analysis.

Power Analysis for Experimental Research

A Practical Guide for the Biological, Medical and Social Sciences

R. BARKER BAUSELL
YU-FANG LI

CAMBRIDGE UNIVERSITY PRESS
Cambridge, New York, Melbourne, Madrid, Cape Town, Singapore, São Paulo

Cambridge University Press
The Edinburgh Building, Cambridge CB2 2RU, UK

Published in the United States of America by Cambridge University Press, New York

www.cambridge.org
Information on this title: www.cambridge.org/9780521809160

First published 2002
This digitally printed first paperback version 2006

A catalogue record for this publication is available from the British Library

ISBN-13 978-0-521-80916-0 hardback
ISBN-10 0-521-80916-9 hardback

ISBN-13 978-0-521-02456-3 paperback
ISBN-10 0-521-02456-0 paperback

This book is dedicated to Jesse Turner Bausell

Contents

Introduction page ix

1 **The conceptual underpinnings of statistical power** 1

2 **Strategies for increasing statistical power** 16

3 **General guidelines for conducting a power analysis** 36

4 **The *t*-test for independent samples** 50

5 **The paired *t*-test** 57

6 **One-way between subjects analysis of variance** 71

7 **One-way between subjects analysis of covariance** 111

8 **One-way repeated measures analysis of variance** 179

9 **Interaction effects for factorial analysis of variance** 239

10 **Power analysis for more complex designs** 302

11 **Other power analytic issues and resources for addressing them** 329

Technical appendix 341
Bibliography 358
Index 362

Introduction

The primary purpose of this book is to provide an *easy-to-use*, unified approach to statistical power analysis in order to enable investigators to avail themselves of the advantages of this powerful tool in the design of their experiments. It is our firm conviction that no other process possesses more potential for increasing the scientific and societal yields accruing from our experiments. It is also our firm belief that the a priori consideration of power is so integral to the entire design process that its consideration should not be delegated to individuals not integrally involved in the conduct of an investigation, hence the present volume has been written to be completely accessible to practicing researchers. For this reason we have studiously avoided the use of technical terms and formulas until the appendix to make it as accessible (and hopefully interesting) to individuals without advanced statistical training as possible.

This is not to say that statistical collaboration in the conduct of most experiments is not desirable. It is, in fact, often absolutely essential and we have written this work to make it as helpful as possible to statisticians charged with the task of performing a power or sample size analysis. It has been our experience, however, that while principal investigators are well versed in formulating research hypotheses, they often conceptualize the determination of power (or the sample size necessary to achieve a desired value thereof) as a technical exercise better delegated to someone with the appropriate expertise. Our purpose in writing this book is to simplify the power analytic process to the point that it can become the integral component of experimental design in *practice* that it occupies in *theory*. The *true* value of the concept of statistical power, in fact, lies in the fact that its consideration forces investigators to think in terms of the strength of the effects their experiments are likely to produce, which is absolutely crucial to the design process itself. It is for this reason that it is not wise or fair to delegate the power analytic process to a statistician, no matter how skilled, who is not immersed in the subject matter and previous research surrounding the experiment being designed. With this principle in mind, we hope that this work will facilitate the statistician's role both with respect to computing the power analysis for a wide range of experimental designs and to involving his/her non-statistical collaborators in the process.

What makes this collaboration so important is the fact that the types of hypotheses that most scientists have been trained to write are not nearly as scientifically (or clinically) relevant as they could be. In designing most experiments, for example, it is almost trivial to posit something to the effect that "Patients exposed to intervention X will experience significantly fewer occurrences of symptom Y at the end of Z weeks than patients receiving standard care." In designing such an experiment it is almost a foregone conclusion that the investigator (and his/her potential funding agency) *believes* that the proposed intervention will be better than a control. From an epistemological point of view it is tautological that the intervention will not produce exactly the same effects as the control. What the investigator's true job description involves, therefore, is the design of experiments that are capable of *demonstrating* the intervention's effectiveness or, at the very least, of designing experiments that provide an adequate chance of demonstrating the intervention's effectiveness.

Many researchers consider this the province of **statistical significance**, and in one sense they are correct. Statistical significance, however, is only one of two pillars upon which the process of accepting or rejecting scientific hypotheses rests. The other pillar is statistical power, or the probability that statistical significance will be obtained and that probability is determined primarily by the **size of the effect** that an experiment is most likely to produce. Statistical significance, the supreme arbiter of an intervention's effectiveness, is also determined primarily by the size of the effect that is actually obtained after the experiment is conducted and the data are collected. If the experiment is not designed with sufficient power to detect the intervention's true **effect size**, then statistical significance will not be obtained once the data are collected and the intervention will be declared non-effective, even if a clinically relevant difference occurs between it and its control and even if the intervention "truly" is effective and is capable of saving thousands of lives (or of improving their quality) if implemented. It is therefore absolutely incumbent upon investigators to design their experiments in such a way that societally important effect sizes will be statistically significant. This, then, is the essence and true purpose of a power analysis. It also represents the true value of the power analytic process in the sense that it forces the investigator to consider what size of effect must be obtained in order to provide a reasonable chance of obtaining statistical significance. Said another way, statistical power involves forcing the investigator to perform a hypothetical statistical analysis prior to collecting data. This is accomplished by simply substituting the minimum effect the intervention is expected to have upon the outcome within the experimental context being designed. This in turn unequivocally yields a discrete, numerical probability of how likely this result will be to occur in the actual data analysis performed at the end of the experiment.

About the book

This book constitutes the most comprehensive power analytic tool available and this statement is not made lightly because it was written standing upon the shoulders of giants – notably Jacob Cohen's seminal *Statistical power analysis for the behavioral sciences* and Mark Lipsey's delightful *Design sensitivity: statistical power for experimental research.* (We would also like to acknowledge Professor Karen L. Soeken for her contributions made to this project from its inception.) What this book basically does is extend the work of these and numerous others to additional designs via tables (and detailed templates to facilitate their use) that permit a one-step approach to power analysis, while making a few advances of its own including the computation of power for multiple comparison procedures and mixed interactions.

The book itself is organized around those parametric statistical procedures most commonly employed in the analysis of experiments involving continuous outcome data. In a sense the volume need not be read from cover to cover, although we do recommend a perusal of the first three introductory chapters since they lay the conceptual foundation for the use of the tables and templates that follow.

Supplementary software

While we believe the treatment of power for *experimental* research is as comprehensive as is possible within the confines of a single volume, there are relatively rare occasions when an investigator or a statistician does need to compute power for, say, an alpha level other than 0.05 or for a different parameter than our tables permit. We have, therefore, in collaboration with Mikolaj Franaszczuk (who at the time of this writing is a brilliant undergraduate computer science student at Cornell University), prepared a computer program entitled *Power analysis for experimental research* that exactly mirrors each of the procedures covered in this volume, but which permits different (and more exact) parameters to be input. The program may be obtained free of charge to readers of this book by email from the first author (bbausell@compmed.umm.edu).

1 The conceptual underpinnings of statistical power

The importance of statistical power

As currently practiced in the social and health sciences, inferential statistics rest solidly upon two pillars: statistical significance and statistical power. The two concepts, both of which are expressed in terms of probabilities (i.e., how likely events are to occur), are so integrally related to one another that it is almost impossible to consider them separately.

Statistical significance, the first pillar, is a probability level generated as a byproduct of the statistical analytic process. It is computed *after* a study is completed and its data are collected. It is used to estimate how *probable* the study's *obtained* difference or relationship (which is called its **effect size**) would be to occur by chance alone. Based in large part upon Sir Ronald Fisher's (1935) recommendations, this probability level is often interpreted as an absolute standard. If it is 0.05 or below, the results are said to be **statistically significant** and the researcher has, by definition, supported his/her hypothesis. If it is 0.06 or above, then statistical significance is not obtained and the research hypothesis is not supported.

Statistical power, on the other hand, is computed *before* a study's final data are collected. It involves a two-step process: (a) hypothesizing the **effect size** which is most likely to occur based upon the study's theoretical and empirical context and (b) estimating how *probable* the study's results are to result in statistical significance if this hypothesis is correct.

Said another way, statistical significance is used to ascertain whether or not a given effect size can be interpreted as being reliable enough to allow the scientific community to accept a hypothesis once a study is *completed*. Statistical power, in contrast, is used to ascertain how likely a study's data are to result in statistical significance *before* the study is begun. It is, in effect, a hypothetical or projected test of statistical significance conducted before an investigator has access to data.

Although reliance upon this genre of hypothesis testing is not without its critics (Neyman & Pearson, 1933), the convention of determining whether or not a given effect size is statistically significant (hence can be considered reliable enough to support a researcher's hypothesis) is almost universally employed in empirical research. Even when confidence intervals are substituted for statistical significance, research consumers still check to see whether a zero effect size resides within that reported confidence interval – which normally occurs only when statistical significance is not achieved and hence is synonymous therewith.

In effect, then, researchers have no choice other than to subject their data to statistical analysis and report the resulting "significance level" in one form or another. There is, in fact, something rather comforting to many scientists about the definitiveness of a decision-making process in which "truth" is always obtained at the end of a study and questions can always be answered with a simple "yes" or "no" and not prefaced with such qualifiers as "perhaps" or "maybe."[1]

Without delving into the philosophical wisdom of this process, its almost universal acceptance in day-to-day scientific practice has resulted in an ironic twist of fate: almost everyone analyzes their data to ascertain whether it is "statistically significant" but a large number of researchers[2] either do not ascertain how likely they are to obtain this statistical significance prior to conducting these studies or at least conduct them with an insufficient amount of power.

What makes this situation ironic is that, to a very real extent, the achievement of statistical significance has become a prime determinant of scientific success and a primary scientific objective in itself. This is because an investigation which does not achieve statistical significance (a) does not support its authors' hypotheses, (b) is often not considered to be as reliable as research which does, and (c) is not as likely to be published. From both a pragmatic and a scientific point of view, therefore, it behoves anyone interested in conducting empirical research to do everything legitimately possible to design his/her research in such a way that it has a reasonable chance of obtaining statistical significance and to a large extent this is synonymous with designing research with a reasonable amount of power.

In effect, then, this entire volume is dedicated to showing researchers how to determine the probability of obtaining statistical significance when designing their empirical investigations. Said another way, this book is dedicated to the second pillar upon which the scientific inferential process rests: statistical power. Said one final way, *this book is dedicated to enabling scientists to be successful in the conduct of their scientific endeavors.*

Statistical power is therefore of paramount importance, not only because its consideration is a necessary condition of achieving *success* in scientific research, but also because it constitutes a procedural facet which is largely under the individual scientist's *control.* As such, it is not only professionally self-destructive to design research which does not have a high probability of success, it is unethical to do so for the simple reason that all scientific investigations consume scarce societal resources – be these monetary in nature or reflected in terms of the time and effort required of their participating human subjects. Since such investments are made on the premise that scientific investigations have the potential of contributing to society and the quality of the human condition, to conduct these investigations without optimizing their chances of success is tantamount to consuming scarce resources under false pretenses.

The book's approach to power

This book is unique in the sense that it represents the most comprehensive treatise presently available on statistical power, yet the only statistical knowledge it assumes on the part of its users is a conceptual understanding of (a) the arithmetic mean, (b) the standard deviation, and (c) the normal curve. Even this prerequisite knowledge is probably not *essential* for the actual determination of a study's statistical power, but it is necessary in *understanding* the process itself.

The remainder of this chapter, then, is largely dedicated to providing the reader with a conceptual basis for understanding what statistical power *means.* To do this it is necessary to first introduce the concept of a study's **effect size**, after which we will demonstrate how power is actually computed using only the mean, standard deviation, and the normal curve as prerequisite concepts.

Chapter 2 presents 11 key design factors which an investigator may manipulate to increase the statistical power of an empirical investigation and Chapter 3 illustrates the use of the power tables presented in this book and provides a number of guidelines regarding the conduct of a power/sample size analysis. Each of the remaining chapters is dedicated to one of the discrete statistical procedures that can be used to analyze the full spectrum of hypotheses and research designs employed in present day scientific practice.

The effect size concept

The most integral statistical component in the power analytic process is the concept of a study's **effect size**, which is nothing more than a standardized measure of the **size** of the mean difference(s) among the study's groups or of the **strength** of the relationship(s) among its variables. Although this book requires no computation whatever to conduct a power analysis, it is always necessary to be able to conceptualize a study's most likely outcome in terms of a **hypothesized effect size**.

To examine this concept in a little more detail, therefore, let us assume the simplest possible example: a two-group design with (a) one group receiving an experimental treatment (E), (b) a separate group serving as the control (C), and (c) a single continuous dependent variable (i.e., a measure which can be appropriately described by the mean). In such a study, the intuitively most obvious indicator of the experimental treatment's success will be the size of the difference between its mean and the mean of its control group. Unfortunately, this particular indicator is dependent upon a number of factors including (a) the scale with which the dependent variable is measured and (b) how heterogeneous the research sample turns out to be. Since the primary benefit to be derived from a power analysis is at the design stage, which occurs *before* subjects are selected and data are collected, it is obviously advantageous to employ an a priori measure of effect size that is independent of the type of dependent variable(s) and the types of subjects the investigator will be using.

Fortunately, by expressing the **potential** mean difference between any two groups in terms of standard deviation units we are able to achieve an effect size measure which is completely independent of both the dependent variable's measurement scale and the type of sample that happens to be selected for the study. This scale- and distribution-free measure of effect size (ES) is expressed by the following formula:

Formula 1.1. Effect size formula for two independent groups

$$\text{ES} = \frac{M_\text{E} - M_\text{C}}{\text{SD}_\text{pooled}}$$

Although computationally quite simple, this formula produces a descriptive statistic of great power and generality. It allows us, for example, to compare directly the effects resulting from two completely different experimental treatments performed on two different samples employing two completely different dependent variables. This very useful characteristic emanates from the standard deviation concept itself, which it will be remembered is, along with the mean, the key component of one of the most successful of all mathematical models: the **normal** (or **bell-shaped**) **curve**. An ES, then, is nothing more than a mean difference expressed in standard deviation units, which means that it is exactly comparable to a z-score (i.e., a standardized score in which the mean is zero and the standard deviation is 1.0). An ES of 1.0 therefore implies that one group's mean differs from that of another by one standard deviation (which is comparable to a z-score of 1.0). An ES of 0.50 implies that one group's mean differs from that of another by one-half (0.50) of one standard deviation (and can be interpreted exactly the same way as an individual who achieves a z-score of 0.50, meaning that he/she scored one-half of a standard deviation above the mean of his/her reference group).

From a research perspective, then, the ES is a superior concept to either the mean or the standard deviation in the sense that it is completely independent of the dependent variable's scale of measurement. The actual value of the standard deviation, like that of the mean, *is* dependent upon the scale of measurement and the characteristics of the sample. The *number* of standard deviations a score is from its mean, or the number of standard deviations (or standard deviation units) two scores are from one another, however, is *not* specific to the particular attribution being described or compared because the ES is expressed in terms of standard deviation units.

To illustrate, let us consider some hypothetical results emanating from two separate experiments: one designed to influence the quantitative subtest of the Scholastic Aptitude Test (SAT), which we will assume has a mean of 500 and a standard deviation of 100, and one designed to change attitudinal scores measured by a single item Likert scale (mean = 3.00; SD = 1.00). Now obviously we could never directly compare the raw mean differences generated by two studies designed to affect such disparate variables. Regardless of how successful the intervention designed to change attitudes was, for example, the experiment as designed could never result in a mean difference greater than 4.00 (since a Likert scale item can only range between 1 and 5). We would hope, however, that an experimental treatment worth testing would differ from its control by considerably more than 4 SAT points.

This is where standard deviation units become so helpful in estimating power. Suppose the first experiment resulted in an experimental vs. control (E vs. C) difference of 50 SAT points while the second yielded a mean difference of only one-half of a Likert scale point. At first glance there might seem to be a major discrepancy between the relative potency of these two interventions. As indicated in the computation in Figure 1.1, however, dividing by the appropriate standard deviation indicates that the **strength** or **size** of the two interventions is identical when expressed in terms of the ES.

This example demonstrates one of the most useful attributes of the ES concept: it provides a distribution-free statistic by which the results (or hypothesized results) of any experiment can be described and by which any two experiments can be directly compared. To be truly useful to investigators who are interested in estimating the amount of power available at the

Table 1.1. *The ES as an index of the proportion of subjects helped by the intervention*

ES	% of experimental > control subjects
0.0	50
0.1	54
0.2	58
0.3	62
0.4	66
0.5	69
0.6	73
0.7	76
0.8	79
1.0	84
1.5	93
2.0	98

Experiment 1 (SAT)

$$\mathrm{ES} = \frac{M_{\mathrm{E}} - M_{\mathrm{C}}}{\mathrm{SD}} = \frac{550 - 500}{100} = 0.50$$

Experiment 2 (LS)

$$\mathrm{ES} = \frac{M_{\mathrm{E}} - M_{\mathrm{C}}}{\mathrm{SD}} = \frac{3.50 - 3.00}{1.0} = 0.50$$

Figure 1.1. The ES as a distribution-free measure. SAT, Scholastic Aptitude Test; LS, Likert Scale.

design phase of a study, however, it is necessary for them to be able to *hypothesize* what will be the most likely value for the ES they will obtain once their research has been completed. To do this, it is quite helpful for them to be able to *visualize* what an ES "means."

The most straightforward way of doing this is probably to think in terms of mean differences and standard deviations and apply Formula 1.1. For investigators who have trouble doing this, there are other ways of thinking about the ES. One of these is to ignore the size of the mean difference likely to accrue from an experiment and to think only in terms of the percentage of subjects that the intervention is likely to "help" in comparison with the control. This index is far from perfect, since automatically (i.e., due to chance alone) 50% of the experimental subjects will score as well as or better than the mean of the control subjects with no intervention at all. Still, if an investigator thinks that his/her intervention will result in, say, approximately two-thirds of the experimental group scoring higher than the mean of the control group as a function of that intervention, Table 1.1 could be used to translate this estimate to a hypothesized ES of approximately 0.40.

Table 1.2. *The ES as a measure of shared variance (r^2) or the Pearson r*

ES	r^2	r
0.0	0.00	0.00
0.1	0.002	0.05
0.2	0.01	0.10
0.3	0.02	0.15
0.4	0.04	0.20
0.5	0.06	0.24
0.6	0.08	0.29
0.7	0.11	0.33
0.8	0.14	0.37
1.0	0.20	0.45
1.5	0.36	0.60
2.0	0.50	0.71

Other researchers find it more convenient to conceptualize the ES in terms of r^2 or the amount of variance in the dependent variable which is explained in terms of the independent variable. (The independent variable in the present case is defined as whether or not an individual receives the intervention.) Thus, a researcher who "thinks" in terms of "percentage of variance explained" (or in terms of the Pearson *r*, which is a measure of association between two variables) can use Table 1.2 to convert such an estimate directly to an ES, although this index too has its limitations. Most experimental researchers, for example, do not visualize their effects in these terms and have difficulty conceptualizing what it means to say that 6% (a medium ES of 0.50) of the dependent variable is shared by variations in group membership.

Finally, those investigators who prefer to think in terms of **proportions** or **success rates** may use Table 1.3 to conceptualize the ES. Thus, if the E vs. C average "success rate" is 50%, an ES of 0.50 translates to an approximate 25% superiority for the intervention as compared to the control. If the E vs. C *average* "success rate" is either below or above 50%, then the E vs. C difference (E − C) for any given ES is less than the estimates presented in Table 1.3.

Hypothesizing an appropriate effect size. Regardless of how an ES is visualized, it is still necessary for an investigator to hypothesize what the most likely value he/she is to obtain from the study being designed *prior* to conducting a power analysis. This is, if anything, the concept's Achilles heel,

Table 1.3. *The ES as a measure of differences in proportion (E vs. C average = 0.50)*

ES	E vs. C success rate		
	C	E	E − C
0.0	0.50	0.50	0.00
0.1	0.47	0.52	0.05
0.2	0.45	0.55	0.10
0.3	0.42	0.57	0.15
0.4	0.40	0.60	0.20
0.5	0.38	0.62	0.24
0.6	0.35	0.64	0.29
0.7	0.33	0.66	0.33
0.8	0.31	0.68	0.37
1.0	0.27	0.72	0.45
1.5	0.20	0.80	0.60
2.0	0.14	0.85	0.71

because many investigators simply throw up their hands in frustration at this step, asking "How can I know what the ES will be before I conduct the study?" The answer, of course, is that they *cannot* – although they can certainly formulate a **hypothesis** regarding this value. In reality, the process of estimating a study's ES is similar to the process of formulating a classical hypothesis; the only difference is that this hypothesis states *how much* better one group will be than another. There are basically three ways in which this is done:

(1) The first is by conducting a pilot study and using the ES obtained therefrom. In some cases this is not optimal because of the small number of subjects typically available for pilot studies. Preliminary studies involving a study's primary variables should *always* be run as a matter of course, however, and the resulting data can often provide relatively good ES estimates. (This is especially true if a relatively large scale preliminary effort involving the same conditions which will be employed in the final study (e.g., the existence of a randomly assigned control group) are employed.)

(2) Another method of hypothesizing an appropriate ES is by carefully reviewing similar studies conducted by other investigators and ascertaining the types of results obtained therein. The increased availability of meta-analyses (i.e., review articles which actually compute mean ES values for research studies testing the same basic hypotheses) makes this a relatively attractive option.

(3) In the absence of a pilot study or sufficiently detailed information from the literature to hypothesize an ES, a very reasonable strategy involves using Jacob Cohen's (1988) ES recommendations. Professor Cohen suggests that researchers hypothesize a small (0.20 standard deviation units for two-group studies), medium (0.50 SD units), or large (0.80 SD units) ES based upon their knowledge of the variables involved. In the absence of specific insights, he recommends choosing a medium ES of 0.50. This latter advice has historically been adopted by a great many researchers and now has an extremely impressive empirical rationale following Lipsey & Wilson's (1993) seminal meta-analysis of 302 social and behavioral meta-analyses in which they found the average ES of over 10000 individual research studies. Incredibly they found the average ES to be *exactly* 0.50, which leads us to paraphrase Professor Cohen's original advice (at least for social and behavioral research):

When in doubt, hypothesize an ES of 0.50.[3]

The meaning of power

Once an ES is hypothesized, power becomes quite easy both to compute and to conceptualize. To gain a clearer understanding of what power means, let us consider a typical power analysis. Let us assume that an investigator wishes to evaluate a treatment's success in decreasing patients' headache pain as measured by a visual analog scale, that she/he hypothesizes that the most likely ES to accrue would be 0.50, and knows she/he has enough subjects to achieve an *N* per group of 64. *Note that the discussion that follows is equally applicable to determining the sample size needed to achieve a desired level of power or the maximum detectable effect size emanating from a fixed sample size and desired level of power: these are all key, integrally related components and are all subsumed under the concept we are referring to as "power" and the process we are referring to as a "power analysis."*

To ascertain how much power would be available for such an experiment, all the researcher would be required to do is decide what statistical procedure would be appropriate for this particular experiment. Since only two groups are involved and the outcome variable is continuous in nature, an independent samples *t*-test is appropriate. The investigator would therefore turn to Table 4.1 at the end of Chapter 4 and locate the power level at the juncture of the 0.50 ES column and the *N*/group row closest to 64 (i.e., 65) which would yield a power estimate of 0.81, which in turn would lead our investigator to conclude correspondingly that, she/he would have an approximately 80% chance of achieving statistical significance (or of rejecting a false null hypothesis) at the 0.05 level with a sample size of 64 subjects per group if indeed the hypothesized ES of 0.50 were appropriate. To communicate the results of this particular power analysis, our investigator would include a statement something like the following in his/her research report or grant proposal:

> A power analysis (Bausell & Li, 2002) indicated 64 subjects per group would yield an 80% chance (i.e., power = 0.80) of detecting an ES of 0.50 between the experimental and control conditions using an independent samples *t*-test (two-tailed alpha = 0.05).

In non-statistical language, what this is telling us is that, *if everything goes as planned*, our investigator will have an 80% chance of achieving statistical significance. In other words, if the experimental treatment is indeed capable of producing a gain of one-half of the dependent variable's standard deviation as compared to no treatment at all (i.e., if the hypothesized ES is accurate), then *eight out of ten properly performed experiments* using 64 subjects per group will result in statistical significance at the 0.05 level.

Although certainly efficient, there is a decided black box quality to this process. Since the computational procedures for power are not difficult, however, perhaps actually seeing how power could be obtained without the help of tables or computer programs may be helpful in conceptualizing what it really means to say that a study "has an 80% chance of achieving statistical significance."

The calculation of power. Re-examining the above statement indicates that our researcher has hypothesized an ES of one-half of a standard deviation (i.e., $(M_E - M_C)/SD$), expects to employ 64 subjects in both the experimental and control groups, and will use a significance level of 0.05. Actually computing the power available for such a study, then, would require access to only three sources of information:

(1) The **critical value** of the *t*-statistic (i.e., t_{cv}) which would be *necessary* to achieve statistical significance at an alpha level of 0.05 if the *N*/group employed were equal to 64. This information can be found in tables present in almost all elementary statistics texts or from commonly used statistical packages such as SPSS or SAS and would be equal to 1.98. This value, then, is the *necessary t* we must obtain in order to achieve statistical significance. (Note that there are no probabilities here: if a *t* of 1.98 is obtained using 64 subjects per group the researcher will (i.e., $p = 1.00$) obtain statistical significance.)

(2) The t-statistic which would accrue if the hypothesized medium ES of 0.50 were *actually* obtained after the study had been conducted and its data analyzed. This value can be obtained by simply substituting an ES of 0.50 and $N = 64$ into the most simple t-test formula:

Formula 1.2. *t*-test formula using the ES concept

$$t_{hyp} = \frac{ES_{hyp}}{\sqrt{2/N}} = \frac{0.50}{\sqrt{2/64}} = 2.82$$

This value ($t = 2.82$) could be called the *hypothesized t* (t_{hyp}) because it is the exact t-statistic that we would obtain if our *hypothesized* ES of 0.50 were obtained employing 64 subjects per group. (Here again, note that this is not a probabilistic statement: a t of 2.82 *will* be obtained ($p = 1.00$) if an ES of 0.50 is obtained.)

(3) A distribution (or table) presenting areas under the unit normal curve associated with different standard deviation units (also called z-scores), which is also present in practically all elementary statistics texts or computer packages and is really nothing more than the values upon which the normal curve is based.

Once these pieces of readily available information are obtained, the computation of power is simplicity itself because for all practical purposes we can subtract our two t values (i.e., $t_{hyp} - t_{cv}$) and treat the difference as though it were expressed in standard deviation units (i.e., a z-score or an ES). This means that to obtain the power available for a study all we need do is look up the difference between the t which we *must* obtain in order to achieve statistical significance (t_{cv}) and the t which *will* be obtained if our hypothesized ES is correct (t_{hyp}) in a table representing the unit normal curve or a distribution of z-scores (all of which are expressed in standard deviation units, as are ES values).

Thus in our present example, given a hypothesized ES of 0.50 and estimated N of 64, what we have so far is a t_{hyp} of 2.82 that we will obtain if (and only if) we obtain our hypothesized ES of 0.50 and a critical value of t (t_{cv}) of 1.98 (which is necessary if we are to obtain statistical significance with 64 subjects per group). To determine the statistical power for this two-group study (with a projected N of 64 and a projected ES of 0.50), all we basically have to do is answer the following question:

> Given a distribution in which the mean is the t representing the hypothesized ES (i.e., t_{hyp}) for a two-group study, what percentage of the t values in this distribution are greater than the critical value needed for statistical significance (i.e., t_{cv})?

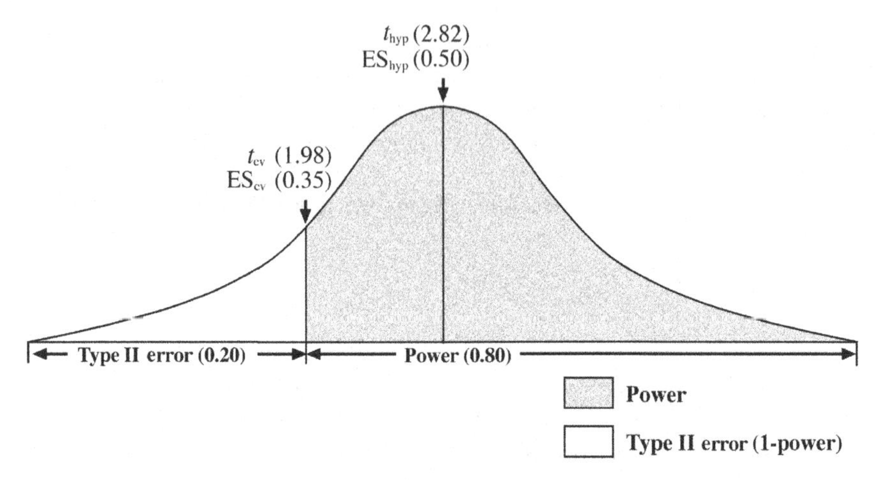

Figure 1.2. Power depicted as the difference between two *t*-statistics.

This specific scenario is illustrated in Figure 1.2.

This figure, then, refers to the theoretical scenario in which an infinite number of experiments are conducted which meet the following criteria:

(1) The "real" ES is exactly the same as the researcher's "hypothesized" ES (i.e., 0.50). Naturally the researcher will never be perfectly accurate in hypothesizing the "real" ES, but hopefully this hypothesis will be a reasonable estimate. (This of course means that a power analysis computed at the design stage is actually an **estimate**, often a very gross estimate, of the amount of power available.)

(2) Each experiment is conducted perfectly, which is another way of saying that the only sources of error are random in nature. At first glance this may seem overly theoretical, but it is really why we need inferential statistics in the first place. To illustrate, suppose a researcher conducted two identical experiments which employed subjects who were randomly drawn from the same population. It would be unreasonable to expect the mean differences resulting from these two studies to be *identical*, since there would be some individual differences among the subjects involved. It follows then, that even if our researcher were perfectly correct in hypothesizing what the "real" ES was for the variables being studied and ran the two studies perfectly, he/she would be extremely unlikely to obtain an ES of exactly 0.50 for both experiments. If enough identical experiments were conducted, however, their distribution of ES values would begin to approximate the curve depicted in

Figure 1.2 in which the *mean ES was the "real" or hypothesized value of 0.50.*

Returning to our numerical example, since the area under the curve represented by Figure 1.2 very nicely *approximates* that of the unit normal curve, the actual area to the right of the t_{cv} can be approximated by using the unit normal distribution. To do this, we would simply subtract the t associated with the critical value (t_{cv} or 1.98) from the t associated with the hypothesized ES of 0.50 (i.e., $t_{ES} = 2.82$) and "pretend" that the resulting difference (i.e., $2.82 - 1.98 = 0.84$) is a z-score representing 0.84 standard deviation units. The unit normal distribution would then inform us that 80% of the area of Figure 1.2 is to the right of t_{cv} (i.e., 50% is to the right of the mean (t_{hyp}) and 30% falls between this mean and t_{cv}), which not coincidentally is the same value we would obtain from the use of a power table.

This method of computing power is very succinctly summarized by the following formula:

Formula 1.3. Power as a function of the difference between *t*-statistics[4]

$$\text{power} = \text{probability of a } z\text{-score} \geqslant (t_{hyp} - t_{cv})$$

An alternative definition of power. Formula 1.3 very nicely encompasses the entire conceptual basis of power. If our hypothesized ES of 0.50 were exactly correct, and if we were to conduct an infinite number of identical experiments, we would obtain a distribution of ES values (i.e., mean differences between E and C) where the **mean** would be 0.50. If we can assume that the only error operating in this process were random in nature, half of these obtained values would be greater than 0.50 and half would be less. We would, in other words, obtain a normal curve in which the hypothesized ES of 0.50 would be the mean and we would be back to the distribution represented in Figure 1.2. (As we will discuss in subsequent chapters, this basic logic holds even when more complex research designs and more sophisticated statistical procedures are employed.)

Said another way, to evoke the statistical theory behind the power concept it is necessary to assume that:

(1) an appropriate intervention is chosen in the first place (i.e., it really was half of a standard deviation "better" than whatever was chosen to compare it to), and
(2) the experiment was conducted *properly*.

Hopefully, then, this discussion has logically led to the following conclusions concerning what power is not:

(1) *Power is not "the probability of obtaining statistical significance if a specified ES is obtained."* If the critical value of *t*, for example, is obtained, then by definition we have a 100% chance of achieving statistical significance (and, also by definition, the null hypothesis will be rejected).

(2) *Power is not "the probability that a test will yield statistical significance" irrespective of whether or not the data upon which it is based were collected appropriately.* If a study is badly run, employs invalid or unreliable measures, utilizes subjects who were inappropriate, or is subject to any number of other glitches, then power is basically irrelevant because any hypothesized ES based upon a true representation of the constructs under study will be inaccurate (and, again, basically irrelevant).

(3) *Power is not "the probability of obtaining statistical significance if the null hypothesis is false."* The null hypothesis can be false, but if the hypothesized ES is not appropriate, then any power analysis based upon it is inappropriate. If the true ES is larger than hypothesized, then power will be underestimated. If smaller, then the computed power will be an overestimate.

It therefore follows that:

> Power is the probability of obtaining statistical significance in a properly run study when the hypothesized ES is correct.

With this definition in mind, then, the computation of statistical power for any projected study involves only (1) hypothesizing the ES most likely to accrue from the study, (2) deciding what statistical procedure will be used to analyze the ES which actually does result, (3) turning to the chapter dedicated to that statistical procedure, and (4) using the appropriate tables (as explained for each test) to estimate (a) the power available for a fixed sample size or (b) the sample size needed to produce a desired level of power. (For more complex designs, additional parameters must be specified, such as the correlation between variables, but these conditions will be clearly specified in the chapters which follow.)

Prior to actually using these tables to conduct a power analysis, however, it is important to consider all of the factors that potentially influence statistical power. Chapter 2 is therefore devoted to those design components under the investigator's control capable of maximizing the chances of a scientific experiment's success.

Endnotes

1 This book will not delve into the myriad criticisms of classic hypothesis testing nor the perceived advantages of employing alternatives thereto. In many ways the present authors' view of the process is closer to that of Neyman & Pearson (1933) than Fisher (1935), but for all practical purposes research practice in the social and health sciences does approximate the latter's dichotomous approach to null hypothesis testing, even though hypotheses themselves are seldom stated in present day published literature.

2 A number of authors have surveyed various social and health related literatures and found that the majority of studies do not have an 80% chance of detecting a medium ES. Examples are Cohen (1962), Brewer (1972), Katzer & Sodt (1973), Haase (1974), Chase & Tucker (1975), Knoll & Chase (1975), Chase & Chase (1976), Reed & Slaichert (1981), Sedlmeier & Gigerenzer (1989), Polit & Sherman (1990), Moyer *et al.* (1994), Clark-Carter (1997). Some disciplines (sociology, marketing, and management), however, were found to have reasonable levels of power (Mazen *et al.*, 1974; Sawyer & Ball, 1981). Freiman *et al.* (1978) present an interesting twist on this literature, finding a decided lack of statistical power in an analysis of clinical trials which did not obtain statistical significance.

3 While excellent advice, no funding agency of which we are aware is likely to accept this as the sole justification for a hypothesized effect size.

4 The Technical appendix presents more detail on the application of this formula plus the following correction factor recommended by Hays (1973) and Scheffe (1959):

$$\text{power} = p\left(z \leq \frac{t_{\text{hyp}} - t_{\text{cv}}}{\sqrt{1 + \dfrac{t_{\text{cv}}^2}{2\text{df}_t}}} \right)$$

2 Strategies for increasing statistical power

The three *basic* parameters that affect power are: (a) the proposed sample size, (b) the significance level which will be used to determine whether or not to accept the study's hypothesis(es) and (c) the hypothesized effect size. Each of these has a number of different facets or applications that potentially affect power. Since power is one of the primary keys to the success of a scientific investigation, we will organize the remainder of the chapter in terms of strategies for *increasing* it. (For ease of exposition, we will assume a two-group experimental design as the basic departure point, although all of the concepts discussed apply to any type of design, experimental or non-experimental.)

The 11 strategies (the mathematical rationale for which is presented in the Technical appendix), along with their relationship to the three basic power parameters, can be summarized as follows:

Parameter I: Sample size

- **Strategy 1: Adding subjects**
- **Strategy 2: Assigning more subjects to groups which are cheaper to run**

Parameter II: Significance level

- **Strategy 3: Choosing a less stringent significance or alpha level**

Parameter III: Effect size

- **Strategy 4: Increasing the size of the hypothesized ES**
- **Strategy 5: Employing as few groups as possible**
- **Strategy 6: Employing covariates and/or blocking variables**
- **Strategy 7: Employing a cross-over or repeated measures/within subject design**
- **Strategy 8: Hypothesizing main effects rather than interactions**
- **Strategy 9: Employing measures which are sensitive to change**
- **Strategy 10: Employing reliable measures**
- **Strategy 11: Using direct rather than indirect dependent variables**

Table 2.1. *The relationship between power and sample size (alpha = 0.05; two-tailed)*

N/group	ES		
	0.20	0.50	0.80
15	0.07	0.26	0.57
25	0.10	0.41	0.79
50	0.16	0.70	0.98
100	0.29	0.94	★[a]
150	0.41	0.99	★[a]

Notes:

[a] Power value is equal to or greater than 0.995.

Eleven strategies for increasing statistical power

Strategy 1: Adding subjects. This is perhaps the most direct (and expensive) means of increasing statistical power available to the researcher. Without exception, the larger the *N* employed, the more power is available for testing a hypothesis. Returning to the two-group example discussed in Chapter 1 (i.e., ES = 0.50; *n* = 64; alpha = 0.05, two-tailed), increasing the sample size by 40 subjects per group would increase the power from 0.80 to 0.95. Alternatively, reducing the group size by 40 subjects would result in a power reduction from 0.80 to 0.39.

This relationship is illustrated quite clearly in Table 2.1 for a two-group design using the types of sample sizes commonly found in experimental research. It is worth repeating that the same relationships hold for more complex experimental and correlational designs as well. They also hold for other significance levels. Thus, assuming an ES of 0.50, increasing the sample size from 50 to 100 subjects per group increases the available power from (a) 0.70 to 0.94 for a two-tailed alpha of 0.05, (b) 0.45 to 0.82 for a two-tailed alpha of 0.01, and (c) 0.80 to 0.97 for a one-tailed alpha of 0.05.

Strategy 2: Assigning more subjects to groups which are cheaper to run. This strategy is really a corollary of the first suggestion, since it is simply a more cost effective means of increasing a study's overall sample size. Equal numbers of subjects per group are always optimal, but if practical constraints limit the number of subjects which can either be run or obtained in, say, the intervention group, then it would be foolish to truncate a study's overall sample size artificially by limiting the number of control subjects if

Table 2.2. *Effects of different N/group ratios (ES = 0.50; alpha = 0.05; two-tailed)*

		N/group			Power	
Group 1	Group 2	Ratio	Harmonic mean	Arithmetic mean	Harmonic mean	Arithmetic mean
40	40	1:1	40	40	0.60	0.60
40	80	1:2	53	60	0.72	0.77
40	120	1:3	60	80	0.77	0.92
40	160	1:4	64	100	0.80	0.94
40	200	1:5	67	120	0.82	0.97

the latter are easily attainable. Instead, the researcher should run as many subjects in his/her control group as necessary to achieve the desired level of power. It is important to keep in mind, however, that it is the harmonic mean (see the Technical appendix) rather than the arithmetic mean which is used to determine how many control subjects would be needed. (As indicated in Table 2.2, then, the relative advantages accruing from this strategy begin to dissipate rapidly after a 1:2 ratio is achieved for two-group studies.)

The same logic holds for non-experimental studies as well. For example, suppose a study were to be conducted comparing smokers to non-smokers with respect to the number of sick days taken during the course of a year's employment. If only 20% of a given company's 100 workers smoke, it would not be wise from a power perspective to compare the 20 smokers with a random sample of 20 non-smokers. Considerably more power would accrue from comparing either the 20 smokers to 40 non-smokers or, if data were easily collected, the 20 smokers to all 80 non-smokers. (The power listed under the "arithmetic mean" refers to the power that would have accrued if the total *N* had been equally distributed in a 1:1 ratio between the two groups.)

Strategy 3: Choosing a less stringent significance or alpha level. As discussed in Chapter 1, the decision regarding what probability level will determine statistical significance is not entirely under the researcher's control unless he/she opts for a more stringent than customary significance level. As indicated in Table 2.3, choosing an alpha level of 0.01 instead of 0.05 significantly *reduces* the chance of obtaining statistical significance from 0.70 to 0.45 for an ES of 0.50 and an *N*/group of 50, while relaxing the significance criterion to 0.10 increases it to 0.80.

Table 2.3. *The relationship between the alpha level (ES= 0.50; two-tailed) and statistical power*

N/group	Alpha			
	0.20	0.10	0.05	0.01
15	0.52	0.37	0.26	0.10
25	0.68	0.54	0.41	0.19
50	0.89	0.80	0.70	0.45
100	0.99	0.97	0.94	0.82
150	*[a]	*[a]	0.99	0.96

Notes:
[a] Power value is equal to or greater than 0.995.

While researchers seldom have the option of increasing their alpha levels to 0.20, there are occasions when an alpha of 0.10 is appropriate and indeed is sometimes the only viable option for a study in which a small ES is the most likely outcome. If, for example, there is *strong* empirical and theoretical evidence regarding the direction of an effect, journal editors and funding agencies will sometimes permit the use of a directional (also called a one-tailed) hypothesis test, which is equivalent to the two-tailed alpha level of 0.10. (Such evidence might include scenarios in which (a) extensive pilot work has been conducted (e.g., in drug trials) or (b) a sufficiently strong theoretical rationale exists (e.g., in mature disciplines involving extensively researched dependent variables) that the researcher can be almost certain that the intervention will not be *harmful*.) In such instances, there is really no reason not to employ a one-tailed significance test if the investigator makes this decision a priori and if he/she is willing to interpret even the strongest negative effect as statistically non-significant. Because of the number of tables required, we have not presented power tables for one-tailed tests. We have, however, included sample size tables for $p=0.10$ (which is equivalent to a one-tailed $p=0.05$) for each statistical procedure covered.

Strategy 4: Increasing the size of the hypothesized ES. It is much easier to achieve statistical significance when the variables involved are strongly related to one another. For experimental research, the ES is synonymous with the strength of the relationship between the independent variable (i.e., the experimental groups) and the dependent variable. For correlational research, stronger relationships can be visualized simply as a larger correlation coefficient (e.g., Pearson r, ϕ) between the independent and dependent variables.

Table 2.1 can also serve to illustrate the relationship between the hypothesized ES and power. Continuing with our two-group example, power increases from 0.10 to 0.41 when the hypothesized ES rises from 0.20 to 0.50 for a fixed *N* of 25 subjects and from 0.41 to 0.79 when the ES rises to 0.80 (assuming a two-tailed alpha level of 0.05).

Increasing the ES likely to accrue from a study can, of course, be a relatively problematic task fraught with difficult trade-offs. If we assume that *different* subjects have been assigned to the two groups and that there are no additional variables that can be used for control purposes (see strategy 6), there are basically three strategies which the investigator can employ. These involve (a) strengthening the intervention, (b) weakening the comparison group, and (for non-manipulated variables) (c) contrasting extreme groups. Let us consider each in turn.

Strengthening the intervention. The strength of an intervention is most directly affected by:

(1) increasing the dose (e.g., literally in the case of a drug trial or by offering more or longer training sessions in a behavioral one), or
(2) adding additional components thereto.

To illustrate, let us assume that an investigator is interested in designing an intervention capable of effecting weight loss in overweight cardiac patients, but has access to only a relatively limited number of such patients. If the literature indicates that a single session of dietary instruction customarily yields a weight loss ES of approximately 0.20, a sample size analysis conducted using Table 4.2 Chart B would indicate that almost 400 subjects per group would be needed to produce an 80% chance of obtaining statistical significance.

Our researcher would thus obviously have little choice but to alter his/her design in some way. To increase the strength of the study's intervention, he/she could (a) design the study to include multiple sessions of instruction and/or (b) add a component (e.g., exercise) to the intervention for which there was adequate theoretical/empirical evidence supporting its effectiveness with respect to inducing weight loss.

Naturally, other considerations would need to be balanced before making such a decision. Adding multiple sessions, for example, might increase experimental attrition as well as make the intervention too costly to implement clinically. Coupling an exercise component with the original instructional intervention, on the other hand, would cloud the ultimate etiology of the effect being demonstrated (i.e., the investigator would have no way of knowing whether the resulting experimental vs. control (E vs. C) difference was *primarily* attributable to dietary instruction, exercise, or some combination/interaction between the two). Decisions such as this must therefore be made by the individual investigator based upon the ultimate scientific objective of the study. One of the purposes of the tables in this book is to alert investigators to the fact that if the ES is likely to be small (e.g., ≤0.20), and if a relatively small number of subjects is available for the study, then the chances of achieving statistical significance at the 0.05 level are exceedingly low.

Weakening the control (comparison) group. Since the ES is determined by the mean difference between the experimental group and its control, weakening the control group can be as effective as strengthening the experimental group. One method of doing this is to use a pure, no-treatment control rather than a treatment-as-usual group or to use the latter in lieu of a placebo. As with altering the intervention, scientific and clinical considerations must always take precedence here over purely statistical ones, *but it should be noted that statistical considerations can effectively prohibit a trial from ever being mounted.* In those cases in which no contact with subjects at all can be justified on epistemological (and ethical) grounds, however, such a strategy will usually result in a larger ES. (Based upon the results of over 300 social and behavioral meta-analyses, Lipsey & Wilson (1993) have estimated that the use of a no treatment control is likely to increase an ES by approximately 0.20 as opposed to use of an attention placebo group. Definitive estimates such as this for health related placebo ES values unfortunately do not yet exist.)

Employing extreme groups for non-manipulated variables. When power is an issue for non-manipulated variables (i.e., in correlational research or for blocking variables in experimental designs), the ES can be increased by "throwing out" the central part of a distribution (Feldt, 1961). As an example, suppose a researcher needed to study the effects of an intervention designed to increase productivity in the work place, but suspected that his/her treatment would be more effective for non-smokers than smokers (because of the number of breaks taken by the latter). Under normal circumstances the investigator would employ smoking as a blocking variable (i.e., identify employees who smoked vs. those who did not and randomly assign each group to experimental vs. control groups). Assuming that only a limited number of employees could be included because of economic constraints, however, the investigator might be wise to increase the smoking vs. non-smoking ES in his/her study by contrasting, say, individuals who smoked two or more packs per day with those who did not smoke at all. The ES for this comparison would probably be considerably greater than for one which included some individuals who only smoked occasionally or who were very light smokers. (Of course the omnipresent trade-off here would involve not knowing anything about the differential effectiveness of the intervention for the remainder of the smoking continuum and productivity.)

Strategy 5: Employing as few groups as possible. As will be discussed in Chapter 6, the use of more than two groups in experimental research necessitates the use of post hoc tests for statistically significant effects. When the sample size is limited, this has the potential of greatly reducing the probability of achieving statistical significance for all resulting pairwise comparisons. As with all the factors affecting power, decisions regarding whether or not to employ multiple groups must ultimately be made by weighing statistical vs. substantive issues.

Returning to the hypothetical clinical weight loss study posited above, suppose our investigator had decided that it was important to determine the etiology of any potential effect resulting from the combined dietary/exercise intervention. Procedurally, he/she would therefore have little choice other than to add two additional intervention groups to the original design: one employing dietary information alone and one employing exercise alone. If he/she wanted to ensure that any observed difference was not due to a placebo effect involving contacts between patients and clinicians, a fifth group would be required as depicted below:

(1) E1: Exercise only.
(2) E2: Dietary information only.
(3) E3: Exercise and dietary information.
(4) C1: No treatment (or treatment as usual).
(5) C2: Attention placebo.

To illustrate the potential adverse effect upon power of such a decision, let us assume that the maximum number of subjects available for this study was 120. Table 2.4 illustrates the relative *loss* in power associated with the addition of groups to the original two-group design proposed for a represented range of ES values. (We will assume a two-tailed alpha of 0.05 and that Tukey's honestly significant difference (HSD) post hoc test will be employed.) As an example, for an ES of 0.50 the power drops from a marginally acceptable 0.77 for a two-group study employing a total *N* of 120 to an abysmal 0.15 for our five-group example above for this specific pairwise comparison (i.e., the 0.50 ES).

Obviously a multiple group study such as this is feasible only for a

Table 2.4. *The relationship between number of groups and statistical power for pairwise comparisons (total N of 120; alpha = 0.05; two-tailed; multiple comparison procedure (MCP), Tukey's HSD)*

ES	Number of groups			
	2 (*N*/group = 60)	3 (*N*/group = 40)	4 (*N*/group = 30)	5 (*N*/group = 24)
0.20	0.19	0.07	0.04	0.02
0.50	0.77	0.45	0.25	0.15
0.80	0.99	0.88	0.69	0.50
1.10	*[a]	0.99	0.95	0.85

Notes:
[a] Power value is equal to or greater than 0.995.

Table 2.5. *The relationship between number of groups and statistical power for pairwise comparisons (N/group = 60; alpha = 0.05, two-tailed; MCP, HSD)*

ES	Number of groups			
	2 (Total *N* = 120)	3 (Total *N* = 180)	4 (Total *N* = 240)	5 (Total *N* = 300)
0.20	0.19	0.10	0.07	0.05
0.50	0.77	0.65	0.56	0.50
0.80	0.99	0.98	0.96	0.95
1.10	*[a]	*[a]	*[a]	*[a]

Notes:
[a] Power value is equal to or greater than 0.995.

relatively large hypothesized ES. As illustrated in Table 2.5, however, the situation improves dramatically if the *N* per group remains unchanged (i.e., power decreases only from 0.77 to 0.50 and the power for the overall *F*-ratio, the statistic used to test the difference between three or more groups, increases dramatically), although a considerable amount of power is still lost *for the individual group-by-group contrasts* (even though the total *N* more than doubles from 120 to 300). This leads us to suggest that if a limited number of subjects is available, the researcher should give serious consideration to employing as few groups as scientifically defensible. (Two groups are always optimal, since this avoids the power reducing necessities of decreasing the number of subjects per group and of employing post hoc tests to evaluate the statistical significance of individual pairwise comparisons.)

Pattern A. ES pattern associated with low F power	①		②③④		⑤
Pattern B. ES pattern associated with medium F power	①	②	③	④	⑤
Pattern C. ES pattern associated with high F power	①②③				④⑤
or	①②				③④⑤

Figure 2.1. The relationship between spread and power of pairwise ES values.

In many ways Tables 2.4 and 2.5 are counter-intuitive, since we have repeatedly stressed the direct relationship between sample size and power, yet the principle holds both for multiple comparison procedures *and* the power of the overall F-ratio. It also holds across different types of designs, such as those employing covariates and repeated measures. The only exception of which we are aware is the investigator's choice of multiple comparison procedures (MCP). This is the primary reason that we have included sample size tables for the Newman–Keuls approach, since it is considerably more liberal with respect to power than the Tukey HSD (as well as most other MCPs) for those contrasts whose pairwise ES values are nearest to one another (see Chapter 6).

Another potential exception involves the power of the overall F-ratio when the spacing or pattern of the hypothesized pairwise ES values for the various groups is allowed to vary. While this does not affect power estimates emanating from MCPs, generally speaking patterns of means that cluster together away from the midpoint of the distribution will result in more power for the overall F-ratio as illustrated in Figure 2.1.

These principles, then, lead us to offer two related, supplementary strategies for those instances in which the number of groups cannot be reduced:

Supplementary strategy 5a. When the number of groups is fixed, consider employing the Newman–Keuls MCP in preference to the Tukey HSD or an equally conservative procedure. The powers of the two procedures are identical for the two largest pairwise ES values of an experiment, but the former is considerably more powerful than the latter for groups that are adjacent to one another (see Figure 2.1).

Supplementary strategy 5b. When scientific considerations permit, design treatments with the greatest possible ES spreads. In other words, an overall hypothesized spread of pairwise ES values such as depicted in pattern C of Figure 2.1 is superior (from the perspective of the power of the overall F-ratio) to a study in which a number of the group means are expected to fall half-way between the extreme groups (pattern A) or are equally spread between them (pattern B).

Strategy 6: Employing covariates and/or blocking variables. Few present day studies employ a single independent variable as depicted in our generic two-group example. Almost without exception investigators collect additional data on their subjects which they have reason to suspect may have a bearing on their experimental outcomes.

These additional variables normally have one of three potential uses: (a) they may be used simply to describe the sample to give the research consumer a feel for the types of subjects used (and hence make an informed decision about how far the results may be generalized), (b) they may be used for statistical control purposes (i.e., as a covariate/blocking variable in experimental research or as a control variable in a regression study), and/or (c) they may be used to test additional hypotheses regarding potential differential effects of the treatment (e.g., as the interaction between a blocking variable and the experimental intervention). Using this additional information either for statistical control purposes, or to test differential treatment effects, has the potential of increasing a study's statistical power, depending upon the degree to which the additional covariate(s) or blocking variable(s) is/are related to the study's dependent variable. (A more thorough discussion on covariates and blocking variables will be provided in Chapter 7; interaction effects will be discussed in Chapter 9.)

Fortunately, when the additional variables do not increase the overall variability among the subjects chosen to participate in the study (e.g., when they are used as covariates or employed as post hoc blocking variables), the resulting increase in statistical power is relatively easily quantified. This is due to the fact that the resulting effect upon power is related to the size of the correlation coefficient between the covariate/blocking variable and the dependent variable, which serves to increase the study's ES.

As illustrated in Table 2.6, the increment to power increases directly with the size of this hypothesized correlation. Thus if an r of 0.60 exists between the covariate and the dependent variable, a study's power would increase from 0.80 to 0.94 assuming an ES of 0.50 and an N/group of 64. Alternatively, if a power of 0.80 were deemed sufficient, the researcher could decrease his/her sample size by 22 subjects per group.

When a blocking variable is employed, an additional increment to statistical power may be provided over and above the increase illustrated in Table 2.6 if this new variable interacts with the treatment and if it can be assumed that the addition of the blocking variable does not increase the overall within-group difference between subjects. Since this latter point is a

Table 2.6. *The relationship between the use of a control variable and power (ES = 0.50; N/group = 64; alpha = 0.05)*

Expected *r*	Adjusted ES	Power	N/group for power of 0.80
0.00	0.50	0.80	64
0.20	0.51	0.82	62
0.40	0.55	0.86	54
0.50	0.58	0.90	49
0.60	0.63	0.94	42
0.70	0.70	0.98	34
0.80	0.83	*[a]	24
0.90	1.15	*[a]	14

Notes:
[a] Power value is equal to or greater than 0.995.

relatively tenuous assumption except for post hoc blocking, we will not bother to quantify the potential increase in power due to a statistically significant interaction in this book. If there is a strong theoretical reason to suspect such an interaction, however, and if it is deemed *not* to be appropriate to use the interaction variable as an inclusion/exclusion criterion, then Chapter 10 presents modeling algorithms to estimate this power increment.

Strategy 7: Employing a cross-over or repeated measures/within subject design. Although often contraindicated (such as when an intervention produces a permanent effect on the dependent variable, hence precluding a return to baseline), allowing each subject to receive all experimental treatments can greatly increase a study's available power. Even when these criteria cannot be met, a randomized matched design (in which subjects are rank ordered via a blocking variable of some sort and randomly assigned to treatments in blocks) can be used to simulate the effects of employing a true repeated measures design. Alternatively, a single group pretest–posttest design, while not a particularly good choice from a methodological perspective, does greatly increase power if the correlation between subject pretest (baseline) and posttest scores is relatively strong.

The increase in power accruing from both types of designs (i.e., those contrasting mean differences involving the same subjects and those involving matched subjects) is directly related to the correlation between the dependent observations (i.e., between each subjects' repeated scores or between matched subjects' scores). As illustrated in Table 2.7, this effect can be considerably more dramatic than was the case for covariates. This is due to the fact that the ES is adjusted (prior to being tested for statistical

Table 2.7. *Power increases as a function of r in a two-group (or single group pretest–posttest) within subjects design (ES = 0.50; alpha = 0.05; N = 50)*

Size of r	Adjusted ES	Power	N/group for power of 0.80
0.00	0.50	0.70	64
0.20	0.56	0.79	52
0.40	0.65	0.89	39
0.50	0.71	0.94	33
0.60	0.79	0.97	27
0.70	0.91	0.99	21
0.80	1.12	*[a]	14
0.90	1.58	*[a]	8

Notes:
[a] Power value is equal to or greater than 0.995.

significance) more directly based upon the relationship between the repeated measures as explained in the Technical appendix.

In Table 2.7, the second column illustrates the increment to the ES that occurs for varying correlations between the dependent observations, assuming an originally hypothesized value of 0.50. This effect becomes quite dramatic when the correlation can be assumed to be as high as 0.60 (which is not an unreasonable expectation for reliable measures), with the originally hypothesized ES rising to almost 0.80 and the power rising to 0.97 (assuming 50 subjects per group). Alternatively, the final column illustrates the decreases in required sample size necessary to produce a desired power level of 0.80, which decreases from 64 assuming no correlation to less than half ($N = 27$) assuming an r of 0.60.

Strategy 8: Hypothesizing main effects rather than interactions. Our rationale for this strategy is more conceptual than statistical and is based primarily upon the difficulties involved in hypothesizing interaction ES values. To illustrate the difference between a main effect ES and its interactional counterpart, let us begin with an extremely potent (and unlikely) hypothesized interaction emanating from the following pattern of means for a 2 (experimental vs. control) × 2 (males vs. females) design in which the standard deviation is held at one (Table 2.8).

Basically Table 2.8 reflects a scenario in which both the treatment and gender ES values as well as the treatment × gender interaction ES are all 0.50, although the hypothesized means themselves reflect a scenario in which the treatment is really only effective for male subjects. A closer examination of these hypothesized means, however, indicates that for this scenario

Table 2.8. *A hypothetical 2 × 2 interaction*

	B1 Males	B2 Females	Treatment means
A1 Treatment	1.00	0.00	0.50
A2 Control	0.00	0.00	0.00
Gender means	0.50	0.00	

Table 2.9. *A more veridical hypothetical 2 × 2 interaction*

	B1 Males	B2 Females	Treatment means
A1 Treatment	0.80	0.40	0.60
A2 Control	0.00	0.00	0.00
Gender means	0.40	0.20	

to occur in an experiment the treatment would need to be extremely potent for males, the documentation of which is difficult in practice because the *N* for the A1B1 cell is only one-half the *N* for the entire experimental group. (Obviously if this were a 3 × 3 design, the *N* for any single cell would be only one-third the *N* for the main effect.)

Since the pattern of means reflected in Table 2.8 would surely almost never occur in an actual experiment, let us hypothesize a more veridical interaction ES in which the treatment was expected to be effective for both genders, but more so for one than the other (Table 2.9).

A close examination of this study indicates that the investigator is still hypothesizing an extremely potent interaction in the sense that he/she is expecting the intervention to be twice as effective for males as it is for females. (In real world clinical research, such dramatic treatment × aptitude interactions are extremely rare.)

For this design, however, the ES for the treatment is considerably larger (0.60, since we are still assuming a standard deviation of 1.0) than the treatment × gender interaction, which is only 0.20 (see Chapter 9 or the Technical appendix for a description of how an interaction ES is calculated). Thus, for a study with a total *N* of 100 subjects, the power for the treatment vs. control contrast would be 0.84 while the power for the treatment × gender interaction would be only 0.16.

This is not to say that factorial studies should not be designed to test interactions. We are simply cautioning the reader to be quite careful in designing an experiment in which the primary hypothesis is tested via an

Figure 2.2. A single group, pretest–posttest design.

interaction *unless this hypothesis is guided by a strong theory*. (Actually we tender even this advice only for experiments involving between subject factors since it often does not apply to designs employing repeated measures.)

Strategy 9: Employing measures which are sensitive to change. The ES, as we have employed it to this point, is a descriptive statistic which indicates the expected improvement that an experimental group is capable of producing *over and above* that of a comparison group. Conceptually, this is no different from estimating how likely the same subjects are to change following the introduction of an intervention via a single group, pretest–posttest design (see Figure 2.2). The existence of a separate control group simply provides a procedurally much cleaner estimate of this effect by "controlling" for a number of experimental artifacts.

The purpose of an experiment, therefore, is to produce *changes* in a dependent variable by introducing an intervention that is theoretically capable of causing such a change. If the dependent variable is a relatively stable attribute, however, this purpose is not likely to be realized. Thus it behoves the investigator to select a dependent variable which has been shown in previous research to be sensitive to change and then to **pilot test** the specific measure, perhaps employing a small single group pretest–posttest design as depicted above to ensure that the dependent variable does indeed change as a function of the intervention.

As an example, suppose a researcher hypothesized that using a focused health education intervention targeted specifically at individuals with high serum cholesterol levels would result in substantive learning gains as opposed to the type of hit-or-miss educational advice normally afforded to these patients by their primary care providers. Our hypothetical investigator would have several options for choosing an appropriate learning measure. He/she might, for example, opt to use a comprehensive knowledge test developed by a national dietary association that possessed extremely high reliability and content/criterion-related validity due to its extensive use both in research and in patient education. Ironically, however, such a measure might be far less likely to document an effect for the tested intervention than a briefer, less reliable assessment tool constructed especially to measure the specific content which would be taught during the

Table 2.10. *The relationship between dependent variable (DV) sensitivity and power (N/group = 50; alpha = 0.05; two-tailed)*

Hypothesized ES	Increased DV sensitivity					
	0%	20%	40%	60%	80%	100%
0.20	0.16	0.22	0.28	0.35	0.43	0.51
0.50	0.70	0.84	0.93	0.98	0.99	>0.99

course of the experiment itself. (This even takes into consideration, as discussed in the next strategy, that everything else being equal, more reliable measures produce greater power than less reliable ones.) The reason is probably obvious: the nationally validated achievement test would contain large numbers of items basically irrelevant to the experimental curriculum, which would have the effect of "watering down" the study's ES.

What our researcher would be wise to do, therefore, is to construct an instrument tailored to his/her study that would be as sensitive as possible to the specific instruction being offered. One method of achieving this objective would be (a) to pretest a group of comparable subjects on as many items as possible which were relevant to the experimental curriculum, (b) to teach this group of students the said curriculum (perhaps using standard instructional methods as opposed to the experimental teaching method), and (c) to construct the final achievement measure based upon those items upon which performance changed most dramatically as a function of instruction. (While this is an extreme case, the bottom line here is that an investigator should employ dependent variables shown to be amenable to experimental change.) Table 2.10 illustrates the rather dramatic increases in power (especially for larger hypothesized ES values) attributable to increases in the sensitivity of a study's dependent variable.

This general principle even applies to correlational research even though no actual change in dependent variable is produced, since the investigator normally is still interested in whether or not the dependent variable is *capable* of being changed by the independent variables of interest.

Strategy 10: Employing reliable measures. Assuming that the dependent variable is sensitive to change (and theoretically linked to the independent variable), the more reliable this dependent variable is the greater the statistical power will be. To illustrate this relationship, let us assume that an ES was hypothesized based upon previous research in which the dependent variables employed had a reliability averaging 0.80 (which means that 20% of any given score generated by the measure in question

Table 2.11 *Changes in power as a function of changes in the reliability of the dependent variable (ES = 0.50, alpha = 0.05, N/group = 50)*

	Reliability		
ES	Hypothesized	Obtained	Power
0.50	0.90	0.90	0.70
0.47		0.80	0.64
0.44		0.70	0.58
0.40		0.60	0.51
0.37		0.50	0.45
0.53	0.80	0.90	0.75
0.50		0.80	0.70
0.47		0.70	0.64
0.43		0.60	0.57
0.39		0.50	0.49
0.57	0.70	0.90	0.89
0.54		0.80	0.76
0.50		0.70	0.70
0.46		0.60	0.62
0.42		0.50	0.55
0.62	0.60	0.90	0.87
0.58		0.80	0.82
0.54		0.70	0.76
0.50		0.60	0.70
0.45		0.50	0.60
0.69	0.50	0.90	0.93
0.65		0.80	0.89
0.60		0.70	0.84
0.55		0.60	0.78
0.50		0.50	0.70

was error and 80% was systematic in nature). If the reliability of the dependent variable measure actually employed in a study turns out to substantially lower than 0.80, then the study's power will be lower as well. Conceptually, this is because the standard deviation will be inflated by the concomitant increase in error and this, in turn, will lower the study's ES (since the mean difference between the two groups, which is systematic in nature, will not increase as a function of an error-inflated standard deviation). This relationship is illustrated in Table 2.11 in which the reliability obtained is contrasted with the one assumed when hypothesizing the study's ES. As indicated therein, power drops off relatively dramatically as the reliability of the dependent variable is reduced below the levels assumed in constructing the study's hypothesized ES. Thus a measure which the investigator expected to have a reliability of 0.80 (based upon either pilot work or the literature), and upon which assumption he/she based a hypothesized ES of 0.50, would produce an estimated power of 0.70 for 50 subjects per group. Should the actual reliability obtained in the study be considerably less than this, however, the "true" hypothesized ES would be lower as would the power of the study. If, for example, the actual reliability turned out to be only 0.50, then the power would be effectively reduced from 0.70 to 0.49 because of the concomitant reduction in the ES.

This principle also holds for correlational research, where it can be directly applied to *both* the independent and dependent variables. The relationship further holds for covariates and blocking variables in experimental designs, although power is not affected as dramatically by an unreliable covariate as it is by an unreliable *dependent* variable.

Strategy 11: Using direct rather than indirect dependent variables (or proximal rather than distal variables (Lipsey, 1990)). In both social and health research, power considerations often preclude employing what investigators consider to be the most important outcomes as a study's primary dependent variable. There are a number of reasons for this, including the length of time it takes for certain outcomes to manifest themselves and their relative rarity. Another factor that must be considered, however, is the inevitable loss of power associated with each step that a study's dependent variable is removed from the actual intervention employed. Let us return to our study designed to evaluate the effects of a dietary education intervention for individuals with high serum cholesterol levels. The "true" purpose of such an intervention would probably be to avoid adverse health consequences such as, say, the incidence of stroke and/or heart attacks. Ignoring the obvious procedural difficulties of waiting the requisite time for such outcomes to manifest themselves, this study would probably not have a great deal of power to detect an experimental effect upon such an outcome given the distance along the causal chain on which the occurrence of a stroke was removed from the educational intervention. The most appropriate way of estimating how many steps are involved in such a chain is to construct a theoretical model of the study itself. One possible model for this experiment is depicted in Figure 2.3, where the development of a stroke is six steps removed from the educational intervention itself.

Examination of this model indicates that the most direct (and hence the greatest accruing ES) would probably occur for a dependent variable measuring *learning* of the content of the experimental curriculum.

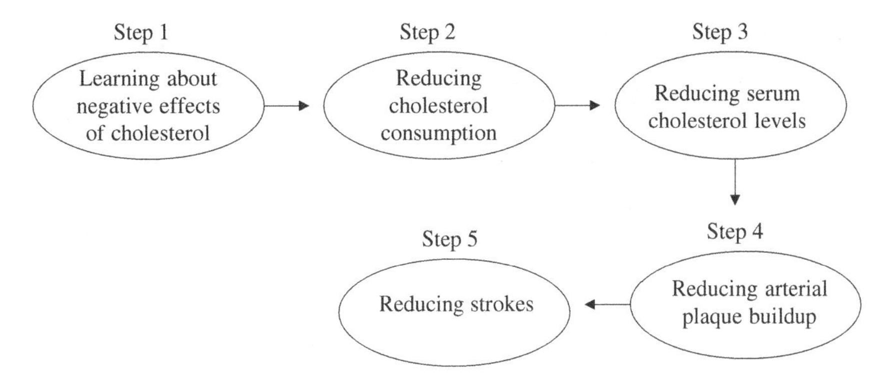

Figure 2.3. Dietary instruction designed to decrease cholesterol consumption.

Unfortunately this would not be a particularly interesting variable. Many health researchers would probably settle for changes in dietary behavior, however, while more physiologically oriented colleagues would consider changes in serum levels to be more meaningful.

Where the selected dependent variable falls along the above continuum is quite important because a study's ES is reduced at each point in this chain to the extent that the correlation between these successive outcomes is not perfect. (For example, some people may learn a great deal about the perils of a high cholesterol diet but not be able to change their behavior; some individuals may change their cholesterol intake as a function of the intervention but other physiological parameters exist which mitigate against the reduction of their serum cholesterol levels; some people may have high serum cholesterol levels but never develop clogged arteries due to protective genetic factors; some may have terribly blocked arteries but never suffer a stroke due to genetic or other as yet undiscovered factors.)

Table 2.12 illustrates the relationship between the number of steps two variables are separated from one another in a causal chain and the subsequent reduction in power based upon less than perfect correlations between the sequential measures representing these steps. Thus, if a medium ES were hypothesized for the intervention's effect upon knowledge and the researcher had access to 64 subjects per group, we already know that the researcher would have an 80% chance of obtaining statistical significance using this specific outcome variable. Table 2.12, however, can be used to estimate the concomitant loss in power should serum cholesterol levels be selected as the dependent variable rather than education if the correlation between these various dependent variables can be estimated. As an example, let us assume an average *r* of 0.90 between these steps (which is actually quite

Table 2.12. *The relationship between dependent variable "distance" from an intervention (N/group = 64; alpha = 0.05; two-tailed)*

r_{hyp} between steps	Two steps		Three steps		Four steps	
	ES	Power	ES	Power	ES	Power
1.00	0.50	0.80	0.50	0.80	0.50	0.80
0.90	0.45	0.71	0.40	0.61	0.36	0.52
0.80	0.40	0.61	0.31	0.41	0.41	0.29
0.70	0.35	0.50	0.24	0.27	0.27	0.16
0.60	0.20	0.37	0.18	0.17	0.17	0.09
0.50	0.24	0.27	0.12	0.10	0.10	0.05
0.40	0.20	0.20	0.08	0.06	0.06	0.04

high). Using Table 2.12 we would observe the ES reducing from 0.50 to 0.45 for the second step (changes in dietary behavior) with a subsequent drop in power from 0.80 to 0.71 (assuming an *N* per group of 64). Moving to reducing serum cholesterol levels (step 3), the power drops to 0.61, all of which should serve to illustrate the importance of selecting dependent variables as closely matched to the intervention as possible. (As always, this relationship holds for correlational as well as experimental research (e.g., a study interested in assessing the relationship between dietary knowledge and the incidence of stroke).) Naturally, these adjustments are extremely tenuous in the absence of good estimates of *r* between steps, but Table 2.12 can at least be used to alert the researcher to the perils of moving too far down the causal chain in the absence of large sample sizes or extremely high correlations between these steps. (It should be noted that these adjustments apply only to the scenario in which the originally hypothesized ES is based upon a dependent variable other than the one ultimately used. Thus, if a study were originally planned around the use of serum levels and the ES based upon previous literature involving that variable, then obviously Table 2.12 does not apply.)

Summary

All 11 strategies presented in this chapter have the capability of increasing a study's statistical power. All 11 also involve trade-offs of one sort or another which must be evaluated based upon scientific considerations. Still, each of these strategies should be considered at the design stage of any experiment because they all have the potential of increasing the sensitivity and validity of the hypothesis test itself.

Keeping the principles discussed in this chapter and the conceptual underpinning of the power analytic concept itself in mind, it is now time to discuss the steps actually required in computing a power analysis. These are illustrated in Chapter 3 and basically involve nothing more than (a) specifying the hypothesized ES, (b) choosing an alpha level (which we will assume to be 0.05), (c) choosing the statistical test which will be used to evaluate the hypothesis in question, and (d) turning to the chapter which presents power and sample size tables for that test.

3 General guidelines for conducting a power analysis

The purpose of this chapter is to provide some very brief guidelines for using the power tables in this book. We will illustrate their application within the context of the *t*-test tables presented at the end of Chapter 4 and a number of common sense principles.

Basically the power tables provided in the chapters which follow may be used to estimate (a) power given an available sample size and hypothesized ES, (b) the sample size required to achieve a given level of power for a hypothesized ES, or (c) the minimum detectable ES for a given level of power and sample size. (For convenience, exact sample size tables are also provided for the most common targeted power levels, ES values, and significance levels.) The following examples illustrate how the power tables may be used for each of these purposes.

Example 1: Calculating power. Suppose that an investigator (a) "knew" (we seldom know anything with any degree of certainty *before* we conduct an experiment) that 45 subjects per group would be available for his/her two-arm study, (b) estimated that his/her ES would be approximately 0.60 (i.e., the experimental group would differ from the control group by slightly more than one-half of a standard deviation), and (c) wished to know how much power would be available for the experiment. Since the *t*-test for independent samples would be the statistic of choice for such an experiment, the investigator would turn to Table 4.1 and simply locate the intersection of the *N*/group = 45 row and the ES = 0.60 column. This, as illustrated in Figure 3.1, would yield a power estimate of 0.80.

Example 2: Determining the *N*/group. Suppose instead our investigator (a) wished to know how many subjects per group he/she would need for (b) an estimated ES of 0.60 and (c) a desired power level of 0.80. The easiest way to accomplish this would be to use Table 4.2 Chart B in Chapter 4 and read the *N*/group value at the intersection of the 0.80 power row and the 0.60 ES column, which would indicate that 45 subjects per group would be needed. Alternatively, he/she could use Table 4.1 and locate the closest

n	Hypothesized ES									
	0.10	0.20	0.30	0.35	0.40	0.45	0.50	0.55	0.60	0.65
5	3	4	6	6	7	8	9	11	12	13
10	4	6	9	11	13	15	18	21	24	27
15	4	7	12	15	18	22	26	30	35	40
20	5	9	15	19	23	28	33	39	45	51
25	5	10	18	22	28	34	41	47	54	61
30	6	11	20	26	33	40	47	55	62	69
35	6	13	23	30	38	46	54	62	69	76
40	6	14	26	34	42	51	60	68	75	82
45	7	15	29	37	40	56	65	73	80	88
50	7	16	32	41	51	60	70	78	84	89

Figure 3.1. Determining power using a typical power table.

n	Hypothesized ES									
	0.10	0.20	0.30	0.35	0.40	0.45	0.50	0.55	0.60	0.65
5	3	4	6	6	7	8	9	11	12	13
10	4	6	9	11	13	15	18	21	24	27
15	4	7	12	15	18	22	26	30	35	40
20	5	9	15	19	23	28	33	39	45	51
25	5	10	18	22	28	34	41	47	54	61
30	6	11	20	26	33	40	47	55	62	69
35	6	13	23	30	38	46	54	62	69	76
40	6	14	26	34	42	51	60	68	75	82
45	7	15	29	37	40	56	65	73	80	88
50	7	16	32	41	51	60	70	78	84	89

Figure 3.2. Determining the *N*/group.

value (see our suggestions regarding estimation/interpolation below when exact values are not obtained) to 0.80 in the ES = 0.60 column. The necessary *N*/group could then be read in the left-most column as illustrated in Figure 3.2.

Example 3: Determining the minimum detectable ES. Finally, let us suppose that an investigator (a) wanted to know the smallest ES he/she could reasonably expect to produce statistical significance if (b) 45 subjects per group were used and (c) he/she desired at least an 80% chance of obtaining statistical significance. Here, our investigator would find the nearest value to

n	Hypothesized ES									
	0.10	0.20	0.30	0.35	0.40	0.45	0.50	0.55	(0.60)	0.65
5	3	4	6	6	7	8	9	11	12	13
10	4	6	9	11	13	15	18	21	24	27
15	4	7	12	15	18	22	26	30	35	40
20	5	9	15	19	23	28	33	39	45	51
25	5	10	18	22	28	34	41	47	54	61
30	6	11	20	26	33	40	47	55	62	69
35	6	13	23	30	38	46	54	62	69	76
40	6	14	26	34	42	51	60	68	75	82
(45)	7	15	29	37	40	56	65	73	(80)	88
50	7	16	32	41	51	60	70	78	84	89

Figure 3.3. Determining the minimum detectable ES.

0.80 in the *N*/group = 45 row in Table 4.1, and then read the minimum detectable ES at the top of the column associated with this value, which of course is 0.60 (Figure 3.3).

Estimating power values

The examples provided above all conveniently used exact values contained in the power table. In actual practice, of course, some degree of estimation or interpolation will often be required. At the risk of being redundant, it is important to remember that the primary purpose of a power analysis is to estimate one of the following three parameters: (a) the number of subjects needed, (b) the maximum detectable effect size, *or* (c) the available power at the design phase of an experiment based upon a fixed number of subjects and the hypothesized effect size. (This assumes that the alpha level is fixed; if it is not, it becomes a fourth parameter.) *Since the hypothesized ES is always an estimate (or guess), however, it is really quite spurious to attempt to predict (a) the necessary N per group to the nearest single subject or (b) the available power to the nearest hundredth.* What this means for the practicing researcher is that it is not necessary to have access to (a) a computer program that produces this level of precision or (b) power/sample size tables that contain all possible ES or *N*/group values. Simple estimation or rounding techniques are perfectly acceptable and may even be preferable if they remind their users that what is being generated is an *estimate*.

To illustrate, let us again use the independent samples *t*-test table as an example. Let us suppose that our investigator estimated that he/she would have 42 subjects per group for the above mentioned experiment and

n	Hypothesized ES									
	0.10	0.20	0.30	0.35	0.40	0.45	0.50	0.55	0.60	0.65
5	3	4	6	6	7	8	9	11	12	13
10	4	6	9	11	13	15	18	21	24	27
15	4	7	12	15	18	22	26	30	35	40
20	5	9	15	19	23	28	33	39	45	51
25	5	10	18	22	28	34	41	47	54	61
30	6	11	20	26	33	40	47	55	62	69
35	6	13	23	30	38	46	54	62	69	76
40	6	14	26	34	42	51	60	68	75	82
45	7	15	29	37	40	56	65	73	80	88
50	7	16	32	41	51	60	70	78	84	89

Figure 3.4. Interpolations for an *N*/group.

estimated the most likely ES to accrue would be 0.50. Unfortunately, as indicated in Figure 3.4, there is an *N*/group of 40 and 45 in Table 4.1, but no 42.

Hence the table "only" produces an exact estimate for the available power for values of *N*/group of 40 and 45, which are 0.60 and 0.65 respectively. The truth of the matter is, however, that either estimate (i.e., 0.60 or 0.65) is sufficiently accurate since the "true" ES is almost certainly not *exactly* 0.50 (e.g., it is surely almost as likely to be, say, 0.52 or 0.47 as it is to be 0.50). Hence we suggest that the users of these tables consistently employ *one* of three rounding strategies:

(1) Use the more conservative tabled value (in this case an *N* value of 40) which would yield an estimated power of 0.60.
(2) Use a linear or estimated interpolation of some sort. This could involve "splitting the difference" and estimating the available power to be $(0.60+0.65)/2=0.63$ (rounded) or an actual linear interpolation (e.g., 42 is 2/5 or 0.40 of the distance between 40 and 45, hence the estimated power is equal to $0.40(0.65-0.60)=0.02$, which when added to the power of 0.60 obtained for the *N*/group of 40 yields 0.62 $(0.60+0.02)$), or
(3) Simply estimate the closest tabled value (which in this case would be a power level of 0.60 since the *N*/group of 42 is closer to 40 than it is to the *N*/group of 45).

Any of these estimates (i.e., 0.60, 0.62, 0.63, or even 0.65) is perfectly acceptable. There are relatively rare situations, it is true, in which reasonably large differences do exist between tabled values. If an investigator feels relatively confident in his/her estimated parameters in such an instance, then option (2) will provide more precision. Alternatively, the formulas provided in the Technical appendix (in this case Formula 1.4) may be used to provide *exact* values, although this is seldom if ever warranted. In the examples that follow we will usually employ a linear interpolation (i.e., strategy (2)) for the sake of consistency.

Recommended power and alpha values

Two of the parameters that must be considered in conducting a power analysis, the acceptable level of power and the significance criterion (alpha), are often set by conventions within the investigator's discipline. Almost without exception the alpha level is set at $p \leq 0.05$ and the minimum acceptable power level is most often considered to be 0.80.

Certainly there are legitimate reasons to use different values for both of these parameters, but there is no question that these are the most commonly employed. A power level of 0.80, for example, means that, if everything goes as planned (i.e., if the hypothesized ES is correct and the study is run properly), the experiment has an 80% chance of achieving statistical significance and a 20% chance of not achieving statistical significance. There are occasions, however, when the stakes are too high to accept odds such as this and the investigator (or his/her funding agency) may not accept a failure rate of anything more than 10% (power = 0.90) or even 5% (power = 0.95).

There is usually less latitude with respect to the alpha level, but occasions exist in which more lenient or more stringent values are justified. (An alpha level of 0.05, it will be remembered, means that 95% of the time a statistically significant difference will indeed be "real.") Here again it is conceivable that this 5% error rate might be considered excessive, hence an investigator might opt for a somewhat lower value – such as 1%. (Similarly, if an experiment involves multiple statistical evaluations of many outcome variables a more stringent alpha level might be employed.) Researchers seldom have the resources to employ alpha levels of 0.01 or below, however, because they require such larger sample sizes (see Chapter 2). A far more likely scenario, therefore, is to employ a one-tailed alpha level of 0.05 which requires fewer subjects *if* the investigator (and his/her audience) are convinced that there is only one reasonable direction for the study outcomes to take *and* if there is no practical or theoretical reason to evaluate statistical significance when the results are not in this direction.[1] There are occasions on which this is completely justifiable (e.g., in the presence of strong theory or prior research), although the investigator must be willing to consider any

Chart A. Independent sample t-test at alpha = 0.05																			
Power	Hypothesized ES																		
	0.20	0.30	0.35	0.40	0.45	0.50	0.55	0.60	0.65	0.70	0.75	0.80	1.00	1.25	1.50	1.75	2.00	2.50	3.00
0.80	394	176	130	100	79	64	54	45	39	34	30	26	17	12	9	7	6	4	4
0.90	527	235	173	133	105	86	71	60	51	45	39	34	23	15	11	9	7	5	4
Chart B. Independent sample t-test at alpha = 0.10																			
Power	Hypothesized ES																		
	0.20	0.30	0.35	0.40	0.45	0.50	0.55	0.60	0.65	0.70	0.75	0.80	1.00	1.25	1.50	1.75	2.00	2.50	3.00
0.80	310	139	102	79	62	51	42	36	31	27	23	21	14	9	7	6	5	4	3
0.90	430	192	141	108	86	70	58	49	42	35	32	28	18	12	9	7	6	4	4

Figure 3.5. Using the sample size tables.

difference in the non-hypothesized direction, no matter how great, as non-significant. This is a heavy price to pay, however, hence this option is used relatively rarely. *The bottom line, then, is that under all but the most unusual circumstances, experiments should be designed to achieve a power level of 0.80 and use a two-tailed alpha of 0.05.*

Since providing power tables for both a one-tailed alpha of 0.05 and a two-tailed alpha of 0.01 would triple the number of tables required for this book, we have opted to provide exact sample size tables for the two most commonly employed levels of desired power (0.80 and 0.90) and alphas of both 0.05 and 0.10 (which is equivalent to a one-tailed 0.05 level).

To illustrate the use of these tables, let us assume that our investigator wanted to know exactly how many subjects would be required to achieve a power level of 0.80 for all three of these different levels of statistical significance. Let us further assume that the hypothesized ES was 0.50 and that an independent samples t-test was again to be employed.

Figure 3.5 presents a portion of the sample size table that would be used for this purpose. (One advantage of these tables is that, if the desired power level is indeed 0.80 or 0.90, there is usually no need to interpolate with respect to the necessary number of subjects needed in each group.) The N/group, then, that would be necessary to assure a power level of 0.80 at a two-tailed alpha of 0.05 would be found at the intersection of the 0.80 power row and the 0.50 ES column in Chart A (alpha = 0.05), which would be 64. For an alpha of 0.10, Chart B would be employed with the resulting N/group being 51, obviously a considerable saving in sample size.

Hypothesizing the effect size

If the power analytic process has an Achilles heel, it is undoubtedly the process of hypothesizing an effect size. It is far easier to hypothesize which group(s) will be superior to which other group(s) than it is to predict the *magnitude* of this difference. From a power analytic perspective, however, the hypothesis being tested is not simply that "the experimental group mean will be significantly greater than that of the control group." Instead the hypothesis is in effect that "the experimental group mean will be a certain number of standard deviation units greater than that of the control group." (Note that if multiple groups or multiple dependent variables are employed, an experiment is almost certain to have more than one ES and hence more than one power analysis will need to be conducted.)

We have discussed defensible methods of estimating an ES (e.g., meta-analyses or primary trials involving similar interventions and outcome variables and/or pilot studies), although all too often investigators determine their ES values based upon how many subjects will be available or how many they can afford to run in order to achieve a specified level of power. An ES generated in this manner, or a power analysis performed based upon an unrealistic ES, is completely worthless at best and fraudulent at worst. Carefully estimating a realistically obtainable ES is an integral part of the scientific process and as such should be justified and documented with great care and explicitness. When there is truly no information that can be brought to bear with respect to hypothesizing an effect size, then for the social and behavioral sciences there is considerable justification for employing an effect size of 0.50 as discussed earlier (e.g., Cohen, 1988; Lipsey, 1990; Lipsey & Wilson, 1993). It is doubtful, however, that many major funding agencies would accept this guideline and most certainly agencies such as the National Insititutes of Health (NIH) or National Science Foundation (NSF) would require a more explicit justification. Regardless of the method used to arrive at a hypothesized ES, the investigator should always present the rationale for his/her final decision(s) clearly and explicitly. If published research is employed, the citation should be provided, the actual summary statistics should be reproduced, and the reasons that these values may and may not apply to the proposed study should be explicated. If a pilot study has been used, the methods (including sample size and summary statistics) should also be presented in the study's Institutional Review Board (IRB) and/or grant application.

Selecting an appropriate statistical procedure

There are always numerous options for analyzing the same data. Generally speaking, the most powerful, classic parametric procedures available have been presented in this book based upon the philosophy (generally supported by Monte Carlo studies) that violations of their assumptions do not produce spurious values or incorrect inferences. We do not, therefore, quibble regarding whether or not a dependent variable is measured at the ordinal or interval level. Instead, if a variable is continuous (as opposed to categorical) we assume that the data will be subjected to one of these procedures, regardless of whether we can justify, say, that the intervals between the possible values are strictly equal to one another.

This is not a universally accepted view, of course, hence some statisticians insist on the use of non-parametric or other options for which convenient power tables are not generally available. What we suggest, therefore, is that an investigator who plans to employ, say, a Mann–Whitney U test based upon the belief that the data are more ordinal than interval (or that they are not normally distributed) still employ the tables presented in Chapter 4 as an estimate of the power available (or sample size needed). Alternatively, should a slightly more sophisticated design (e.g., one employing a nested factor) be employed, or a new statistical approach used for which power calculation procedures have not yet been developed, we would suggest that the closest analog present in the chapters that follow be employed. As we have emphasized before, a power/sample size analysis is an estimate at best and is not exact because the hypothesized effect size is far from an exact quantity under even the best circumstances. (Should the investigator or his/her statistician feel especially insecure about such an estimate, the computation can be based upon a desired power of 0.90 instead of 0.80 with the reason for this change being explicitly stated when the power analysis is reported.) Also, as a general rule of thumb, if an investigator estimates power based upon a slightly less sophisticated model than the one actually employed (and also states this clearly when the results of the power analysis are described), then he/she is not likely to be criticized because the resulting power/sample size values would err only *slightly* on the conservative side.

Using pairwise (1 df) contrasts

When more than two groups or arms are employed in an experiment, the primary comparisons of interest are the individual comparisons between the various groups involved. Thus if an experiment were to employ four groups, our statistical textbooks tell us that the most appropriate way to analyze the data resulting from this experiment would be an analysis of variance procedure that would produce an *F*-ratio with three degrees of freedom in the numerator. If this statistic were found to be statistically significant, the next step would be to determine which of the four groups differed significantly from one another – thereby producing six pairwise contrasts.

From a power analytic perspective, the power present to produce a statistically significant *F*-ratio is largely irrelevant, however, since the primary purpose of the experiment lies embedded in these individual contrasts. (Determining the power of an *F*-ratio is analogous to determining the power of the multiple *R* in a regression study where the primary question of interest is whether or not a particular independent variable is *independently* related to an outcome variable. In the latter case, the statistic that is of interest is the individual beta weight associated with the independent variable, not the multiple *R* – which addresses the question of whether or not the *set* of included independent variables are significantly related to the dependent variable of interest.)

In the case of the omnibus *F*-ratio for a four-group study, then, it is not the pattern or spread of all of the mean differences involved that is of interest, but the power for each of the six individual contrasts that is crucial (most of which will incidentally have completely different hypothesized ES values and hence be associated with completely different power/sample size estimates). For this reason, the tables provided in Chapters 6 through 8 (which deal with between subject, analysis of covariance, and repeated measures designs respectively) allow the reader to compute the power of individual pairwise mean differences as well as the overall *F*-ratio.

The importance of modeling different power parameters

Since the ES and available *N* are usually only estimates, it is wise to model different values for these parameters as well as different analytic/design strategies when performing a power analysis. In other words, since one never knows for sure exactly what a study's final ES will be (and we seldom know how many subjects will be eligible or will agree to participate), it makes sense to produce multiple power/sample size estimates based upon different experimental scenarios. When these scenarios are equally probable, the final power/sample size estimates should be presented either as a range or as the most conservative estimate generated (i.e., the largest required sample size, the smallest maximum detectable ES, or the largest *N*/group).

The primary benefit generated from modeling such as this,

however, is to determine those design components that could potentially be added to a study to assure the availability of sufficient power – which happens to be far and away the most important contribution the power analytic process makes to research itself.

As one example of this principle, suppose that an investigator's best estimate for an ES was indeed 0.50, but he/she knew that only 40 subjects per group for a two-group trial would be available (or that resources existed only to recruit and run 40 subjects per group). Use of Table 4.1 would indicate that the power available for such a study if it were to be analyzed via an independent samples *t*-test would be 0.60, and Table 4.2 Chart B would indicate that another 24 subjects per group would be needed to achieve a desired power level of 0.80.

For anything but a pilot study, a 60% chance of achieving statistical significance is really not adequate, hence (assuming that the sample size is indeed fixed), our hypothetical investigator needs to explore his/her options to see whether indeed a more reasonable power estimate can be obtained given these constraints. Referral to Chapter 2 provides the full range of options available, but from the perspective of the tables presented in this book there are basically three genres of variable that can be modeled.

Modeling genre 1: Relaxing the significance level from 0.05 to 0.10 (in the relatively rare environment in which this would be permitted). To ascertain the effect of this modification, Table 4.2 Chart C could be employed. Locating the intersection between the power = 0.80 row and the 0.50 ES column, producing a sample size requirement of 51 subjects per group, which in this particular experiment does not particularly help the investigator.

Modeling genre 2: Changing the design from a between subject to a within subject trial (in the equally rare situations in which the same or a priori matched subjects can be employed in both intervention and control groups). As will be described in more detail in Chapter 5, an additional parameter must be hypothesized for this scenario: the correlation between subjects' dependent variable scores under the two different treatments (e.g., the experimental vs. control groups). Table 5.6 allows the investigator to model several options, here, which indicate that when the *r* is as low as 0.40, only 40 total patients are required to achieve a power level of 0.80. This is indicative of how efficient this strategy is, although it is an option for only a relatively limited range of studies (e.g., those for which the treatment effect is transitory and subjects' dependent variable scores return to baseline values).

Modeling genre 3: Employing a baseline (pretest) measure of the dependent variable (or other comparable variable) collected on all subjects prior to the introduction of the intervention (which usually is feasible). For this scenario, different subjects are still randomly assigned to the two groups, but all subjects are measured prior to treatment implementation. As with the within subject option above, a correlation between the baseline and end-of-treatment measures must be hypothesized (as described in Chapter 7) and, depending upon the proposed *r* (which may be ascertained via a pilot study), either Table 7.1 (for $r-0.40$) or Tablc 7.2 ($r=0.60$) can bc employed if the baseline measure is conceptualized as a covariate. In the first case, for an ES of 0.50 and 40 subjects per group, the power would rise to 0.68 and in the second it would become an acceptable 0.80, which the investigator might select as the most viable approach; of course he/she might also prefer one of the options presented in Chapter 2 (e.g., changing the intervention or control group to increase the ES, increasing the reliability of sensitivity of the dependent variable, or combining strategies such as relaxing the alpha level and using a covariate).

An algorithm for selecting among analytic options available in later chapters

Basically, three of the next five chapters are devoted to options 2 and 3 above. Although each chapter begins with an explicit description of the types of scenarios in which its tables are appropriate, we also offer the following algorithm for facilitating the process of determining which analytic procedure is most appropriate for any given study design.

For within subject (repeated measures) designs

- For two-group studies involving only a single administration of the dependent variable or one-group studies employing baseline and end of treatment (EOT) measures, see Chapter 5.
- For three- to five-group studies involving only a single administration of the dependent variable, see Chapter 8.
- For two- to five-group studies employing another between subject grouping variable (or for two- to five-group studies employing different subjects in the treatment groups but that employ multiple administrations of the dependent variable (e.g., baseline, EOT, and follow-up)), see Chapter 9.

For between subject designs

- For two-group studies involving only a single administration of the dependent variable, see Chapter 4.
- For three- to five-group studies involving only a single administration of the dependent variable, see Chapter 6.
- For two- to five-group studies involving a baseline measure or other variable that will be employed as a covariate, see Chapter 7.
- For two- to five-group studies that also contain a second grouping variable (with or without a covariate), see Chapter 9.

For other designs

- See Chapter 10, or simply employ the nearest analog discussed therein or in one of the previous chapters (always explaining this compromise when reporting the results of the power analysis).

Reporting the results of a power analysis

Since even the most simple power analysis involves specifying three parameters (i.e., the *N*/group, the ES, and the alpha level) in addition to the statistical procedure that will be employed to ascertain statistical significance or the lack thereof, it follows that all of these factors should be described when the results are presented. In addition, the ES should be justified by telling the reviewer of the research protocol, in a few succinct paragraphs, exactly how it was derived – such as via a pilot test, a meta-analysis, or a very similar experiment with appropriate citations. (Normally the other parameters involved in a power analysis do not need justification unless an alpha level other than 0.05 is to be employed. The statistical procedure and the alpha level should be mentioned, however, if they are not specified elsewhere in the proposal.)

Most journals do not require the explicit ES justification that we are describing here (although they probably should), but almost all funding agencies do. The important point to remember in describing and justifying one's hypothesized ES, however, is to be honest and realistic. To be otherwise increases the probability of the proposed study's *failure*, and the most immediate victim of this failure is the investigator who will spend a significant portion of his/her life conducting an unsuccessful trial. (Ultimately, of course, those members of society who might have profited from the discovery of a successful intervention suffer perhaps the greatest loss.)

The power analysis itself, along with a reference to alert the reader to the specific method employed in arriving at the final results, may be described in a single sentence, such as:

> A power analysis (Bausell & Li, 2002) indicated that 64 subjects per group would result in an 80% chance of obtaining statistical significance via an independent samples *t*-test ($p \leq 0.05$) assuming the ES of 0.50 observed in our previously described pilot study.

Alternatively, if the necessary sample size were derived for a desired level of power, this sentence might read as follows:

> Assuming the use of an independent samples *t*-test, a power analysis (Bausell & Li, 2002) indicated that 64 subjects per group would be needed in order to produce an 80% chance of obtaining statistical significance at the 0.05 level for the 0.50 ES observed in our previously reported pilot study (Jones & Smith, 2000).

As the complexity of experimental designs increases, additional parameters are involved in the calculation of power (e.g., the correlation between the independent variable and the covariate in an analysis of covariance), hence they too must be specified and justified. (Examples of how the results of these power/sample size analyses can be reported are presented in the chapters that follow.) Often it is difficult to justify adequately all of the values needed in some of these designs, especially in the absence of a pilot study, since they are seldom found in published reports. In these instances, modeling different parameter values can be especially helpful in arriving at reasonable power or sample size estimates. Additional suggestions for this modeling process (for which the provided templates are especially helpful) are presented in the sections devoted to the relevant statistical procedures.

Summary

This chapter has provided guidelines for using the tables presented in the next six chapters to estimate power, sample size, and the maximum detectable ES. A case is made for the importance of explicitly stating one's assumptions and rationales made during the course of a power analysis, as well as being extremely forthright and honest in the reporting of this information. Modeling suggestions are provided and an algorithm presented that directs the reader to the appropriate chapter for more information regarding the analytic options available.

Endnote

1 A one-tailed test assumes that any outcome in the opposite direction than the one hypothesized is spurious and therefore cannot be statistically significant. Thus if an investigator uses a one-tailed test to evaluate the effect of an intervention and the control group is actually superior, the inferential conclusion must be that there is no statistically significant difference between the two treatments no matter how large this difference turns out to be. By implication, therefore, use of a one-tailed test is inappropriate for an experiment in which it is also important to evaluate potential *harmful* effects of an intervention.

4 The *t*-test for independent samples

Purpose of the statistic

The *t*-test for independent samples is used to ascertain how likely an observed mean difference between two groups would be to occur by chance alone. The groups may be experimental conditions to which subjects have been randomly assigned (e.g., an intervention vs. a control), a naturally occurring dichotomy (e.g., a comparison among males and females), or a binary comparison of any sort (e.g., smokers vs. non-smokers) as long as the two groups are *not* made up of the same or matched[1] subjects.

The independent samples *t*-test, then, is used when:

(1) there is a single, dichotomous independent variable (e.g., two discrete groups),
(2) the dependent variable is continuous in nature,
(3) the hypothesis to be tested is expressed in terms of a *mean difference*, and
(4) the subjects or observations contained in the two groups are independent of one another (e.g., they have been randomly assigned or assigned by some non-matched manner).

The independent samples *t*-test is *not* used when:

(1) there is more than one independent variable such as a covariate (which is used strictly for statistical control purposes) or a blocking variable (which is used to ascertain if one *type* of subject responds differently to the intervention),
(2) this single independent variable is continuous in nature or contains *more* than two groups,
(3) the hypothesis is expressed in terms of a relationship between variables rather than a difference between groups,
(4) the dependent variable is categorical in nature, or
(5) the two groups contain the same or matched subjects.

The *t*-test tables

Table 4.1 presents a power table that can be used for most two-group experiments that fit the criteria listed above. As described in Chapter 3, this table may be used (with appropriate estimating/interpolating steps) to estimate (a) the power that will be available for a wide range of hypothesized ES values or (b) the minimum detectable ES given a fixed *N*/group and a desired level of power.

Table 4.2 presents the exact sample size needed to achieve two commonly employed levels of power, 0.80 and 0.90 and three alpha levels: 0.01, 0.05, and 0.10 (which is equivalent to a one-tailed test of significance at the 0.05 level). To use this table the investigator need only locate the intersection between the row associated with the level of power and the column associated with the hypothesized ES in the appropriate chart (which in most cases will be Chart B). Thus, should an investigator wish to determine how many subjects per group would be needed to provide an 80% chance of achieving statistical significance for a hypothesized ES of 0.50 at the 0.05 level of significance, he/she would simply locate the intersection of the 0.80 power row and the ES = 0.50 column in Chart B (yielding a required *N*/group of 64). Alternatively, if a power level of 0.90 were desired, this *N*/group would rise to 86 subjects. Should a one-tailed test be desired, Chart C would provide respective values of *N*/group of 51 and 70. (Note that the necessary *N*/group becomes quite small for large ES values, which are sometimes applicable to laboratory research involving animals.)

Templates for the independent samples *t*-test

Conducting a power analysis for an experiment that can be analyzed via a *t*-test for independent samples involves so few parameters that a template is hardly necessary except for modeling purposes. For the latter, we have provided a template that allows the use of (a) three different hypothesized ES values, each with three different values of *N*/group, to be modeled or (b) three different hypothesized ES values to be modeled for the desired power level. Obviously modeling such as this is only indicated when different values of ES or *N*/group are feasible.

Although the use of the power and sample size tables in this chapter has already been illustrated in Chapter 3, we will provide a further example to demonstrate the use of Template 4.1. Let us assume that an investigator wished to evaluate an educational program developed by the Arthritis Foundation with respect to its effect upon increasing functional ability in

Chart 4.1. A modeling strategy employing Template 4.1 (*t*-test for independent samples)

Hypothesized ES =		0.40			0.50			0.60	
N/group =	50	60	70	50	60	70	50	60	70
Power = from Table 4.1 or 4.2	0.51	0.58	0.65	0.70	0.77	0.83	0.84	0.90	0.94

individuals recently diagnosed with rheumatoid arthritis. Based upon a pilot study employing 12 subjects who took the course over a six-week period, the investigator's best estimate for this program's ES was hypothesized to be four-tenths of a standard deviation. Let us further assume that he/she estimated that 50 patients per group would probably be available for this study and that logistic issues precluded the possibility of obtaining baseline measures on any of the patients.

Following steps 1 and 2 in Template 4.1, then, would produce the following results:

Step 1. To estimate the power, specify one or more hypothesized ES values in the blanks below as well as the *N*/group available for the experiment. (To estimate the required *N*/group for a desired level of power of 0.80 or 0.90, go to step 3.)

Hypothesized ES = 0.40

N/group = 50

Step 2. For each set of parameters listed in step 1, find the available power at the intersection of the ES column and the *N*/group row in Table 4.1. Interpolate as necessary.

Power (Table 4.1) = 0.51

Not being particularly impressed by this value, our investigator's next step might be to consult Table 4.2 (steps 3 and 4) to ascertain how many patients would be necessary to achieve a reasonable level of power (say, 0.80). Locating the intersection of the 0.80 power row and the ES = 0.40 column indicates this value to be 100 subjects per group, which was deemed to be too large an experiment to be feasible.

Faced with this problem, the investigator decided that the best approach would be to model different ES values and more realistic values of *N*/group, since it might be possible to increase the ES by increasing the length of the course and possibly by selecting individuals who had been diagnosed with the disease for a longer period of time. Chart 4.1 gives the results of this modeling exercise. The results indicated to the investigator that a more feasible approach might be to increase the strength (duration) of the intervention to the point at which the hypothesized ES would be 50% greater than the originally projected 0.40, thereby allowing him/her to conduct the experiment with the originally projected *N*/group.

As indicated, our investigator finds that an acceptable power level only begins to manifest itself if an ES of 0.50 can be achieved and even then a slightly increased sample size is indicated. It would be necessary to double the group size to achieve a power of 0.80 if the originally hypothesized ES of 0.40 is still deemed the most realistic. (Alternatively, our investigator could ascertain what type of ES would be required to yield a desired power of 0.80 for his/her fixed *N*/group of 50 by locating the nearest value to 50 in the appropriate chart in Table 4.2 and finding its associated ES column. Here, he/she would find that 50 subjects fell about half-way between the 0.55 and 0.60 ES columns, indicating that the study would need to be designed to obtain an ES within this interval if nothing else varied.)

Usually, of course, an investigator would have additional options, the most obvious being to measure the subjects at baseline on the dependent variable, which changes the analytic model from an independent samples *t*-test to, among other options, a two-group ANCOVA. By simply turning to Chapter 7 he/she learns that the trial's power increases to 0.71, 0.79, and 0.84 for the different modeled values of *N*/group (i.e., 50, 60, and 70) for an ES of 0.40, assuming that the covariate–dependent variable relationship is hypothesized to be approximately 0.60 (Table 7.2).

Alternatively some of the additional options listed in Chapter 2 might be appropriate, but the point to be emphasized here is that (a) modeling is almost always indicated since the input parameters for a power analysis are almost always estimates and (b) the most powerful statistical model available should be employed (although methodological and scientific constraints always take precedence in this decision). It should also be noted that an additional benefit of modeling power and sample size estimates is to warn the investigator what will occur if his/her parameter estimates tend to be overestimates, hence it is always wise to build in some room for error in this direction if this is deemed likely.

Summary

Power and sample size tables are presented for a wide range of value of *N* and hypothesized ES. In addition, a power/sample size template is used to allow the modeling of different ES, *N*/group, and power parameters.

Endnote

1 An example of matching within an experimental context is the scenario in which a continuous variable exists on the subjects prior to random assignment. Subjects are rank ordered on this variable and then randomly assigned in blocks of two to either the experimental or the control group, beginning with the highest two subjects and continuing through to the subjects with the lowest two scores.

Template 4.1. Power and sample size template for the independent samples *t*-test

---

Step 1. To estimate the power, specify one or more hypothesized ES values in the blanks below as well as the *N*/group available for the experiment. (To estimate the required *N*/group for a desired level of power of 0.80 or 0.90, go to step 3.)

Hypothesized ES = ___ ___ ___

N/group = ___ ___ ___ ___ ___ ___ ___ ___ ___

Step 2. For each set of parameters listed in step 1, find the available power at the intersection of the ES column and the *N*/group row in Table 4.1. Interpolate as necessary.

Power (Table 4.1) = ___ ___ ___ ___ ___ ___ ___ ___ ___

Step 3. For the estimated *N*/group necessary for a fixed level of power, enter the desired power level and the hypothesized ES.

Desired power = ___

Hypothesized ES = ___ ___ ___

Step 4. For desired powers other than 0.80 or 0.90, locate the closet value to the targeted power in the ES column of Table 4.1 and read the necessary *N*/group in the left-most column. (Interpolate as desired.) For a power of 0.80 or 0.90, employ the chart in Table 4.2 that corresponds to the desired alpha level (in most cases this will be Chart B, alpha = 0.05). Locate the required *N*/group at the intersection of the desired power row and the ES column. Interpolate as desired.

N/group = ___ ___ ___

Table 4.1. Power table for the independent samples t-test at alpha = 0.05

n	Hypothesized ES																		
	0.20	0.30	0.35	0.40	0.45	0.50	0.55	0.60	0.65	0.70	0.75	0.80	1.00	1.25	1.50	1.75	2.00	2.50	3.00
5	4	6	6	7	8	9	11	12	13	15	17	18	27	39	52	66	77	92	98
6	5	6	7	8	10	11	13	14	16	18	20	23	33	48	63	76	87	97	.
7	5	7	8	10	11	13	15	17	19	21	24	27	39	56	72	84	92	99	.
8	5	8	9	11	12	14	17	19	22	25	28	31	45	63	79	90	96	.	.
9	6	8	10	12	14	16	19	21	24	28	31	35	50	69	84	93	98	.	.
10	6	9	11	13	15	18	21	24	27	31	34	38	55	74	88	96	99	.	.
11	6	9	11	14	16	19	22	26	30	34	38	42	60	79	91	97	99	.	.
12	7	10	12	15	18	21	24	28	32	37	41	46	64	83	94	98	.	.	.
13	7	11	13	16	19	22	26	30	35	39	44	49	68	86	95	99	.	.	.
14	7	11	14	17	20	24	28	33	37	42	47	52	71	89	97	99	.	.	.
15	7	12	15	18	22	26	30	35	40	45	50	55	75	91	98	.	.	.	.
20	9	15	19	23	28	33	39	45	51	57	63	69	87	97	.	.	.	.	.
25	10	18	22	28	34	41	47	54	61	68	73	79	93	99	.	.	.	.	.
30	11	20	26	33	40	47	55	62	69	76	81	86	97	.	.	.	.	.	.
35	13	23	30	38	46	54	62	69	76	82	87	91	98	.	.	.	.	.	.
40	14	26	34	42	51	60	68	75	82	87	91	94	99	.	.	.	.	.	.
45	15	29	37	46	56	65	73	80	86	91	94	96	.	.	.	.	.	.	.
50	16	32	41	51	60	70	78	84	89	93	96	98	.	.	.	.	.	.	.
55	18	34	44	55	65	74	81	88	92	95	97	99	.	.	.	.	.	.	.
60	19	37	47	58	68	77	85	90	94	97	98	99	.	.	.	.	.	.	.
65	20	39	51	62	72	81	87	92	96	98	99	99	.	.	.	.	.	.	.
70	22	42	54	65	75	83	90	94	97	98	99	.	.	.	.	.	.	.	.
75	23	45	57	68	78	86	92	95	98	99	.	.	.	.	.	.	.	.	.
80	24	47	59	71	81	88	93	96	98	99	.	.	.	.	.	.	.	.	.
90	26	52	65	76	85	92	96	98	99	.	.	.	.	.	.	.	.	.	.
100	29	56	69	80	89	94	97	99	.	.	.	.	.	.	.	.	.	.	.
110	31	60	73	84	91	96	98	99	.	.	.	.	.	.	.	.	.	.	.
120	34	64	77	87	93	97	99	.	.	.	.	.	.	.	.	.	.	.	.
130	36	67	80	89	95	98	99	.	.	.	.	.	.	.	.	.	.	.	.
140	38	71	83	92	96	99	.	.	.	.	.	.	.	.	.	.	.	.	.
150	41	74	86	93	97	99	.	.	.	.	.	.	.	.	.	.	.	.	.
175	46	80	90	96	99	.	.	.	.	.	.	.	.	.	.	.	.	.	.
200	51	85	94	98	99	.	.	.	.	.	.	.	.	.	.	.	.	.	.
225	56	89	96	99	.	.	.	.	.	.	.	.	.	.	.	.	.	.	.
250	61	92	97	99	.	.	.	.	.	.	.	.	.	.	.	.	.	.	.
300	69	96	99	.	.	.	.	.	.	.	.	.	.	.	.	.	.	.	.
400	81	99	.	.	.	.	.	.	.	.	.	.	.	.	.	.	.	.	.
500	88	.	.	.	.	.	.	.	.	.	.	.	.	.	.	.	.	.	.

Table 4.2. Sample size table for the independent sample *t*-test

Chart A. Independent sample *t*-test at alpha = 0.01																			
Power	Hypothesized ES																		
	0.20	0.30	0.35	0.40	0.45	0.50	0.55	0.60	0.65	0.70	0.75	0.80	1.00	1.25	1.50	1.75	2.00	2.50	3.00
0.80	586	262	193	148	118	96	80	67	58	50	44	39	26	17	13	10	8	6	5
0.90	746	333	245	188	149	121	101	85	73	63	55	49	32	21	16	12	10	7	6
Chart B. Independent sample *t*-test at alpha = 0.05																			
Power	Hypothesized ES																		
	0.20	0.30	0.35	0.40	0.45	0.50	0.55	0.60	0.65	0.70	0.75	0.80	1.00	1.25	1.50	1.75	2.00	2.50	3.00
0.80	394	176	130	100	79	64	54	45	39	34	30	26	17	12	9	7	6	4	4
0.90	527	235	173	133	105	86	71	60	51	45	39	34	23	15	11	9	7	5	4
Chart C. Independent sample *t*-test at alpha = 0.10																			
Power	Hypothesized ES																		
	0.20	0.30	0.35	0.40	0.45	0.50	0.55	0.60	0.65	0.70	0.75	0.80	1.00	1.25	1.50	1.75	2.00	2.50	3.00
0.80	310	139	102	79	62	51	42	36	31	27	23	21	14	9	7	6	5	4	3
0.90	430	192	141	108	86	70	58	49	42	35	32	28	18	12	9	7	6	4	4

5 The paired *t*-test

Purpose of the statistic

The paired *t*-test (also called the correlated *t*-test and the *t*-test for dependent means) is used to ascertain how likely the difference between two means that contain the same (or matched) observations is to occur by chance alone. These means may represent pretest–posttest differences involving the same group of subjects, posttest differences when subjects are randomly assigned to two groups in pairs based upon a pre-existing variable (or a pretest), or differences between two scores available on the same group of subjects in non-experimental research.

The paired *t*-test, then, is used when:

(1) there are two continuous sets of numbers, and
(2) the hypothesis to be tested is expressed in terms of a *mean difference* between these two sets of numbers.

The paired *t*-test is *not* used when:

(1) the hypothesis to be tested is expressed in terms of whether or not these two sets of continuous numbers are related to one another,
(2) there are more than two continuous sets of numbers (e.g., when there are pretest and posttest scores available on two or more groups),
(3) there is another independent variable of interest besides the contrast between paired observations (e.g., it is desired to contrast a single group of subjects in a second manner, such as males vs. females), or
(4) the two sets of continuous numbers are independent of one another (i.e., are not generated from the same group of subjects or matched pairs of subjects).

The paired *t*-test normally yields significantly more power than an independent samples *t*-test, especially when the correlation between the two paired sets of numbers is relatively high.

The paired *t*-test tables

More power and sample size tables are required for the paired *t*-test than was the case for the independent samples *t*-test because the former requires the estimation of one additional parameter: the most likely correlation between two sets of continuous numbers. Thus if an investigator wishes to estimate the power available for a fixed number of available subjects who will be measured at baseline, administered an intervention, and then measured again, he/she must hypothesize *both* an effect size (in the same manner that would be done if two independent groups were being contrasted) *and* the correlation that is likely to be obtained between subjects' baseline and follow-up scores.

Guidelines for estimating the correlation between paired observations. While the effect size is hypothesized in the same manner as described previously (e.g., from a pilot study or from similar research conducted using similar interventions/dependent variables), estimating the Pearson *r* between the paired observations involved (i.e., the set of two continuous numbers alluded to above) is sometimes a slightly more tenuous proposition since many studies do not report this value. It can be relatively easily calculated from studies involving the paired *t*-test that report both the *t* and summary statistics (which most do), however. Alternatively the authors of previous studies that employed the independent variable of interest in a pre–post design (it is not particularly important that the same intervention be employed) may be contacted to ascertain whether they still have access to this information.

If this is not feasible, the next best options are either a small scale pilot study or, depending upon the dependent variable's reliability, for the investigator simply to assume that this correlation will be somewhere within the 0.40 to 0.60 range. Specifically, if the dependent variable is reasonably reliable (e.g., has a test–retest or internal consistency reliability $\geq$0.75) it is probably safe to assume that the correlation between baseline and follow-up administrations of the measure will approach 0.60. If the reliability of the measure is less than this, but still fairly high, then 0.40 may be a reasonable estimate. In the absence of good information regarding the stability of a measure, we recommend (as always) that the investigator (a) err on the conservative side and/or (b) model different values in order to be aware of the effects that discrepant values may play on power/sample size estimates. In summary, then, we suggest the following strategies:

(1) Attempt to derive the Pearson *r* from previous studies.
(2) When this is not possible, contact the authors of those studies.

(3) Conduct a small pilot (which should always be done anyway before launching the final study) to calculate the Pearson *r*.
(4) For stable measures (i.e., whose reliabilities tend to be relatively high), assume that the Pearson *r* may be as high as 0.60.
(5) For less stable but still reasonably reliable measures, assume an *r* of 0.40.
(6) Model several values of *r*, since this parameter has a relatively dramatic effect on power/sample size estimates.

Using the paired *t*-test power/sample size tables. Tables 5.1 to 5.5 present power tables for five different estimated correlation coefficients, $r =$ 0.40 through 0.80. Basically these tables are used identically to those in Chapter 4 (and as described in Chapter 3) except that the choice of tables is dictated by the estimated *r* between paired observations. Thus if this value is estimated to be 0.40 using the guidelines presented above, then Table 5.1 is used to determine power. As described in Chapter 3, these tables may be used to estimate (a) the power that will be available for a wide range of hypothesized ES values or (b) the minimum detectable ES given a fixed *N*/group and a desired level of power. The same interpolation advice holds for ES and *N* values that fall between those provided, although such precision is a bit more elusive here because of the additional parameter (i.e., *r*) that must be estimated.

Tables 5.6 through 5.8 present sample size tables for desired power levels of 0.80 and 0.90. Table 5.6 presents these required *N* values for the five correlation coefficients for $p = 0.05$, while Tables 5.7 and 5.8 present the comparable values for $p = 0.01$ and 0.10.

Example. To illustrate the use of these tables, let us return to the hypothetical study posited in Chpater 4 in which the Arthritis Foundation's educational program is to be evaluated with respect to increasing the functional ability of rheumatoid arthritis patients. Let us assume, however, that all of the power calculations involving the independent samples *t*-test (which employed two groups, one that received the intervention and one that did not) were deemed moot for the simple reason that the investigator only had access to a total of 30 patients (which would produce a power estimate of only 0.18 if this sample size had to be divided between two experimental groups).

One option that our investigator would have, however, would be to employ a single group design in which his/her 30 patients were administered the test of functional ability at baseline, then exposed to the educational program, and measured again eight weeks later. While this design is obviously not optimal (given the lack of a control group), it has the potential of increasing the available statistical power, depending upon the size of the correlation coefficient that could be expected to accrue by administering the same measure to the same group of individuals twice. Let us assume that previous research had shown this particular dependent variable to be quite stable and a Pearson r of 0.60 could be expected to accrue over the relatively brief time period chosen, although the investigator wished to model an r of 0.50 as well to be safe. (Generally speaking, the longer the interval between the repeated measures, the lower the correlation coefficient would be expected to be.)

Assuming, then, that research with previous educational interventions had found that the difference between the baseline and posttest means divided by the pooled standard deviation generally resulted in an increase of 0.50 standard deviation units for dependent variables such as functional ability, the investigator decided to proceed with his/her power analysis based upon this information.

To facilitate this process, Template 5.1 is provided. Following steps 1 through 4 would produce the following results:

Step 1. Hypothesize the most likely ES to be obtained between pre and post administrations of the dependent variable. This is done by dividing the difference between the pretest (or baseline) and posttest means by their pooled standard deviation (not the standard deviation of the differences scores).

Hypothesized ES = 0.50

Step 2. Hypothesize one or more likely values for the Pearson r between the repeated measures.

Hypothesized r = 0.50 0.60

Step 3. For power, specify the total N available. For sample size, go to step 5.

N = 30

Step 4. Locate the power available at the intersection of the ES column and the N/row in the table specified below. Interpolate as necessary.

For r = 0.40, access Table 5.1.
For r = 0.50, access Table 5.2.
For r = 0.60, access Table 5.3.
For r = 0.70, access Table 5.4.
For r = 0.80, access Table 5.5.

Power = 0.75 (r = 0.50, Table 5.2) 0.84 (r = 0.60, Table 5.3)

Based upon these results, the investigator might decide to go ahead with the experiment, in effect trading a superior design for a reasonable chance of obtaining statistical significance. The analysis itself could be described to a funding agency or IRB as follows:

> A power analysis indicated that the experiment would have an 84% chance of achieving statistical significance assuming an ES of 0.50 between baseline and post-intervention means and an estimated correlation of 0.60 between the two repeated measures (Bausell & Li, 2002).

Prior to this statement, of course, the investigator would need to describe where he/she obtained the estimates for both the ES and the Pearson r. This justification, since it was based upon previous research, should include citations regarding the actual studies employed and possibly a table including summary statistics gleaned therefrom.

Should our investigator wish to estimate more exactly what type of sample size would be available to achieve, say, a power level of 0.80, he/she could follow steps 5 and 6 of Template 5.1. Assuming that the alpha level was set at 0.05, Charts B ($r=0.50$) and C ($r=0.60$) of Table 5.6 would be used for this purpose, resulting in estimates of 34 and 28 participants respectively for the two modeled r values (for sample size requirements for an alpha level of 0.01, see Table 5.7; for 0.10, see Table 5.8):

Step 5. For required sample size, specify the desired level of power.

Desired power = 0.80

Step 6. Locate the chart corresponding to the hypothesized r in Table 5.6. For powers of 0.80 and 0.90, the required sample size is located at the intersection of the power row and the hypothesized ES column.

Sample size = 34 ($r=0.50$, Chart B, Table 5.6)   28 ($r=0.60$, Chart C, Table 5.6)

Following a description of the ES and correlational rationale (the latter of which could cite our guidelines presented in this chapter), the results of this analysis might be described as follows:

> Employing the tables in Bausell & Li (2002) it was estimated that 28 participants would be needed to enable the detection of an ES of 0.50 between baseline and post-intervention means, assuming that the two measures were correlated 0.60. Should the correlation be as low as 0.50, 34 participants would be needed.

Summary

The paired t-test is an extremely powerful statistical procedure that can be used when the same subjects are measured before and after an intervention.

Following guidelines for employing the procedure as well as for estimating the most likely correlation across the dual administrations of the dependent variable, power and sample size tables are presented for a wide range of r, ES, and N values across these repeated observations.

Template 5.1. Power and sample size template for the paired t-test

Step 1. Hypothesize the most likely ES to be obtained between pre and post administrations of the dependent variable. This is done by dividing the difference between the pretest (or baseline) and posttest means by their pooled standard deviation.

Hypothesized ES = ___

Step 2. Hypothesize one or more likely values for the Pearson r between the repeated measures.

Hypothesized r= ___ ___

Step 3. For power, specify the total N available. For sample size, go to step 5.

N= ___ ___ ___ ___ ___ ___

Step 4. Locate the power available at the intersection of the ES column and the N/row in the table specified below. Interpolate as necessary.

For r=0.40, access Table 5.1.
For r=0.50, access Table 5.2.
For r=0.60, access Table 5.3.
For r=0.70, access Table 5.4.
For r=0.80, access Table 5.5.

Power= ___ ___ ___ ___ ___ ___

Step 5. For required sample size, specify the desired level of power.

Desired power= ___

Step 6. Locate the chart corresponding to the hypothesized r in Table 5.6. For powers of 0.80 and 0.90, the required sample size is located at the intersection of the power row and the hypothesized ES column. For p-values of 0.01, see Table 5.7. For p-values of 0.10, use Table 5.8.

Sample size= ___ ___ ___ ___ ___ ___

Table 5.1. Power table for correlated t-test; $r = 0.40$ at alpha $= 0.05$

n	Hypothesized ES																		
	0.20	0.30	0.35	0.40	0.45	0.50	0.55	0.60	0.65	0.70	0.75	0.80	1.00	1.25	1.50	1.75	2.00	2.50	3.00
5	5	6	7	8	9	11	12	13	15	17	19	21	30	44	58	71	82	95	99
6	5	7	8	10	11	13	15	17	19	22	24	27	40	57	73	85	93	99	.
7	5	8	10	11	13	16	18	21	24	27	30	34	49	68	83	93	97	.	.
8	6	9	11	13	15	18	21	25	28	32	36	40	57	77	90	97	99	.	.
9	6	10	12	15	18	21	24	28	32	37	41	46	65	83	94	98	.	.	.
10	7	11	13	16	20	23	28	32	37	42	47	52	71	88	97	99	.	.	.
11	7	12	15	18	22	26	31	36	41	46	52	57	76	92	98	.	.	.	.
12	8	13	16	20	24	29	34	39	45	50	56	62	81	94	99	.	.	.	.
13	8	14	17	22	26	31	37	43	49	55	60	66	85	96	99	.	.	.	.
14	9	15	19	23	28	34	40	46	52	58	64	70	88	97	.	.	.	.	.
15	9	16	20	25	30	36	43	49	56	62	68	74	90	98	.	.	.	.	.
20	11	21	26	33	40	48	56	63	70	77	82	87	97	.	.	.	.	.	.
25	13	25	33	41	50	58	67	74	81	86	90	94	99	.	.	.	.	.	.
30	16	30	39	48	58	67	75	82	88	92	95	97	.	.	.	.	.	.	.
35	18	34	45	55	65	74	82	88	92	96	97	99	.	.	.	.	.	.	.
40	20	39	50	61	71	80	87	92	95	98	99	99	.	.	.	.	.	.	.
45	22	43	55	66	77	85	91	95	97	99	99	.	.	.	.	.	.	.	.
50	24	47	60	71	81	88	93	97	98	99	.	.	.	.	.	.	.	.	.
55	26	51	64	76	85	91	95	98	99	.	.	.	.	.	.	.	.	.	.
60	28	55	68	79	88	93	97	99	99	.	.	.	.	.	.	.	.	.	.
65	30	58	72	82	90	95	98	99	.	.	.	.	.	.	.	.	.	.	.
70	32	61	75	85	92	96	99	99	.	.	.	.	.	.	.	.	.	.	.
75	34	65	78	88	94	97	99	.	.	.	.	.	.	.	.	.	.	.	.
80	36	67	80	90	95	98	99	.	.	.	.	.	.	.	.	.	.	.	.
90	40	73	85	93	97	99	.	.	.	.	.	.	.	.	.	.	.	.	.
100	44	77	88	95	98	99	.	.	.	.	.	.	.	.	.	.	.	.	.
110	47	81	91	97	99	.	.	.	.	.	.	.	.	.	.	.	.	.	.
120	51	84	93	98	99	.	.	.	.	.	.	.	.	.	.	.	.	.	.
130	54	87	95	98	.	.	.	.	.	.	.	.	.	.	.	.	.	.	.
140	57	90	96	99	.	.	.	.	.	.	.	.	.	.	.	.	.	.	.
150	60	91	97	99	.	.	.	.	.	.	.	.	.	.	.	.	.	.	.
175	67	95	99	.	.	.	.	.	.	.	.	.	.	.	.	.	.	.	.
200	73	97	99	.	.	.	.	.	.	.	.	.	.	.	.	.	.	.	.
225	78	98	.	.	.	.	.	.	.	.	.	.	.	.	.	.	.	.	.
250	82	99	.	.	.	.	.	.	.	.	.	.	.	.	.	.	.	.	.
300	88	.	.	.	.	.	.	.	.	.	.	.	.	.	.	.	.	.	.
400	95	.	.	.	.	.	.	.	.	.	.	.	.	.	.	.	.	.	.
500	98	.	.	.	.	.	.	.	.	.	.	.	.	.	.	.	.	.	.

Table 5.2. Power table for correlated t-test; $r = 0.50$ at alpha $= 0.05$

n	Hypothesized ES																		
	0.20	0.30	0.35	0.40	0.45	0.50	0.55	0.60	0.65	0.70	0.75	0.80	1.00	1.25	1.50	1.75	2.00	2.50	3.00
5	5	7	8	9	10	12	13	15	17	19	22	24	35	51	66	79	89	98	.
6	5	8	9	11	13	15	17	20	22	25	28	32	46	65	80	91	96	.	.
7	6	9	11	13	15	18	21	24	28	31	35	39	56	76	89	96	99	.	.
8	6	10	12	15	18	21	25	29	33	37	42	47	65	84	94	99	.	.	.
9	7	11	14	17	20	24	28	33	38	43	48	53	73	89	97	99	.	.	.
10	8	12	15	19	23	27	32	37	43	48	54	59	79	93	99	.	.	.	.
11	8	13	17	21	26	31	36	42	47	53	59	65	84	96	99	.	.	.	.
12	9	15	19	23	28	34	39	46	52	58	64	70	87	97	.	.	.	.	.
13	9	16	20	25	31	37	43	49	56	62	68	74	90	98	.	.	.	.	.
14	10	17	22	27	33	39	46	53	60	66	72	78	93	99	.	.	.	.	.
15	10	18	23	29	35	42	49	57	64	70	76	81	95	99	.	.	.	.	.
20	13	24	31	39	47	55	64	71	78	84	88	92	99	.	.	.	.	.	.
25	15	29	38	48	57	66	74	82	87	92	95	97	.	.	.	.	.	.	.
30	18	35	45	56	66	75	82	88	93	96	98	99	.	.	.	.	.	.	.
35	20	40	51	63	73	82	88	93	96	98	99	.	.	.	.	.	.	.	.
40	23	45	57	69	79	87	92	96	98	99	.	.	.	.	.	.	.	.	.
45	26	50	63	74	84	90	95	98	99	.	.	.	.	.	.	.	.	.	.
50	28	54	68	79	87	93	97	99	99	.	.	.	.	.	.	.	.	.	.
55	30	59	72	83	90	95	98	99	.	.	.	.	.	.	.	.	.	.	.
60	33	62	76	86	93	97	99	.	.	.	.	.	.	.	.	.	.	.	.
65	35	66	79	89	95	98	99	.	.	.	.	.	.	.	.	.	.	.	.
70	38	69	82	91	96	98	99	.	.	.	.	.	.	.	.	.	.	.	.
75	40	72	85	93	97	99	.	.	.	.	.	.	.	.	.	.	.	.	.
80	42	75	87	94	98	99	.	.	.	.	.	.	.	.	.	.	.	.	.
90	46	80	91	96	99	.	.	.	.	.	.	.	.	.	.	.	.	.	.
100	51	84	93	98	99	.	.	.	.	.	.	.	.	.	.	.	.	.	.
110	55	88	95	99	.	.	.	.	.	.	.	.	.	.	.	.	.	.	.
120	58	90	97	99	.	.	.	.	.	.	.	.	.	.	.	.	.	.	.
130	62	92	98	99	.	.	.	.	.	.	.	.	.	.	.	.	.	.	.
140	65	94	98	.	.	.	.	.	.	.	.	.	.	.	.	.	.	.	.
150	68	95	99	.	.	.	.	.	.	.	.	.	.	.	.	.	.	.	.
175	75	98	.	.	.	.	.	.	.	.	.	.	.	.	.	.	.	.	.
200	80	99	.	.	.	.	.	.	.	.	.	.	.	.	.	.	.	.	.
225	85	99	.	.	.	.	.	.	.	.	.	.	.	.	.	.	.	.	.
250	88	.	.	.	.	.	.	.	.	.	.	.	.	.	.	.	.	.	.
300	93	.	.	.	.	.	.	.	.	.	.	.	.	.	.	.	.	.	.
400	98	.	.	.	.	.	.	.	.	.	.	.	.	.	.	.	.	.	.
500	99	.	.	.	.	.	.	.	.	.	.	.	.	.	.	.	.	.	.

Table 5.3. Power table for correlated *t*-test; $r = 0.60$ at alpha $= 0.05$

n	Hypothesized ES																		
	0.20	0.30	0.35	0.40	0.45	0.50	0.55	0.60	0.65	0.70	0.75	0.80	1.00	1.25	1.50	1.75	2.00	2.50	3.00
5	5	7	9	10	12	14	16	18	21	23	26	29	42	60	76	87	94	99	.
6	6	9	11	13	15	18	20	24	27	31	34	38	55	75	88	96	99	.	.
7	6	10	12	15	18	21	25	29	33	38	43	47	66	85	95	99	.	.	.
8	7	12	14	18	21	25	30	35	40	45	50	56	75	91	98	.	.	.	.
9	8	13	16	20	25	29	34	40	46	51	57	63	82	95	99	.	.	.	.
10	9	14	18	23	28	33	39	45	51	57	63	69	87	97	.	.	.	.	.
11	9	16	20	25	31	37	43	50	56	63	69	75	91	98	.	.	.	.	.
12	10	17	22	28	34	41	47	54	61	68	74	79	93	99	.	.	.	.	.
13	10	19	24	30	37	44	51	59	66	72	78	83	95	.	.	.	.	.	.
14	11	20	26	33	40	47	55	63	70	76	82	86	97	.	.	.	.	.	.
15	12	22	28	35	43	51	59	66	73	79	85	89	98	.	.	.	.	.	.
20	15	29	37	46	56	65	73	80	86	91	94	96	.	.	.	.	.	.	.
25	18	36	46	57	67	76	83	89	93	96	98	99	.	.	.	.	.	.	.
30	21	42	54	65	75	84	90	94	97	98	99	.	.	.	.	.	.	.	.
35	25	48	61	72	82	89	94	97	99	99	.	.	.	.	.	.	.	.	.
40	28	51	67	78	87	93	97	98	99	.	.	.	.	.	.	.	.	.	.
45	31	59	72	83	91	96	98	99	.	.	.	.	.	.	.	.	.	.	.
50	34	64	77	87	94	97	99	.	.	.	.	.	.	.	.	.	.	.	.
55	37	68	81	90	95	98	99	.	.	.	.	.	.	.	.	.	.	.	.
60	40	72	84	92	97	99	.	.	.	.	.	.	.	.	.	.	.	.	.
65	42	76	87	94	98	99	.	.	.	.	.	.	.	.	.	.	.	.	.
70	45	79	90	96	99	.	.	.	.	.	.	.	.	.	.	.	.	.	.
75	48	82	92	97	99	.	.	.	.	.	.	.	.	.	.	.	.	.	.
80	50	84	93	98	99	.	.	.	.	.	.	.	.	.	.	.	.	.	.
90	55	88	96	99	.	.	.	.	.	.	.	.	.	.	.	.	.	.	.
100	60	91	97	99	.	.	.	.	.	.	.	.	.	.	.	.	.	.	.
110	64	94	98	.	.	.	.	.	.	.	.	.	.	.	.	.	.	.	.
120	68	95	99	.	.	.	.	.	.	.	.	.	.	.	.	.	.	.	.
130	71	97	99	.	.	.	.	.	.	.	.	.	.	.	.	.	.	.	.
140	75	98	.	.	.	.	.	.	.	.	.	.	.	.	.	.	.	.	.
150	78	98	.	.	.	.	.	.	.	.	.	.	.	.	.	.	.	.	.
175	84	99	.	.	.	.	.	.	.	.	.	.	.	.	.	.	.	.	.
200	88	.	.	.	.	.	.	.	.	.	.	.	.	.	.	.	.	.	.
225	92	.	.	.	.	.	.	.	.	.	.	.	.	.	.	.	.	.	.
250	94	.	.	.	.	.	.	.	.	.	.	.	.	.	.	.	.	.	.
300	97	.	.	.	.	.	.	.	.	.	.	.	.	.	.	.	.	.	.
400	99	.	.	.	.	.	.	.	.	.	.	.	.	.	.	.	.	.	.
500	.	.	.	.	.	.	.	.	.	.	.	.	.	.	.	.	.	.	.

Table 5.4. Power table for correlated t-test; $r = 0.70$ at alpha $= 0.05$

n	Hypothesized ES																		
	0.20	0.30	0.35	0.40	0.45	0.50	0.55	0.60	0.65	0.70	0.75	0.80	1.00	1.25	1.50	1.75	2.00	2.50	3.00
5	6	9	10	12	15	17	20	23	26	29	33	37	53	72	87	95	98	.	.
6	7	10	13	16	19	22	26	30	34	39	44	49	68	86	95	99	.	.	.
7	7	12	15	19	23	27	32	37	43	48	54	59	79	93	99	.	.	.	.
8	8	14	18	22	27	32	38	44	50	56	62	68	86	97	.	.	.	.	.
9	9	16	21	26	31	37	44	51	57	64	70	75	91	99	.	.	.	.	.
10	10	18	23	29	35	42	49	57	64	70	76	81	95	99	.	.	.	.	.
11	11	20	26	32	39	47	55	62	69	75	81	86	97	.	.	.	.	.	.
12	12	22	28	35	43	51	59	67	74	80	85	89	98	.	.	.	.	.	.
13	13	24	31	39	47	55	64	71	78	84	88	92	99	.	.	.	.	.	.
14	14	26	33	42	50	59	68	75	82	87	91	94	99	.	.	.	.	.	.
15	14	28	36	45	54	63	71	79	85	90	93	96	.	.	.	.	.	.	.
20	19	37	47	58	68	77	85	90	94	97	98	99	.	.	.	.	.	.	.
25	23	45	57	69	79	87	92	96	98	99	.	.	.	.	.	.	.	.	.
30	27	53	66	78	86	92	96	98	99	.	.	.	.	.	.	.	.	.	.
35	31	60	73	84	91	96	98	99	.	.	.	.	.	.	.	.	.	.	.
40	35	66	79	89	95	98	99	.	.	.	.	.	.	.	.	.	.	.	.
45	39	72	84	92	97	99	.	.	.	.	.	.	.	.	.	.	.	.	.
50	43	76	88	95	98	99	.	.	.	.	.	.	.	.	.	.	.	.	.
55	46	80	91	96	99	.	.	.	.	.	.	.	.	.	.	.	.	.	.
60	50	84	93	98	99	.	.	.	.	.	.	.	.	.	.	.	.	.	.
65	53	87	95	98	.	.	.	.	.	.	.	.	.	.	.	.	.	.	.
70	56	89	96	99	.	.	.	.	.	.	.	.	.	.	.	.	.	.	.
75	59	91	97	99	.	.	.	.	.	.	.	.	.	.	.	.	.	.	.
80	62	93	98	.	.	.	.	.	.	.	.	.	.	.	.	.	.	.	.
90	68	95	99	.	.	.	.	.	.	.	.	.	.	.	.	.	.	.	.
100	72	97	99	.	.	.	.	.	.	.	.	.	.	.	.	.	.	.	.
110	76	98	.	.	.	.	.	.	.	.	.	.	.	.	.	.	.	.	.
120	80	99	.	.	.	.	.	.	.	.	.	.	.	.	.	.	.	.	.
130	83	99	.	.	.	.	.	.	.	.	.	.	.	.	.	.	.	.	.
140	86	.	.	.	.	.	.	.	.	.	.	.	.	.	.	.	.	.	.
150	88	.	.	.	.	.	.	.	.	.	.	.	.	.	.	.	.	.	.
175	92	.	.	.	.	.	.	.	.	.	.	.	.	.	.	.	.	.	.
200	95	.	.	.	.	.	.	.	.	.	.	.	.	.	.	.	.	.	.
225	97	.	.	.	.	.	.	.	.	.	.	.	.	.	.	.	.	.	.
250	98	.	.	.	.	.	.	.	.	.	.	.	.	.	.	.	.	.	.
300	99	.	.	.	.	.	.	.	.	.	.	.	.	.	.	.	.	.	.
400	.	.	.	.	.	.	.	.	.	.	.	.	.	.	.	.	.	.	.
500	.	.	.	.	.	.	.	.	.	.	.	.	.	.	.	.	.	.	.

Table 5.5. Power table for correlated t-test; $r = 0.80$ at alpha $= 0.05$

n	Hypothesized ES																		
	0.20	0.30	0.35	0.40	0.45	0.50	0.55	0.60	0.65	0.70	0.75	0.80	1.00	1.25	1.50	1.75	2.00	2.50	3.00
5	7	11	14	17	20	24	28	32	37	41	46	51	71	88	96	99	.	.	.
6	8	14	17	21	26	31	37	42	48	54	60	66	84	96	99	.	.	.	.
7	9	17	21	26	32	39	45	52	59	65	71	77	92	99	.	.	.	.	.
8	11	19	25	31	38	46	53	61	68	74	80	85	96	.	.	.	.	.	.
9	12	22	29	36	44	52	60	68	75	81	86	90	98	.	.	.	.	.	.
10	13	25	33	41	50	58	67	74	81	86	91	94	99	.	.	.	.	.	.
11	15	28	36	45	55	64	72	79	85	90	94	96	.	.	.	.	.	.	.
12	16	31	40	50	59	69	77	84	89	93	96	98	.	.	.	.	.	.	.
13	17	33	43	54	64	73	81	87	92	95	97	99	.	.	.	.	.	.	.
14	18	36	47	58	68	77	84	90	94	97	98	99	.	.	.	.	.	.	.
15	20	39	50	61	71	80	87	92	96	98	99	99	.	.	.	.	.	.	.
20	26	51	64	76	85	91	96	98	99	.	.	.	.	.	.	.	.	.	.
25	32	62	75	85	92	96	99	99	.	.	.	.	.	.	.	.	.	.	.
30	38	70	83	91	96	99	.	.	.	.	.	.	.	.	.	.	.	.	.
35	44	77	89	95	98	99	.	.	.	.	.	.	.	.	.	.	.	.	.
40	49	83	93	97	99	.	.	.	.	.	.	.	.	.	.	.	.	.	.
45	54	87	95	99	.	.	.	.	.	.	.	.	.	.	.	.	.	.	.
50	59	91	97	99	.	.	.	.	.	.	.	.	.	.	.	.	.	.	.
55	63	93	98	.	.	.	.	.	.	.	.	.	.	.	.	.	.	.	.
60	67	95	99	.	.	.	.	.	.	.	.	.	.	.	.	.	.	.	.
65	71	96	99	.	.	.	.	.	.	.	.	.	.	.	.	.	.	.	.
70	74	97	.	.	.	.	.	.	.	.	.	.	.	.	.	.	.	.	.
75	77	98	.	.	.	.	.	.	.	.	.	.	.	.	.	.	.	.	.
80	80	99	.	.	.	.	.	.	.	.	.	.	.	.	.	.	.	.	.
90	84	99	.	.	.	.	.	.	.	.	.	.	.	.	.	.	.	.	.
100	88	.	.	.	.	.	.	.	.	.	.	.	.	.	.	.	.	.	.
110	91	.	.	.	.	.	.	.	.	.	.	.	.	.	.	.	.	.	.
120	93	.	.	.	.	.	.	.	.	.	.	.	.	.	.	.	.	.	.
130	95	.	.	.	.	.	.	.	.	.	.	.	.	.	.	.	.	.	.
140	96	.	.	.	.	.	.	.	.	.	.	.	.	.	.	.	.	.	.
150	97	.	.	.	.	.	.	.	.	.	.	.	.	.	.	.	.	.	.
175	99	.	.	.	.	.	.	.	.	.	.	.	.	.	.	.	.	.	.
200	99	.	.	.	.	.	.	.	.	.	.	.	.	.	.	.	.	.	.
225	.	.	.	.	.	.	.	.	.	.	.	.	.	.	.	.	.	.	.
250	.	.	.	.	.	.	.	.	.	.	.	.	.	.	.	.	.	.	.
300	.	.	.	.	.	.	.	.	.	.	.	.	.	.	.	.	.	.	.
400	.	.	.	.	.	.	.	.	.	.	.	.	.	.	.	.	.	.	.
500	.	.	.	.	.	.	.	.	.	.	.	.	.	.	.	.	.	.	.

Table 5.6. Sample size table for the paired t-test (alpha = 0.05)

Chart A. Paired t-test; $r = 0.4$ at alpha = 0.05																			
Power	Hypothesized ES																		
	0.20	0.30	0.35	0.40	0.45	0.50	0.55	0.60	0.65	0.70	0.75	0.80	1.00	1.25	1.50	1.75	2.00	2.50	3.00
0.80	238	107	80	62	49	40	34	29	25	22	20	18	12	9	7	6	5	4	4
0.90	318	143	106	82	65	53	44	38	33	29	25	23	15	11	9	7	6	5	4
Chart B. Paired t-test; $r = 0.5$ at alpha = 0.05																			
Power	Hypothesized ES																		
	0.20	0.30	0.35	0.40	0.45	0.50	0.55	0.60	0.65	0.70	0.75	0.80	1.00	1.25	1.50	1.75	2.00	2.50	3.00
0.80	199	90	67	52	42	34	29	25	21	19	17	15	11	8	6	6	5	4	4
0.90	265	119	89	68	55	45	38	32	28	24	21	19	13	10	8	6	6	5	4
Chart C. Paired t-test; $r = 0.6$ at alpha = 0.05																			
Power	Hypothesized ES																		
	0.20	0.30	0.35	0.40	0.45	0.50	0.55	0.60	0.65	0.70	0.75	0.80	1.00	1.25	1.50	1.75	2.00	2.50	3.00
0.80	160	73	54	42	34	28	24	20	18	16	14	13	9	7	6	5	5	4	4
0.90	213	96	71	55	44	36	31	26	23	20	18	16	11	8	7	6	5	4	4
Chart D. Paired t-test; $r = 0.7$ at alpha = 0.05																			
Power	Hypothesized ES																		
	0.20	0.30	0.35	0.40	0.45	0.50	0.55	0.60	0.65	0.70	0.75	0.80	1.00	1.25	1.50	1.75	2.00	2.50	3.00
0.80	121	55	41	32	26	22	18	16	14	13	11	10	8	6	5	5	4	4	3
0.90	160	73	54	42	34	28	24	20	18	16	14	13	9	7	6	5	5	4	4
Chart E. Paired t-test; $r = 0.8$ at alpha = 0.05																			
Power	Hypothesized ES																		
	0.20	0.30	0.35	0.40	0.45	0.50	0.55	0.60	0.65	0.70	0.75	0.80	1.00	1.25	1.50	1.75	2.00	2.50	3.00
0.80	81	38	28	22	18	15	13	12	10	9	9	8	6	5	4	4	4	3	3
0.90	108	49	37	29	24	20	17	15	13	11	10	9	7	6	5	4	4	4	3

Table 5.7. Sample size table for the paired *t*-test (alpha = 0.01)

Chart A. Paired *t*-test; $r = 0.4$ at alpha = 0.01

Power	Hypothesized ES																		
	0.20	0.30	0.35	0.40	0.45	0.50	0.55	0.60	0.65	0.70	0.75	0.80	1.00	1.25	1.50	1.75	2.00	2.50	3.00
0.80	355	160	119	92	73	60	51	43	37	33	29	26	18	13	10	9	8	6	6
0.90	451	203	150	116	92	76	63	54	46	41	36	32	22	16	12	10	9	7	6

Chart B. Paired *t*-test; $r = 0.5$ at alpha = 0.01

Power	Hypothesized ES																		
	0.20	0.30	0.35	0.40	0.45	0.50	0.55	0.60	0.65	0.70	0.75	0.80	1.00	1.25	1.50	1.75	2.00	2.50	3.00
0.80	296	134	100	77	62	51	43	37	32	28	25	22	16	12	9	8	7	6	5
0.90	376	169	126	97	78	64	53	46	39	35	31	27	19	14	11	9	8	6	6

Chart C. Paired *t*-test; $r = 0.6$ at alpha = 0.01

Power	Hypothesized ES																		
	0.20	0.30	0.35	0.40	0.45	0.50	0.55	0.60	0.65	0.70	0.75	0.80	1.00	1.25	1.50	1.75	2.00	2.50	3.00
0.80	238	108	80	63	50	42	35	30	26	23	21	19	14	10	8	7	6	5	5
0.90	302	136	101	79	63	52	44	37	32	28	25	23	16	12	9	8	7	6	5

Chart D. Paired *t*-test; $r = 0.7$ at alpha = 0.01

Power	Hypothesized ES																		
	0.20	0.30	0.35	0.40	0.45	0.50	0.55	0.60	0.65	0.70	0.75	0.80	1.00	1.25	1.50	1.75	2.00	2.50	3.00
0.80	179	82	61	48	39	32	27	24	21	19	17	15	11	9	7	6	6	5	5
0.90	227	103	77	60	48	40	34	29	25	22	20	18	13	10	8	7	6	5	5

Chart E. Paired *t*-test; $r = 0.8$ at alpha = 0.01

Power	Hypothesized ES																		
	0.20	0.30	0.35	0.40	0.45	0.50	0.55	0.60	0.65	0.70	0.75	0.80	1.00	1.25	1.50	1.75	2.00	2.50	3.00
0.80	121	56	42	33	27	23	20	17	15	14	13	12	9	7	6	6	5	5	4
0.90	153	70	53	41	34	28	24	21	18	16	15	14	10	8	7	6	5	5	4

Table 5.8. Sample size table for the paired *t*-test (alpha = 0.10)

Chart A. Paired *t*-test; *r* = 0.4 at alpha = 0.10																			
Power	Hypothesized ES																		
	0.20	0.30	0.35	0.40	0.45	0.50	0.55	0.60	0.65	0.70	0.75	0.80	1.00	1.25	1.50	1.75	2.00	2.50	3.00
0.80	188	85	63	49	39	32	27	23	20	17	15	14	10	7	6	5	4	4	3
0.90	259	116	86	66	53	43	36	31	27	23	20	18	13	9	7	6	5	4	4
Chart B. Paired *t*-test; *r* = 0.5 at alpha = 0.10																			
Power	Hypothesized ES																		
	0.20	0.30	0.35	0.40	0.45	0.50	0.55	0.60	0.65	0.70	0.75	0.80	1.00	1.25	1.50	1.75	2.00	2.50	3.00
0.80	157	71	53	41	33	27	23	19	17	15	13	12	9	6	5	5	4	4	3
0.90	216	97	72	56	44	36	30	26	22	20	17	16	11	8	6	5	5	4	3
Chart C. Paired *t*-test; *r* = 0.6 at alpha = 0.10																			
Power	Hypothesized ES																		
	0.20	0.30	0.35	0.40	0.45	0.50	0.55	0.60	0.65	0.70	0.75	0.80	1.00	1.25	1.50	1.75	2.00	2.50	3.00
0.80	126	57	43	33	27	22	19	16	14	12	11	10	7	6	5	4	4	3	3
0.90	173	78	58	45	36	30	25	21	18	16	14	13	9	7	5	5	4	4	3
Chart D. Paired *t*-test; *r* = 0.7 at alpha = 0.10																			
Power	Hypothesized ES																		
	0.20	0.30	0.35	0.40	0.45	0.50	0.55	0.60	0.65	0.70	0.75	0.80	1.00	1.25	1.50	1.75	2.00	2.50	3.00
0.80	95	43	33	25	21	17	15	13	11	10	9	8	6	5	4	4	3	3	3
0.90	131	59	44	34	28	23	19	17	14	13	11	10	7	6	5	4	4	3	3
Chart E. Paired *t*-test; *r* = 0.8 at alpha = 0.10																			
Power	Hypothesized ES																		
	0.20	0.30	0.35	0.40	0.45	0.50	0.55	0.60	0.65	0.70	0.75	0.80	1.00	1.25	1.50	1.75	2.00	2.50	3.00
0.80	64	30	22	18	14	12	10	9	8	7	7	6	5	4	4	3	3	3	3
0.90	88	40	30	24	19	16	14	12	10	9	8	8	6	5	4	4	3	3	3

6 One-way between subjects analysis of variance

Purpose of the statistic

The one-way between subjects analysis of variance (ANOVA) is used to ascertain how likely the differences among *three* or more groups would be to occur by chance alone. It is a direct extension of the *t*-test for independent samples, which assesses only differences between two groups. The groups may be experimental conditions to which subjects have been randomly assigned (e.g., two experimental treatments vs. a control) or they may be defined by a naturally occurring phenomenon (e.g., a comparison of several different diagnoses with respect to the amount of chronic pain experienced). As with the independent samples *t*-test (other than the standard assumptions of the procedure itself), the only stipulations for its use are that the hypotheses being tested involve group means (which implies that the dependent variable is continuous in nature) and the groups are *not* made up of the same or matched subjects. Also, as with the *t*-test for independent samples, from a power analytic perspective, *this procedure is not recommended if covariates, baseline values on the dependent variable, or blocking variables are available.*

To review then, a one-way between subjects ANOVA is used when:

(1) there is a *single*, independent variable which is defined as group membership in three or more groups,
(2) the dependent variable is measured in such a way that it can be described by a mean (i.e., it is continuous in nature and not categorical),
(3) the hypothesis being tested is expressed in terms of a mean difference, and
(4) the subjects assigned or contained in the groups are statistically independent of one another (i.e., they are different individuals and are not specifically matched in some way).

The statistic is *not* used when:

(1) there is more than one independent variable (such as covariates, which are used to control statistically for initial differences between

groups, or blocking variables, which are used to ascertain whether one type of subject responds differently to the interventions(s) than another),

(2) this single independent variable is continuous in nature (unless a decision is made to divide it into three or more arbitrary categories),

(3) the hypothesis is expressed in terms of a relationship between variables rather than a difference among groups,

(4) the dependent variable is categorical in nature, or

(5) the multiple groups involved contain the same or matched subjects.

A one-way ANOVA is really a two-step process. The *first* step entails computing an F-ratio, which is analogous to a t (for two groups, in fact, $t^2 = F$). Unlike the t-test, however, the process does not stop at this point. If the F-ratio is statistically significant, then the *second* step involves ascertaining which groups differ significantly from which other groups.

As mentioned previously, we are of the opinion that statistics involving more than one degree of freedom for the primary contrast of interest have very limited utility from a scientific perspective. It is therefore very rare for an overall F-ratio to be used directly to test a hypothesis. In most day-to-day research practice, all the F-ratio does is serve in a sort of gate-keeping function to tell the scientist when he/she is "permitted" to conduct a multiple comparison procedure (MCP) to ascertain which experimental groups differ from which others. *This approach helps to guard against the production of false positive results by providing a degree of protection against inflating the alpha level when performing multiple significance tests.*

Thus, if the F-ratio is not statistically significant, many statisticians advise their clients not to proceed further but to declare the entire study as "non-significant" unless prior, orthogonal contrasts[1] have been specified. We consider this approach too rigid, however, given that the appropriate use of an MCP should itself provide adequate protection against false positive findings (Miller, 1966; Petrinovich & Hardyck, 1969; Davis & Gaito, 1984), especially with respect to the *primary* hypothesized contrasts of interest. Since most journals expect an overall F-ratio to be reported prior to considering individual differences among groups, however, we provide tables for computing power and sample size requirements for this statistic.

The remainder of this chapter will therefore be divided into two parts. The first will present power and sample size tables for the overall F-ratio. The second will present tables that allow the investigator to estimate the power (or sample size requirements) for individual contrasts using two common multiple comparison procedures

Part I. The *F*-ratio tables

Tables 6.1 through 6.11 present estimated power values and required sample sizes for a wide range of values of ES and *N*/group for *F*-ratios involving three through five groups. In addition to the parameters encountered in Chapters 4 and 5, the use of these tables involves one additional step: estimating what the pattern of the various group means is most likely to look like. This latter step, of course, necessitates hypothesizing the ES values involved in all of the pairwise contrasts among each of the groups employed in the proposed trial.

Power. To illustrate how these tables are employed, let us assume the use of a three-group study contrasting the effects of an intervention consisting of a combination of dietary education and exercise, an attention placebo comprising information about cardiovascular disease, and treatment-as-usual with the outcome variable being the amount of weight loss registered by overweight post-surgical cardiovascular patients. Computing the power of the overall *F*-ratio for such a study involves specifying the following parameters:

(1) the ES for the largest mean difference (remembering that there are three in this particular study – the weight loss intervention vs. the attention placebo, the weight loss intervention vs. treatment-as-usual, and the attention placebo vs. treatment-as-usual),
(2) the projected *N*/group, and
(3) whether the spread of means is likely to reflect a low to medium, or a high dispersion pattern.

To facilitate this process, two templates (6.1 and 6.2) are provided. Let us therefore illustrate how an investigator would estimate the power for the overall *F*-ratio for this three-group study employing these two templates, the first of which is basically designed to facilitate the estimation of the three power parameters just listed. Variations in any of these parameters exhibit a dramatic effect upon the power of a study. The type of hypothesized pattern alone, for example, can more than double the required sample size to achieve a desired level of power. Thus in a five-group study with a hypothesized ES of 0.50, the estimated *N*/group necessary to achieve a power level of 0.80 is 96 if three of the five means are expected to fall half-way between the two extreme groups (a low dispersion pattern). Should two of the five means be expected to fall at one end of the ES continuum and three at the other (a high dispersion pattern), however, only 41 subjects per group would be needed.

The first step in preliminary Template 6.1 simply requests that the groups involved are ordered from weakest to strongest with respect to their likely effect sizes. In the current example we would obviously expect the usual treatment group to experience the least amount of weight loss, followed by the attention placebo (since this group would also receive treatment-as-usual for ethical reasons), with patients receiving the experimental intervention hypothesized to lose the most weight:

Step 1. Write in the names/codes of the groups in the chart to the right in ascending order based upon their expected means (i.e., the name of the group expected to have the lowest mean or the weakest effect will be written next to ①, followed by the next strongest treatment and so forth).	Group definitions ① Treatment-as-usual ② Attention placebo ③ Exercise + education

The next step in the process is the most difficult, because here the investigator must have access to preliminary data, either from his/her own pilot work or from the empirical literature in order to be able to estimate the effect sizes (i.e., the average amount of weight loss divided by the standard deviation) for each group. Let us assume, therefore, that our investigator has indeed conducted such a pilot study in which his/her exercise/education intervention was found to produce an ES of 0.50 compared to treatment-as-usual. Group 3 is correspondingly plotted on the ES line.

Let us further assume that while the investigator did not pilot the planned attention placebo, other studies upon which it was based had observed an ES of approximately 0.20 with comparable samples. Armed with this information the middle group is plotted on the ES line, completing the most difficult (and crucial) step in the power analytic process as illustrated below. Note that the treatment-as-usual group ES has been arbitrarily set at zero. When beginning with raw data, this is done by dividing each hypothesized mean by the pooled standard deviation and then simply subtracting the lowest resulting standardized mean from each of the other standardized group means (including itself). In the present case this will produce a zero value for the treatment-as-usual group, which in no way implies that this group will lose or gain no weight at all.

Converting multiple raw means to ES values. In the current example, let us assume that our investigator based his/her weight loss estimates upon both a previous pilot study and the empirical literature employing similar populations. From these sources, he/she found that the standard deviation for weight loss averaged 10 lb with even treatment-as-usual controls normally losing approximately 5 lb. An attention placebo was further hypothesized to result in an additional weight loss of at least 2 lb, while the experimental intervention was expected to result in twice as much weight loss as treatment-as-usual. This produces the following scenario:

Condition	Weight loss (standard deviation)
Treatment-as-usual	5 lb (10.0)
Placebo	7 lb (10.0)
Intervention	10 lb (10.0)

Since use of all of the power tables in this book involve converting raw means such as this to ES values, the next step involves dividing each of these means by the pooled standard deviation and then subtracting the lowest standardized group mean from all of the others (including itself), producing the following results:

Condition	Standardized mean (with the lowest mean set at 0)
Treatment-as-usual	5 lb/10.0 = 0.5 − 0.5 = 0.0
Placebo	7 lb/10.0 = 0.7 − 0.5 = 0.2
Intervention	10 lb/10.0 = 1.0 − 0.5 = 0.5

Step 2. Plot these groups, using the circled numbers rather than their codes, on the ES line below. In the case of a tie between two groups with respect to their actual ES value, place one to the right of the other based upon the best theoretical evidence for which might be slightly superior.

ES line

① ② ③

0 0.1 0.2 0.3 0.4 0.5 0.6 0.7 0.8 0.9 1.0 1.1 1.2 1.3 1.4 1.5

The next step involves either specifying (a) how many subjects are expected to be available per group (if power is desired) or (b) what the desired power level is if the investigator wishes to calculate the number of subjects needed to achieve this desired level of power. For present purposes we will assume the former, although a sample size analysis will also be illustrated later in the chapter. Let us therefore assume that our investigator wishes to ascertain how much power would be available if 50 subjects per group were available. (Note that the *N*/group posited here should take attrition into account, thus if 50 subjects/group was the maximum sample size that was likely to be available and the investigator expected that approximately 10% of his/her sample would drop out of the study, then 45 would be entered in step 3.)

The final parameter called for by preliminary Template 6.1 is the hypothesized pattern of means that will accrue from the experiment. Since this pattern has already been graphed in step 2, all that is necessary is to contrast it to the three-group options presented under step 4(a). While none of the three options matches this pattern perfectly, the first pattern is closer than the second two, hence our investigator would conclude that the low/medium dispersion pattern is the best model for estimating power for this particular study. Formulas exist (see the Technical appendix) that exactly express all of the possible patterns of ES values that can accrue from different numbers of groups. Given our belief that such precision is in many ways specious, we have opted to present graphically three patterns that reflect (a) the largest possible spread among multiple means (i.e., the scenario in which the means produce the largest number of maximum ES values, which implies that half will be located at one end of the line and the other half will be clustered at the other end), (b) the smallest possible spread (the scenario in which the extra means are located at the middle of the continuum), and (c) a medium pattern, in which the means are evenly spaced along the continuum.[2] (Note that for a three-group experiment, and only for a three-group experiment, the patterns for a low and medium spread of means are identical, hence only two patterns need be considered for this type of experiment.)

For all practical purposes, now, all that is necessary to compute the power available for the overall *F*-ratio for this hypothetical experiment is to access the appropriate power table, which is dictated by Template 6.2. Here, step 5 requests the largest ES from step 2 above, which is the 0.5 difference between group 3 and group 1. Tables 6.1 and 6.2 are both relevant to *F*-ratios based upon three groups, but as indicated in step 6, it is Table 6.1 that is specifically designed for the low/medium dispersion pattern. (Table 6.2 would have been used if the high dispersion pattern had been selected.)

Step 7 correspondingly informs the reader that the power for his/her study will be located at the intersection of the 0.50 ES column and the *N*/group = 50 row of Table 6.1, which is 0.60. This means, in effect, that our investigator will have a 60% chance of obtaining statistical significance if the assumptions he/she has made are correct. (Note that the power would have been only 0.73 if the high ES pattern had been hypothesized, which would have indicated that the three group means would have been hypothesized to be located at the extremes of the ES line (e.g., treatment as usual 0, attention placebo 0.10, and true acupuncture 0.50).)

The entire process could be communicated as follows:

> Based upon our preliminary work contrasting the exercise/education intervention with a treatment as usual control, an ES (*d*) of 0.50 was hypothesized between these two groups. In addition, we used the weight loss study conducted by Jones & Smith (2000) to estimate that the attention placebo condition would result in an ES of 0.20 as compared to cardiovascular patients who received routine clinical care only. Using the tables prepared by Bausell & Li (2002), we therefore estimated that 50 subjects per group would yield a 60% chance of the one-way ANOVA *F*-ratio reaching statistical significance at the 0.05 level.

Modeling the results

Since all of these parameters (*N*/group, largest pairwise ES, and the ES pattern) are obviously estimates prior to conducting an experiment, it is always a good idea to model different values for each. This is not a time consuming endeavor once the first power value has been computed and simply involves duplicating multiple copies of the templates and modeling as many different values of the above parameters as relevant. Should a 60% chance of achieving statistical significance not be deemed sufficient, as it probably would not, our investigator would have the option of changing the study design such as (a) employing only one control group, thereby producing an *N*/group of 75 and increasing power to 0.86 (Table 4.1), (b) adding covariates (see Chapter 7), or for certain types of studies, (c) employing a within subject design of some sort (see Chapter 8).

Estimating sample size

Tables 6.9 to 6.11 allow the investigator to calculate the minimum sample size needed to achieve two desired power levels (0.80 and 0.90) for experiments involving three, four, and five groups respectively. Basically the steps are the same here as they were for computing power, except that step 3 of the preliminary Template 6.1 requires that the investigator specify the desired level of power instead of the *N*/group. For present purposes, let us assume the minimum level of power our investigator is willing to accept is 0.80. Once this decision is made and the pattern of means is hypothesized (step 4), the last step in the preliminary template directs the investigator to the sample size template (6.3). Here, the first task (step 5) is the same as in Template 6.2, specifying the ES associated with the most powerful group (the exercise + dietary education intervention), which again is 0.50.

Step 6 then indicates that the *N*/group will be found at the intersection of the 0.50 ES column and the power = 0.80 row of Table 6.9, Chart C (i.e., *N*/group = 77). Should this value be deemed unrealistic, the researcher has the option of either (a) adopting one of the strategies listed in Chapter 2 (e.g., increasing the sample size, decreasing the number of groups, making the intervention stronger and/or the control groups weaker, employing a covariate or blocking variable, and so forth) or (b) accepting a lower power level. Let us therefore assume that this latter option was chosen and that the investigator was specifically interested in determining how many subjects per group would be necessary to achieve a power level of 0.70 instead of 0.80.

Following the instructions in step 7 of the sample size template, the investigator is instructed to access Table 6.1 from step 6 of the previous template and locate the closest approximation (or interpolated value) to the desired power of 0.70 in the 0.50 ES column. This value is 0.69, which corresponds to an *N*/group of 60 (or 61 if interpolation is used) located at the extreme left of the row containing said value. Thus, an *N*/group of approximately 60 (or a total sample size of 180) would be needed to yield a 70% chance of achieving statistical significance if the specified parameters were appropriate. The results of such a sample size analysis could be written in the same manner as suggested for the power analysis above, with a few relatively obvious changes:

> Based upon our preliminary work contrasting the exercise/education intervention with a treatment as usual control, an ES (*d*) of 0.50 was hypothesized between these two groups. In addition, we used the weight loss study conducted by Jones & Smith (2000) to estimate that the attention placebo condition would result in an ES of 0.20 as compared to cardiovascular patients who received routine clinical care only. Using the tables prepared by Bausell & Li (2002), we therefore estimated that 60 subjects per group would be necessary to provide a 70% chance of the one-way ANOVA *F*-ratio reaching statistical significance at the 0.05 level.

Part II. The multiple comparison tables

In many ways it is the second step in the evaluation of multiple group experiments that is crucial, since it is usually of little scientific interest to know that, say, statistically significant differences exist *somewhere* among three different experimental groups. What the researcher really needs to know is which specific groups differ from which other groups and this involves the computation of what are termed **multiple comparison**

procedures (MCPs). These are conceptually quite similar to *t*-tests in the sense that they test one mean difference at a time. They differ from the *t*-test, however, in that they avoid the inflation of false positive errors associated with repeated tests involving an alpha level of 0.05.[3]

A wide variety of MCPs are presently computed by most of the major statistical packages. Basically these procedures differ from one another with respect to how they go about protecting the study against false positive results. The method selected is quite relevant, however, because it can have a major impact upon the statistical power available for the pairwise comparisons among means (Li, 1997).

Some procedures are quite conservative, such as the Bonferroni *t* (the original developer of whom is unknown, although Dunn (1961) has perhaps done the most to explicate its use) which basically adjusts the alpha level by dividing it by the number of pairwise comparisons, while other Bonferroni-type adaptations such as Sidak's multiplicative inequality are slightly more powerful (Sidak, 1967). We have chosen to present two commonly employed MCPs based upon the Studentized range statistic: the relatively liberal Newman–Keuls (Newman, 1939; Keuls, 1952) which was the first multiple range test developed and Tukey's relatively conservative honestly significant difference (HSD) procedure (1984). For a more thorough treatment of additional options and their relative power, see Li (1997).

Power. Unlike estimating the power of the overall *F*-ratio, the hypothesized dispersion pattern among the means involved in an experiment does not enter into the estimation of power for pairwise contrasts. Only three power tables (Tables 6.12 through 6.14) and three sample size charts (Table 6.15 through 6.17) are necessary, therefore, to represent experiments involving three through five groups.

The same three templates, using the same parameter values (with the exception of the hypothesized dispersion pattern which is not needed) that were used to calculate the power/required sample size for the *F*-ratio are also employed for these analyses. To compute the power of MCPs for a three-group experiment, then, the preliminary template (6.1) is used in exactly the same way as described for estimating the power available for the overall *F*-ratio (Part I above) except that step 4 is superfluous.

The last instructions in Template 6.1 direct the researcher to Template 6.2 for determining power. Entering this template at step 8 under the heading "Power of the pairwise contrasts," the reader is instructed to perform the indicated subtractions based upon the ES line already completed. Since only three groups are involved in the present example, only three differences need to be calculated:

Step 8. Using the hypothesized values from step 2 in the preliminary template (6.1), fill out the hypothesized ES values by performing the indicated subtractions listed under the "Contrasts" column in the chart below.

ES ② − ① = 0.2 − 0 = 0.2
ES ③ − ① = 0.5 − 0 = 0.5
ES ③ − ② = 0.5 − 0.2 = 0.3

Step 9 next instructs the investigator to choose either the Tukey HSD or the Newman–Keuls procedure. The latter is a more liberal (i.e., powerful) test than is the Tukey procedure for all individual pairwise contrasts except for the one representing the largest possible ES. Once this decision has been made, the actual tables to be employed are provided in step 9.

These tables are set up and employed identically to the *t*-test tables in Chapters 4 and 5. For each of the three contrasts, the intersection between the ES column and the *N*/group row yields the power estimate for that particular ES. Thus, the Tukey HSD, which employs Table 6.12 for all three comparisons, yields power values of 0.09, 0.20, and 0.55 while Newman–Keuls yields higher estimates for two of the three contrasts:

ES	Power (Tukey HSD)	Power (Newman–Keuls)
[2 − 1] 0.2	0.09	0.16
[3 − 2] 0.3	0.20	0.32
[3 − 1] 0.5	0.55	0.55

Obviously there is not sufficient power for the treatment-as-usual vs. attention placebo contrast using either (or any) MCP. This lack of power for certain contrasts, in fact, is a characteristic of almost all multiple group experiments and should be taken into consideration at the design stage. There are other reasons for including multiple arms in a study than the power of pairwise contrasts, but if a major reason to include a particular condition does involve ascertaining whether or not it is statistically significant from another condition, and if the likelihood of achieving same is less than 10% as in the example above, the investigator might well reconsider the original design. Unfortunately there is not a great deal of power for any of these pairwise contrasts employing either procedure, hence the investigator might be wise to rethink his/her design at this stage. Assuming that this is not the case, however, the following power statement would be produced, which actually could be tacked onto the results for the overall *F*-ratio presented above:

Based upon our preliminary work contrasting the exercise/education intervention with a treatment-as-usual control, an ES (*d*) of 0.50 was hypothesized between these two groups. In addition, we used the weight loss study conducted by Jones & Smith (2000) to estimate that the attention placebo condition would result in an ES of 0.20 as compared to cardiovascular patients who received routine clinical care only. Using the tables prepared by Bausell & Li (2002), we therefore estimated that 50 subjects per group would yield a 60% chance of the one-way ANOVA *F*-ratio reaching statistical significance at the 0.05 level. These same parameters would produce a power of 0.55 for the exercise/education vs. treatment-as-usual contrast, although only 0.20 and 0.09 respectively for the intervention vs. attention placebo and the contrast (employing the Tukey HSD multiple comparison procedure).

Modeling the power of an MCP

As always, we encourage investigators to model different values (e.g., ES, *N*/group) when computing the power of an MCP. (This can be done by simply duplicating the relevant template and changing the parameter values as desired.) The power values provided by an MCP analysis can be especially helpful in selecting the optimal number and types of groups to be employed at the design phase. Taking our hypothetical example above, let us assume that the primary contrast of interest was the intervention vs. treatment-as-usual control and that the estimated power for this contrast (0.55) was indeed deemed not to be acceptable. Obviously the investigator has the usual options for improving this estimate (e.g., increasing the *N*/group, employing one or more covariates, increasing the ES by increasing the intensity of the treatment), but should none of these strategies prove viable he/she might opt to do away with the education control and redistribute the proposed total sample size ($N = 150$) to the resulting two groups, thereby producing a power level of 0.86 (see Table 4.1). Needless to say, scientific considerations must take precedence for decisions such as this, but ultimately science should not be conducted that does not have a reasonable chance of success. Let us assume, however, that increasing the sample size is a viable option, in which case a sample size analysis would be in order.

Estimating the required sample size to produce a desired power level. The second half of Template 6.3 is provided for this option. Its use predisposes the same preliminary steps as was indicated for estimating the power of an MCP except that the desired power would be inputted in step 3 of preliminary Template 6.1 instead of the *N*/group. The follow-up sample size template instructs the user to calculate the same pairwise ES values as was performed in step 8 of Template 6.2. Step 9 directs the investigator to the sample size table necessary for powers of 0.80 and 0.90 for both the Tukey and Newman–Keuls procedures. For other power levels, the reader is instructed to use the power tables suggested in step 9 of Template 6.2.

In way of illustration, let us assume that a power level of 0.80 was specified in step 3 of Template 6.1. To calculate the required sample sizes for the three ES values represented by the previously illustrated ES line (i.e., 0.2, 0.3, and 0.5), step 8 instructs the reader to access Table 6.15, Chart B for the Tukey HSD procedure. For the Newman–Keuls procedure, Chart A of the same table is mandated for contrasting group 1 vs. group 2 and group 2 vs. group 3 while Chart B is relevant for group 1 vs. group 3. Accessing these charts would produce the following eye-opening sample size requirements for the two procedures:

ES	*N*/group (Tukey HSD)	*N*/group (Newman–Keuls)
[2 − 1] 0.2	509	394
[3 − 2] 0.3	227	176
[3 − 1] 0.5	83	83

From these results it should be fairly obvious that a very large size trial will be required to provide an 80% chance of obtaining statistical significance of ES values of anything much less than 0.50 for individual contrasts. This stands in relatively sharp contrast to the *N*/group requirements for achieving an 80% chance of obtaining statistical significance for the overall *F*-ratio and serves to highlight the difference between omnibus and pairwise post hoc procedures. If, indeed, the investigator is satisfied with providing an 80% chance of obtaining statistical significance only for the single contrast indicated above (i.e., in the present example, the exercise/education intervention vs. treatment-as-usual), then perhaps it would be reasonable to conduct the present study. It probably would be necessary, however, to justify exactly *why* the second control was designed into this particular trial in the funding or IRB proposal.

Summary

The one-way between subjects analysis of variance is an extension of the *t*-test for independent samples to three or more groups. The technique involves a two-step process whereby (a) an overall omnibus *F* statistic is first computed to ascertain whether the multiple groups involved differ significantly from one another and, if they do, (b) a multiple comparison procedure is next computed to ascertain which of the individual groups differ from one another and which do not. Three templates are presented to facilitate the use of the power and sample size tables provided for this purpose. Through the use of examples, it is shown that it is often quite difficult to design a study in which sufficient power is present for all of the pairwise contrasts normally present in a multiple group study. While this is not problematic for comparisons of minimal scientific interest, a power analysis of this sort can force an investigator to make some difficult design decisions.

Endnotes

1 One relatively rare exception exists to the rule that multiple group studies must be analyzed by MCPs (also called a posteriori tests). If an investigator is only interested in a limited number of contrasts among his/her available groups, *and* if hypotheses involving these contrasts are specified prior to collecting data, then no protection of the alpha level is required. This means that, in effect, the investigator may perform multiple *t*-tests on his/her data and need not even perform an initial ANOVA as a gate-keeping step. Additionally, this strategy permits the combining of two or more groups into a single group to compare with another group (or combination of other groups).

While an effective procedure for increasing statistical power, a priori (or orthogonal) contrasts, as they are sometimes called, do not come without a heavy price. In the first place, only the number of groups minus one contrasts are permitted – which means that only two a priori contrasts are permitted for a three-group study, three comparisons for a four-group study, and so forth. The second price that must be paid resides in the requirement that the same group may not be used in more than one contrast. Hence in our three-group weight loss study, the intervention group could not be used in two separate contrasts involving the two control groups. What it could be compared to is the combination (or average) of the *two* control groups (i.e., treatment as usual and placebo), leaving one additional comparison, which in this case could only be between the two controls. (Algorithms exist in most standard texts (e.g., Bausell, 1986; Keppel, 1991) to illustrate which contrasts are and are not orthogonal to one another, although again this option is seldom used in modern day experimentation.)

2 These patterns represent the fact that the largest pairwise ES does not completely determine the size of an *F*-ratio (and therefore the power associated with a multiple group experiment), although certainly the larger this ES is, the more power will be available for the *F*-ratio. The size (and hence the power) of an *F*-ratio is actually proportionate to the average distance each mean (or ES) is from

the grand mean (or mean ES), hence if all of the means are as far from this midpoint as possible, the *F*-ratio will be larger than if some of the means fall very near (or exactly on) the grand mean.

3 If a researcher were to compute the ten *t*-tests necessary to ascertain whether all the mean differences contained within a five-group study are statistically significant, the effective level of significance rises from 0.05 to 0.40 $(1-(1-0.05)^{10}=0.40)$.

Template 6.1. Preliminary one-way between subjects ANOVA power/sample size template

This preliminary template is applicable to all one-way between subjects ANOVA designs employing between three and five groups.

Step 1. Write in the names/codes of the groups in the chart to the right in ascending order based upon their expected means (i.e., the name of the group expected to have the lowest mean or the weakest effect will be written next to ①, followed by the next strongest treatment and so forth).

Group definitions
①
②
③
④
⑤

Step 2. Plot these groups, using the circled numbers rather than their codes, on the ES line below. In the case of a tie between two groups with respect to their actual ES values, place one to the right of the other based upon the best theoretical evidence for which might be slightly superior.

ES line

①
____
0 0.1 0.2 0.3 0.4 0.5 0.6 0.7 0.8 0.9 1.0 1.1 1.2 1.3 1.4 1.5

Step 3. For the estimation of power, specify the *N*/group available. For required sample size, enter the desired power level.

N/group =   or, desired power =

Step 4. Compare the low, medium, and high ES patterns below with the graphed ES line from step 2 above. Choose the pattern which most closely matches the hypothesized pattern of means (step 2). Note that this step does not need to be performed for two-group designs (see Chapter 4), since there is only one possible pattern of means. Note also that three-group studies have only two patterns as we define them.

(a) Three-group designs

ES pattern for low/medium *F* power	①	②		③
ES pattern for high *F* power	①②			③
or	①			②③

(b) Four-group designs

ES pattern for low *F* power	①	②③		④
ES pattern for medium *F* power	①	②	③	④
ES pattern for high *F* power	①②			③④

(c) Five-group designs

ES pattern for low F power	①		②③④		⑤
ES pattern for medium F power	①	②	③	④	⑤
ES pattern for high F power	①②③				④⑤
or	①②				③④⑤

To compute power, turn to Template 6.2. To determine N/group, turn to Template 6.3.

Template 6.2. One-way between subjects ANOVA power template

Power of the overall F-ratio

Steps 1–4, see preliminary Template 6.1.

Step 5. Use the largest hypothesized ES produced in step 2 (Template 6.1). For two-group studies, this will be the ES for group ②, for three-group studies it will be the ES for group ③, and so forth.

Largest ES from preliminary step 2:

Step 6. Use the chart below to find the appropriate power table.

(a) For two groups use Table 4.1 in Chapter 4.

(b) For three groups, choose the appropriate model below:

Pattern of means	Table
L/M pattern	6.1
H pattern	6.2

(c) For four groups, choose the appropriate model below:

Pattern of means	Table
L pattern	6.3
M pattern	6.4
H pattern	6.5

(d) For five groups, choose the appropriate model below:

Pattern of means	Table
L pattern	6.6
M pattern	6.7
H pattern	6.8

Step 7. Turn to the table identified in step 6 and find the power at the intersection between the ES column from step 5 and the N/group specified in step 3 of the preliminary template. Interpolate as desired.

Power =

Power of the pairwise contrasts

Step 8. Using the hypothesized values from step 2 in the preliminary template (6.1), fill out the hypothesized ES values by performing the indicated subtractions listed under the "Contrasts" column in the chart below.

ES ② − ① = ES ④ − ① = ES ⑤ − ① =
ES ③ − ① = ES ④ − ② = ES ⑤ − ② =
ES ③ − ② = ES ④ − ③ = ES ⑤ − ③ =
ES ⑤ − ④ =

Step 9. For power of the above pairwise contrasts (step 8) using the **Tukey HSD** procedure, use the tables indicated in the "Power tables for the Tukey HSD procedure" chart. For the **Newman–Keuls** procedure, only sample size tables (see Template 6.3) are provided. (Note that the tables specified for the Tukey procedure are based *only* upon the number of groups and not the particular contrast involved while the tables used for Newman–Keuls are based *only* upon the contrasts involved and *not* the number of groups.) Locate the power in the indicated table at the intersection of the ES column (step 8) and the *N*/group row (preliminary step 3). Interpolate as desired.

Power tables for the Tukey HSD procedure

Three-group studies	Table 6.12
Four-group studies	Table 6.13
Five-group studies	Table 6.14

Power ② − ① = Power ④ − ① = Power ⑤ − ① =
Power ③ − ① = Power ④ − ② = Power ⑤ − ② =
Power ③ − ② = Power ④ − ③ = Power ⑤ − ③ =
Power ⑤ − ④ =

Template 6.3. One-way between subjects ANOVA sample size template

Required sample size for a statistically significant overall *F*-ratio

Steps 1–4, see the preliminary Template 6.1.

Step 5. Use the largest hypothesized ES produced in preliminary step 2. For two-group studies, this will be the ES for group ②, for three-group studies it will be the ES for group ③, and so forth.

Largest ES from preliminary step 2:

Step 6. For desired power values of 0.80 and 0.90, use the chart below to find the appropriate sample size table. (For desired power values other than 0.80 or 0.90, go to step 7.) Locate the *N*/group at the intersection of the ES column (step 5) and the desired power row. (Note that these tables also provide the necessary *N*/group for $p=0.01$ and $p=0.10$.)

(a) For two groups see Chapter 4.

(b) For three groups, choose the appropriate model below:

Pattern of means	$p=0.05$	$p=0.01$	$p=0.10$
L/M pattern	Table 6.9 Chart C	Table 6.9 Chart A	Table 6.9 Chart E
H pattern	Table 6.9 Chart D	Table 6.9 Chart B	Table 6.9 Chart F

(c) For four groups, choose the appropriate model below:

Pattern of means	$p=0.05$	$p=0.01$	$p=0.10$
L pattern	Table 6.10 Chart D	Table 6.10 Chart A	Table 6.10 Chart G
M pattern	Table 6.10 Chart E	Table 6.10 Chart B	Table 6.10 Chart H
H pattern	Table 6.10 Chart F	Table 6.10 Chart C	Table 6.10 Chart I

(d) For five groups, choose the appropriate model below:

Pattern of means	$p=0.05$	$p=0.01$	$p=0.10$
L pattern	Table 6.11 Chart D	Table 6.11 Chart A	Table 6.11 Chart G
M pattern	Table 6.11 Chart E	Table 6.11 Chart B	Table 6.11 Chart H
H pattern	Table 6.11 Chart F	Table 6.11 Chart C	Table 6.11 Chart I

Step 7. For desired powers other than 0.80 and 0.90, turn to the table identified in step 6 of the power template (6.2) and find the *N*/group to the left of the row in the ES column (from step 5) that most closely matches the desired power level. Interpolate as desired.

N/group =

Sample size for the pairwise contrasts

Step 8. Using the hypothesized values from step 2 in the preliminary template, fill out the hypothesized ES values by performing the indicated subtractions listed under the "Contrasts" column in the chart below.

ES ② − ① =	ES ④ − ① =	ES ⑤ − ① =
ES ③ − ① =	ES ④ − ② =	ES ⑤ − ② =
ES ③ − ② =	ES ④ − ③ =	ES ⑤ − ③ =
		ES ⑤ − ④ =

Step 9. For the required sample size for powers of 0.80 or 0.90 of the pairwise contrasts listed in step 8 for the **Tukey HSD** procedure, use the tables indicated in the "Tukey HSD" chart below. For powers of 0.80 or 0.90 for the **Newman–Keuls** procedure, use the tables indicated in the "Newman–Keuls" chart. (Note that the tables specified for the Tukey procedure are based *only* upon the number of groups and not the particular contrast involved while the tables used for Newman–Keuls are based *only* upon the contrasts involved and *not* the number of groups.) Locate the *N*/group in the indicated table at the intersection of the ES column (step 8) and the desired power row (preliminary step 3). For powers other than 0.80 and 0.90, use the appropriate tables indicated in step 9 of Template 6.2 by finding the nearest power value in the ES column (step 8) and reading the *N*/group associated with that row. Interpolate as desired.

Sample size tables for the Tukey HSD procedure

	$p=0.05$	$p=0.01$	$p=0.10$
Three-group studies	Table 6.15 Chart B	Table 6.16 Chart B	Table 6.17 Chart B
Four-group studies	Table 6.15 Chart E	Table 6.16 Chart E	Table 6.17 Chart E
Five-group studies	Table 6.15 Chart I	Table 6.16 Chart I	Table 6.17 Chart I

Sample size tables for the Newman–Keuls procedure

Determining the sample size requirements for the Newman–Keuls procedure is slightly more complicated since different charts within the Tables 6.15 ($p=0.05$), 6.16 ($p=0.01$), and 6.17 ($p=0.10$) are used to determine sample sizes for both the number of groups and the number of intervening variables.

Three groups

Contrast	Chart
② − ①	A
③ − ①	B
③ − ②	A

Four groups

Contrast	Chart	Contrast	Chart
② − ①	C	④ − ①	E
③ − ①	D	④ − ②	D
③ − ②	C	④ − ③	C

Five groups

Contrast	Chart	Contrast	Chart	Contrast	Chart
② − ①	F	④ − ①	H	⑤ − ①	I
③ − ①	G	③ − ①	G	⑤ − ②	H
③ − ②	F	④ − ③	F	⑤ − ③	G
				⑤ − ④	F

N/group ② − ① =	*N*/group ④ − ① =	*N*/group ⑤ − ① =
N/group ③ − ① =	*N*/group ④ − ② =	*N*/group ⑤ − ② =
N/group ③ − ② =	*N*/group ④ − ③ =	*N*/group ⑤ − ③ =
		N/group ⑤ − ④ =

Table 6.1. Power table for F-test; pattern L/M, 3 groups at alpha = 0.05

n	Hypothesized ES																		
	0.20	0.30	0.35	0.40	0.45	0.50	0.55	0.60	0.65	0.70	0.75	0.80	1.00	1.25	1.50	1.75	2.00	2.50	3.00
5	5	6	7	7	8	9	10	11	12	13	14	16	22	33	45	58	71	89	97
6	5	6	7	8	9	10	11	12	14	15	17	19	27	40	55	69	81	95	99
7	6	7	8	9	10	11	12	14	16	17	20	22	32	48	64	78	88	98	.
8	6	7	8	9	10	12	14	16	18	20	22	25	37	54	71	84	93	99	.
9	6	7	9	10	11	13	15	17	20	22	25	28	42	61	77	89	96	.	.
10	6	8	9	11	12	14	16	19	22	25	28	31	46	66	82	92	97	.	.
11	6	8	10	11	13	15	18	21	24	27	31	34	51	71	86	95	99	.	.
12	6	8	10	12	14	17	19	23	26	30	33	38	55	75	89	97	99	.	.
13	6	9	11	13	15	18	21	24	28	32	36	41	59	79	92	98	.	.	.
14	6	9	11	13	16	19	22	26	30	34	39	44	63	83	94	98	.	.	.
15	7	10	12	14	17	20	24	28	32	37	42	47	66	85	95	99	.	.	.
20	7	12	15	18	22	27	31	37	42	48	54	60	80	94	99	.	.	.	.
25	8	14	18	22	27	33	39	45	52	58	65	71	89	98	.	.	.	.	.
30	9	16	21	26	32	39	46	53	60	67	73	79	94	99	.	.	.	.	.
35	10	18	24	30	37	45	52	60	68	74	80	85	97	.	.	.	.	.	.
40	11	20	27	34	42	50	59	67	74	80	86	90	98	.	.	.	.	.	.
45	12	23	30	38	47	55	64	72	79	85	90	93	99	.	.	.	.	.	.
50	13	25	33	42	51	60	69	77	83	89	93	95	.	.	.	.	.	.	.
55	14	27	36	45	55	65	73	81	87	92	95	97	.	.	.	.	.	.	.
60	15	29	39	49	59	69	77	84	90	94	96	98	.	.	.	.	.	.	.
65	16	32	42	52	63	73	81	87	92	95	98	99	.	.	.	.	.	.	.
70	17	34	44	56	66	76	84	90	94	97	98	99	.	.	.	.	.	.	.
75	18	36	47	59	70	79	86	92	95	98	99	99	.	.	.	.	.	.	.
80	19	38	50	62	73	82	89	93	97	98	99	.	.	.	.	.	.	.	.
90	21	42	55	67	78	86	92	96	98	99	.	.	.	.	.	.	.	.	.
100	23	47	60	72	82	90	95	97	99	.	.	.	.	.	.	.	.	.	.
110	25	50	64	76	86	92	96	98	99	.	.	.	.	.	.	.	.	.	.
120	27	54	68	80	89	94	98	99	.	.	.	.	.	.	.	.	.	.	.
130	29	58	72	83	91	96	98	99	.	.	.	.	.	.	.	.	.	.	.
140	31	61	75	86	93	97	99	.	.	.	.	.	.	.	.	.	.	.	.
150	33	64	78	88	95	98	99	.	.	.	.	.	.	.	.	.	.	.	.
175	37	72	85	93	97	99	.	.	.	.	.	.	.	.	.	.	.	.	.
200	42	78	89	96	99	.	.	.	.	.	.	.	.	.	.	.	.	.	.
225	47	82	93	97	99	.	.	.	.	.	.	.	.	.	.	.	.	.	.
250	51	86	95	99	.	.	.	.	.	.	.	.	.	.	.	.	.	.	.
300	59	92	98	.	.	.	.	.	.	.	.	.	.	.	.	.	.	.	.
400	72	98	.	.	.	.	.	.	.	.	.	.	.	.	.	.	.	.	.
500	82	99	.	.	.	.	.	.	.	.	.	.	.	.	.	.	.	.	.

Table 6.2. Power table for *F*-test; pattern H, 3 groups at alpha = 0.05

n	Hypothesized ES																		
	0.20	0.30	0.35	0.40	0.45	0.50	0.55	0.60	0.65	0.70	0.75	0.80	1.00	1.25	1.50	1.75	2.00	2.50	3.00
5	6	7	7	8	9	10	11	13	14	16	18	20	29	42	57	71	83	96	99
6	6	7	8	9	10	12	13	15	17	19	21	24	35	52	68	82	91	99	.
7	6	7	9	10	11	13	15	17	20	22	25	28	42	60	77	89	95	.	.
8	6	8	9	11	13	15	17	20	22	25	29	32	48	68	83	93	98	.	.
9	6	8	10	12	14	16	19	22	25	29	32	36	53	74	88	96	99	.	.
10	6	9	11	13	15	18	21	24	28	32	36	40	59	79	92	98	.	.	.
11	7	9	11	14	16	19	23	27	31	35	40	44	64	83	94	99	.	.	.
12	7	10	12	15	18	21	25	29	34	38	43	48	68	87	96	99	.	.	.
13	7	10	13	16	19	23	27	31	36	41	47	52	72	90	97	.	.	.	.
14	7	11	14	17	20	24	29	34	39	44	50	56	76	92	98	.	.	.	.
15	7	12	14	18	22	26	31	36	42	47	53	59	79	94	99	.	.	.	.
20	9	14	18	23	28	34	41	47	54	61	67	73	90	98	.	.	.	.	.
25	10	17	22	28	35	42	50	57	65	72	78	83	96	.	.	.	.	.	.
30	11	20	26	34	42	50	58	66	73	80	85	90	98	.	.	.	.	.	.
35	12	23	31	39	48	57	65	73	80	86	91	94	99	.	.	.	.	.	.
40	13	26	35	44	53	63	72	79	86	91	94	96	.	.	.	.	.	.	.
45	15	29	39	49	59	69	77	84	90	94	96	98	.	.	.	.	.	.	.
50	16	32	42	53	64	73	82	88	93	96	98	99	.	.	.	.	.	.	.
55	17	35	46	57	68	78	85	91	95	97	99	99	.	.	.	.	.	.	.
60	19	38	50	61	72	81	88	93	96	98	99	.	.	.	.	.	.	.	.
65	20	41	53	65	76	85	91	95	98	99	.	.	.	.	.	.	.	.	.
70	21	44	56	69	79	87	93	96	98	99	.	.	.	.	.	.	.	.	.
75	23	46	60	72	82	90	94	97	99	.	.	.	.	.	.	.	.	.	.
80	24	49	63	75	85	91	96	98	99	.	.	.	.	.	.	.	.	.	.
90	27	54	68	80	89	94	98	99	.	.	.	.	.	.	.	.	.	.	.
100	29	59	73	84	92	96	99	.	.	.	.	.	.	.	.	.	.	.	.
110	32	63	77	88	94	98	99	.	.	.	.	.	.	.	.	.	.	.	.
120	34	67	81	90	96	99	.	.	.	.	.	.	.	.	.	.	.	.	.
130	37	71	84	93	97	99	.	.	.	.	.	.	.	.	.	.	.	.	.
140	40	74	87	94	98	99	.	.	.	.	.	.	.	.	.	.	.	.	.
150	42	77	89	96	99	.	.	.	.	.	.	.	.	.	.	.	.	.	.
175	48	84	93	98	99	.	.	.	.	.	.	.	.	.	.	.	.	.	.
200	54	89	96	99	.	.	.	.	.	.	.	.	.	.	.	.	.	.	.
225	59	92	98	.	.	.	.	.	.	.	.	.	.	.	.	.	.	.	.
250	64	95	99	.	.	.	.	.	.	.	.	.	.	.	.	.	.	.	.
300	72	97	.	.	.	.	.	.	.	.	.	.	.	.	.	.	.	.	.
400	85	.	.	.	.	.	.	.	.	.	.	.	.	.	.	.	.	.	.
500	92	.	.	.	.	.	.	.	.	.	.	.	.	.	.	.	.	.	.

Table 6.3. Power table for F-test; pattern L, 4 groups at alpha = 0.05

n	Hypothesized ES																		
	0.20	0.30	0.35	0.40	0.45	0.50	0.55	0.60	0.65	0.70	0.75	0.80	1.00	1.25	1.50	1.75	2.00	2.50	3.00
5	5	6	6	7	7	8	9	10	10	11	12	14	19	28	40	52	65	85	95
6	5	6	7	7	8	9	10	11	12	13	15	16	23	35	49	63	76	92	99
7	5	6	7	8	9	10	11	12	13	15	17	19	28	42	57	72	84	96	.
8	5	7	7	8	9	10	12	13	15	17	19	21	32	48	65	79	90	98	.
9	6	7	8	9	10	11	13	15	17	19	21	24	36	54	71	85	93	99	.
10	6	7	8	9	11	12	14	16	19	21	24	27	41	60	77	89	96	.	.
11	6	7	9	10	11	13	15	18	20	23	26	30	45	65	82	92	98	.	.
12	6	8	9	10	12	14	17	19	22	25	29	32	49	69	85	95	99	.	.
13	6	8	9	11	13	15	18	21	24	27	31	35	52	74	89	96	99	.	.
14	6	8	10	12	14	16	19	22	26	30	34	38	56	77	91	97	99	.	.
15	6	9	10	12	15	17	20	24	28	32	36	41	60	81	93	98	.	.	.
20	7	10	13	15	19	23	27	32	37	42	48	53	74	92	98	.	.	.	.
25	8	12	15	19	23	28	33	39	45	52	58	64	84	97	.	.	.	.	.
30	8	14	18	22	27	33	40	47	54	60	67	73	91	99	.	.	.	.	.
35	9	16	20	26	32	39	46	54	61	68	75	80	95	.	.	.	.	.	.
40	10	17	23	29	36	44	52	60	68	75	81	86	97	.	.	.	.	.	.
45	10	19	25	32	40	49	57	66	73	80	86	90	99	.	.	.	.	.	.
50	11	21	28	36	45	54	63	71	78	84	89	93	99	.	.	.	.	.	.
55	12	23	31	39	49	58	67	75	82	88	92	95	.	.	.	.	.	.	.
60	13	25	33	43	52	62	71	79	86	91	94	97	.	.	.	.	.	.	.
65	13	27	36	46	56	66	75	83	89	93	96	98	.	.	.	.	.	.	.
70	14	29	39	49	60	70	79	86	91	95	97	99	.	.	.	.	.	.	.
75	15	31	41	52	63	73	82	88	93	96	98	99	.	.	.	.	.	.	.
80	16	33	44	55	66	76	84	90	95	97	99	99	.	.	.	.	.	.	.
90	18	37	49	61	72	81	89	94	97	98	99	.	.	.	.	.	.	.	.
100	19	40	53	66	77	86	92	96	98	99	.	.	.	.	.	.	.	.	.
110	21	44	58	70	81	89	94	97	99	.	.	.	.	.	.	.	.	.	.
120	23	48	62	74	85	92	96	98	99	.	.	.	.	.	.	.	.	.	.
130	24	51	66	78	88	94	97	99	.	.	.	.	.	.	.	.	.	.	.
140	26	55	69	81	90	95	98	99	.	.	.	.	.	.	.	.	.	.	.
150	28	58	72	84	92	97	99	.	.	.	.	.	.	.	.	.	.	.	.
175	32	65	80	90	96	98	.	.	.	.	.	.	.	.	.	.	.	.	.
200	36	72	85	93	98	99	.	.	.	.	.	.	.	.	.	.	.	.	.
225	41	77	89	96	99	.	.	.	.	.	.	.	.	.	.	.	.	.	.
250	45	82	92	97	99	.	.	.	.	.	.	.	.	.	.	.	.	.	.
300	53	89	96	99	.	.	.	.	.	.	.	.	.	.	.	.	.	.	.
400	66	96	99	.	.	.	.	.	.	.	.	.	.	.	.	.	.	.	.
500	77	99	.	.	.	.	.	.	.	.	.	.	.	.	.	.	.	.	.

Table 6.4. Power table for F-test; pattern M, 4 groups at alpha = 0.05

n	Hypothesized ES																		
	0.20	0.30	0.35	0.40	0.45	0.50	0.55	0.60	0.65	0.70	0.75	0.80	1.00	1.25	1.50	1.75	2.00	2.50	3.00
5	5	6	7	7	8	8	9	10	11	12	13	15	21	31	44	57	70	89	97
6	5	6	7	8	8	9	10	11	13	14	16	17	26	39	54	68	80	95	99
7	5	7	7	8	9	10	12	13	15	16	18	20	31	46	62	77	88	98	.
8	6	7	8	9	10	11	13	14	16	19	21	23	35	53	70	84	93	99	.
9	6	7	8	9	11	12	14	16	18	21	24	26	40	59	76	89	96	.	.
10	6	7	9	10	11	13	15	18	20	23	26	30	45	65	82	92	97	.	.
11	6	8	9	11	12	14	17	19	22	25	29	33	49	70	86	95	99	.	.
12	6	8	10	11	13	16	18	21	24	28	32	36	53	74	89	97	99	.	.
13	6	8	10	12	14	17	20	23	26	30	34	39	57	78	92	98	.	.	.
14	6	9	10	13	15	18	21	24	28	33	37	42	61	82	94	99	.	.	.
15	6	9	11	13	16	19	22	26	30	35	40	45	65	85	95	99	.	.	.
20	7	11	14	17	21	25	30	35	40	46	52	58	79	94	99	.	.	.	.
25	8	13	16	21	25	31	37	43	50	57	63	69	88	98	.	.	.	.	.
30	9	15	19	24	30	37	44	51	58	66	72	78	94	99	.	.	.	.	.
35	9	17	22	28	35	43	51	58	66	73	79	85	97	.	.	.	.	.	.
40	10	19	25	32	40	48	57	65	73	79	85	90	98	.	.	.	.	.	.
45	11	21	28	36	45	54	62	71	78	84	89	93	99	.	.	.	.	.	.
50	12	23	31	40	49	58	68	76	83	88	92	95	.	.	.	.	.	.	.
55	13	25	34	43	53	63	72	80	86	91	95	97	.	.	.	.	.	.	.
60	14	28	37	47	57	67	76	84	90	94	96	98	.	.	.	.	.	.	.
65	15	30	40	50	61	71	80	87	92	95	98	99	.	.	.	.	.	.	.
70	16	32	42	54	65	75	83	89	94	97	98	99	.	.	.	.	.	.	.
75	16	34	45	57	68	78	86	92	95	98	99	.	.	.	.	.	.	.	.
80	17	36	48	60	71	81	88	93	96	98	99	.	.	.	.	.	.	.	.
90	19	40	53	66	77	86	92	96	98	99	.	.	.	.	.	.	.	.	.
100	21	45	58	71	81	89	94	97	99	.	.	.	.	.	.	.	.	.	.
110	23	49	63	75	85	92	96	98	99	.	.	.	.	.	.	.	.	.	.
120	25	52	67	79	88	94	98	99	.	.	.	.	.	.	.	.	.	.	.
130	27	56	71	83	91	96	98	99	.	.	.	.	.	.	.	.	.	.	.
140	29	59	74	86	93	97	99	.	.	.	.	.	.	.	.	.	.	.	.
150	31	63	77	88	95	98	99	.	.	.	.	.	.	.	.	.	.	.	.
175	35	70	84	93	97	99	.	.	.	.	.	.	.	.	.	.	.	.	.
200	40	76	89	96	99	.	.	.	.	.	.	.	.	.	.	.	.	.	.
225	45	82	92	97	99	.	.	.	.	.	.	.	.	.	.	.	.	.	.
250	49	86	95	99	.	.	.	.	.	.	.	.	.	.	.	.	.	.	.
300	57	92	98	.	.	.	.	.	.	.	.	.	.	.	.	.	.	.	.
400	71	98	.	.	.	.	.	.	.	.	.	.	.	.	.	.	.	.	.
500	81	99	.	.	.	.	.	.	.	.	.	.	.	.	.	.	.	.	.

Table 6.5. Power table for F-test; pattern H, 4 groups at alpha = 0.05

n	Hypothesized ES																		
	0.20	0.30	0.35	0.40	0.45	0.50	0.55	0.60	0.65	0.70	0.75	0.80	1.00	1.25	1.50	1.75	2.00	2.50	3.00
5	6	7	8	9	10	12	13	15	17	19	21	24	36	53	70	84	93	99	.
6	6	8	9	10	12	13	15	18	20	23	26	29	44	64	81	92	97	.	.
7	6	8	10	11	13	15	18	21	24	27	31	35	52	73	88	96	99	.	.
8	6	9	10	12	15	17	20	24	27	31	36	40	59	80	93	98	.	.	.
9	7	9	11	14	16	19	23	27	31	36	40	46	66	86	96	99	.	.	.
10	7	10	12	15	18	21	25	30	35	40	45	51	71	90	98	.	.	.	.
11	7	11	13	16	20	24	28	33	38	44	50	55	76	93	99	.	.	.	.
12	7	11	14	17	21	26	31	36	42	48	54	60	81	95	99	.	.	.	.
13	8	12	15	19	23	28	33	39	45	52	58	64	84	97	.	.	.	.	.
14	8	13	16	20	25	30	36	42	49	55	62	68	87	98	.	.	.	.	.
15	8	13	17	21	27	32	38	45	52	59	65	71	90	98	.	.	.	.	.
20	10	17	22	28	35	43	51	59	66	73	80	85	97	.	.	.	.	.	.
25	11	21	27	35	44	53	61	70	77	84	89	93	99	.	.	.	.	.	.
30	13	25	33	42	52	61	71	79	85	90	94	96	.	.	.	.	.	.	.
35	14	29	38	48	59	69	78	85	91	94	97	98	.	.	.	.	.	.	.
40	16	32	43	54	66	76	84	90	94	97	99	99	.	.	.	.	.	.	.
45	17	36	48	60	71	81	88	93	97	98	99	.	.	.	.	.	.	.	.
50	19	40	53	65	76	85	92	96	98	99	.	.	.	.	.	.	.	.	.
55	21	44	57	70	81	89	94	97	99	.	.	.	.	.	.	.	.	.	.
60	22	47	61	74	84	91	96	98	99	.	.	.	.	.	.	.	.	.	.
65	24	51	65	78	87	94	97	99	.	.	.	.	.	.	.	.	.	.	.
70	26	54	69	81	90	95	98	99	.	.	.	.	.	.	.	.	.	.	.
75	28	57	72	84	92	96	99	.	.	.	.	.	.	.	.	.	.	.	.
80	29	61	75	86	94	97	99	.	.	.	.	.	.	.	.	.	.	.	.
90	33	66	80	90	96	99	.	.	.	.	.	.	.	.	.	.	.	.	.
100	36	71	85	93	98	99	.	.	.	.	.	.	.	.	.	.	.	.	.
110	40	76	88	95	99	.	.	.	.	.	.	.	.	.	.	.	.	.	.
120	43	80	91	97	99	.	.	.	.	.	.	.	.	.	.	.	.	.	.
130	46	83	93	98	.	.	.	.	.	.	.	.	.	.	.	.	.	.	.
140	49	86	95	99	.	.	.	.	.	.	.	.	.	.	.	.	.	.	.
150	52	88	96	99	.	.	.	.	.	.	.	.	.	.	.	.	.	.	.
175	60	93	98	.	.	.	.	.	.	.	.	.	.	.	.	.	.	.	.
200	66	96	99	.	.	.	.	.	.	.	.	.	.	.	.	.	.	.	.
225	72	98	.	.	.	.	.	.	.	.	.	.	.	.	.	.	.	.	.
250	77	99	.	.	.	.	.	.	.	.	.	.	.	.	.	.	.	.	.
300	84	.	.	.	.	.	.	.	.	.	.	.	.	.	.	.	.	.	.
400	93	.	.	.	.	.	.	.	.	.	.	.	.	.	.	.	.	.	.
500	98	.	.	.	.	.	.	.	.	.	.	.	.	.	.	.	.	.	.

Table 6.6. Power table for *F*-test; pattern L, 5 groups at alpha = 0.05

n	Hypothesized ES																		
	0.20	0.30	0.35	0.40	0.45	0.50	0.55	0.60	0.65	0.70	0.75	0.80	1.00	1.25	1.50	1.75	2.00	2.50	3.00
5	5	6	6	7	7	8	8	9	10	10	11	12	17	26	36	48	60	82	94
6	5	6	6	7	7	8	9	10	11	12	13	14	21	32	45	59	72	90	98
7	5	6	7	7	8	9	10	11	12	14	15	17	25	38	53	68	80	95	99
8	5	6	7	8	9	10	11	12	14	15	17	19	29	44	60	75	87	98	.
9	5	7	7	8	9	10	12	13	15	17	19	21	33	49	67	81	91	99	.
10	6	7	8	9	10	11	13	15	17	19	21	24	36	55	73	86	94	.	.
11	6	7	8	9	10	12	14	16	18	21	23	26	40	60	78	90	96	.	.
12	6	7	8	10	11	13	15	17	20	23	26	29	44	65	82	93	98	.	.
13	6	7	9	10	12	14	16	18	21	24	28	31	48	69	86	95	99	.	.
14	6	8	9	11	12	15	17	20	23	26	30	34	51	73	88	96	99	.	.
15	6	8	9	11	13	15	18	21	25	28	32	36	55	76	91	97	.	.	.
20	7	9	11	14	17	20	24	28	33	38	43	48	70	89	97	.	.	.	.
25	7	11	13	17	21	25	30	35	41	47	53	59	81	95	99	.	.	.	.
30	8	12	16	20	24	30	36	42	49	56	62	69	88	98	.	.	.	.	.
35	8	14	18	23	28	35	42	49	56	63	70	76	93	99	.	.	.	.	.
40	9	15	20	26	32	40	47	55	63	70	77	82	96	.	.	.	.	.	.
45	9	17	23	29	36	44	53	61	69	76	82	87	98	.	.	.	.	.	.
50	10	19	25	32	40	49	58	66	74	81	86	91	99	.	.	.	.	.	.
55	11	20	27	35	44	53	62	71	78	85	90	93	99	.	.	.	.	.	.
60	11	22	30	38	48	57	67	75	82	88	92	95	.	.	.	.	.	.	.
65	12	24	32	41	51	61	71	79	86	91	94	97	.	.	.	.	.	.	.
70	13	26	34	44	55	65	74	82	88	93	96	98	.	.	.	.	.	.	.
75	14	27	37	47	58	68	78	85	91	95	97	99	.	.	.	.	.	.	.
80	14	29	39	50	61	72	80	87	93	96	98	99	.	.	.	.	.	.	.
90	16	33	44	56	67	77	85	91	95	98	99	.	.	.	.	.	.	.	.
100	17	36	48	61	72	82	89	94	97	99	99	.	.	.	.	.	.	.	.
110	19	40	53	66	77	86	92	96	98	99	.	.	.	.	.	.	.	.	.
120	20	43	57	70	81	89	94	97	99	.	.	.	.	.	.	.	.	.	.
130	22	46	61	74	84	92	96	98	99	.	.	.	.	.	.	.	.	.	.
140	23	50	64	77	87	94	97	99	.	.	.	.	.	.	.	.	.	.	.
150	25	53	68	80	89	95	98	99	.	.	.	.	.	.	.	.	.	.	.
175	29	60	75	87	94	98	99	.	.	.	.	.	.	.	.	.	.	.	.
200	33	67	81	91	97	99	.	.	.	.	.	.	.	.	.	.	.	.	.
225	36	73	86	94	98	99	.	.	.	.	.	.	.	.	.	.	.	.	.
250	40	78	90	96	99	.	.	.	.	.	.	.	.	.	.	.	.	.	.
300	48	85	95	99	.	.	.	.	.	.	.	.	.	.	.	.	.	.	.
400	61	94	99	.	.	.	.	.	.	.	.	.	.	.	.	.	.	.	.
500	72	98	.	.	.	.	.	.	.	.	.	.	.	.	.	.	.	.	.

Table 6.7. Power table for F-test; pattern M, 5 groups at alpha = 0.05

n	Hypothesized ES																		
	0.20	0.30	0.35	0.40	0.45	0.50	0.55	0.60	0.65	0.70	0.75	0.80	1.00	1.25	1.50	1.75	2.00	2.50	3.00
5	5	6	6	7	8	8	9	10	11	12	13	14	21	32	44	58	71	90	98
6	5	6	7	7	8	9	10	11	13	14	16	17	26	39	55	70	82	96	99
7	5	6	7	8	9	10	11	13	14	16	18	20	31	47	64	78	89	98	.
8	6	7	8	9	10	11	13	14	16	18	21	23	35	53	71	85	94	99	.
9	6	7	8	9	10	12	14	16	18	21	23	26	40	60	78	90	96	.	.
10	6	7	8	10	11	13	15	17	20	23	26	29	45	66	83	93	98	.	.
11	6	8	9	10	12	14	17	19	22	25	29	33	50	71	87	96	99	.	.
12	6	8	9	11	13	15	18	21	24	28	32	36	54	76	90	97	99	.	.
13	6	8	10	12	14	16	19	23	26	30	34	39	58	80	93	98	.	.	.
14	6	9	10	12	15	18	21	24	28	33	37	42	62	83	95	99	.	.	.
15	6	9	11	13	16	19	22	26	30	35	40	45	66	86	96	99	.	.	.
20	7	11	13	17	20	25	30	35	41	47	53	59	80	95	99	.	.	.	.
25	8	13	16	20	25	31	37	43	50	57	64	70	89	98	.	.	.	.	.
30	9	15	19	24	30	37	44	52	59	66	73	79	95	.	.	.	.	.	.
35	9	17	22	28	35	43	51	59	67	74	80	86	97	.	.	.	.	.	.
40	10	19	25	32	40	49	57	66	74	80	86	91	99	.	.	.	.	.	.
45	11	21	28	36	45	54	63	72	79	85	90	94	99	.	.	.	.	.	.
50	12	23	31	40	49	59	68	77	84	89	93	96	.	.	.	.	.	.	.
55	13	25	34	43	54	64	73	81	88	92	95	98	.	.	.	.	.	.	.
60	13	27	37	47	58	68	77	85	91	94	97	98	.	.	.	.	.	.	.
65	14	30	40	51	62	72	81	88	93	96	98	99	.	.	.	.	.	.	.
70	15	32	43	54	66	76	84	90	95	97	99	99	.	.	.	.	.	.	.
75	16	34	45	58	69	79	87	92	96	98	99	.	.	.	.	.	.	.	.
80	17	36	48	61	72	82	89	94	97	99	99	.	.	.	.	.	.	.	.
90	19	41	54	67	78	87	93	96	98	99	.	.	.	.	.	.	.	.	.
100	21	45	59	72	83	90	95	98	99	.	.	.	.	.	.	.	.	.	.
110	23	49	63	76	86	93	97	99	.	.	.	.	.	.	.	.	.	.	.
120	25	53	68	80	89	95	98	99	.	.	.	.	.	.	.	.	.	.	.
130	27	57	72	84	92	97	99	.	.	.	.	.	.	.	.	.	.	.	.
140	29	60	75	87	94	98	99	.	.	.	.	.	.	.	.	.	.	.	.
150	31	64	78	89	95	98	.	.	.	.	.	.	.	.	.	.	.	.	.
175	35	71	85	94	98	99	.	.	.	.	.	.	.	.	.	.	.	.	.
200	40	78	90	96	99	.	.	.	.	.	.	.	.	.	.	.	.	.	.
225	45	83	93	98	.	.	.	.	.	.	.	.	.	.	.	.	.	.	.
250	50	87	96	99	.	.	.	.	.	.	.	.	.	.	.	.	.	.	.
300	58	93	98	.	.	.	.	.	.	.	.	.	.	.	.	.	.	.	.
400	72	98	.	.	.	.	.	.	.	.	.	.	.	.	.	.	.	.	.
500	82	.	.	.	.	.	.	.	.	.	.	.	.	.	.	.	.	.	.

Table 6.8. Power table for *F*-test; pattern H, 5 groups at alpha = 0.05

n	Hypothesized ES																		
	0.20	0.30	0.35	0.40	0.45	0.50	0.55	0.60	0.65	0.70	0.75	0.80	1.00	1.25	1.50	1.75	2.00	2.50	3.00
5	6	7	8	9	10	12	13	15	17	20	22	25	38	57	75	88	95	.	.
6	6	8	9	10	12	14	16	18	21	24	28	31	47	69	85	95	99	.	.
7	6	8	10	11	13	16	19	22	25	29	33	37	56	77	92	98	.	.	.
8	6	9	11	13	15	18	21	25	29	33	38	43	63	84	95	99	.	.	.
9	7	10	12	14	17	20	24	28	33	38	43	49	70	89	98	.	.	.	.
10	7	10	13	15	19	23	27	32	37	43	48	54	76	93	99	.	.	.	.
11	7	11	14	17	20	25	30	35	41	47	53	59	81	95	99	.	.	.	.
12	7	12	15	18	22	27	33	38	45	51	58	64	85	97	.	.	.	.	.
13	8	12	16	20	24	30	35	42	48	55	62	68	88	98	.	.	.	.	.
14	8	13	17	21	26	32	38	45	52	59	66	72	91	99	.	.	.	.	.
15	8	14	18	22	28	34	41	48	55	63	69	76	93	99	.	.	.	.	.
20	10	18	23	30	37	46	54	63	70	77	83	88	98	.	.	.	.	.	.
25	11	22	29	37	47	56	65	74	81	87	92	95	.	.	.	.	.	.	.
30	13	26	35	45	55	65	75	83	89	93	96	98	.	.	.	.	.	.	.
35	15	30	40	52	63	73	82	89	93	96	98	99	.	.	.	.	.	.	.
40	16	34	46	58	70	80	87	93	96	98	99	.	.	.	.	.	.	.	.
45	18	39	51	64	75	85	91	96	98	99	.	.	.	.	.	.	.	.	.
50	20	43	56	69	80	89	94	97	99	.	.	.	.	.	.	.	.	.	.
55	22	47	61	74	84	92	96	98	99	.	.	.	.	.	.	.	.	.	.
60	24	51	65	78	88	94	97	99	.	.	.	.	.	.	.	.	.	.	.
65	25	54	69	82	90	96	98	99	.	.	.	.	.	.	.	.	.	.	.
70	27	58	73	85	93	97	99	.	.	.	.	.	.	.	.	.	.	.	.
75	29	61	76	87	94	98	99	.	.	.	.	.	.	.	.	.	.	.	.
80	31	64	79	90	96	99	.	.	.	.	.	.	.	.	.	.	.	.	.
90	35	70	84	93	98	99	.	.	.	.	.	.	.	.	.	.	.	.	.
100	39	75	88	96	99	.	.	.	.	.	.	.	.	.	.	.	.	.	.
110	42	80	91	97	99	.	.	.	.	.	.	.	.	.	.	.	.	.	.
120	46	84	94	98	.	.	.	.	.	.	.	.	.	.	.	.	.	.	.
130	49	87	95	99	.	.	.	.	.	.	.	.	.	.	.	.	.	.	.
140	53	89	97	99	.	.	.	.	.	.	.	.	.	.	.	.	.	.	.
150	56	91	98	.	.	.	.	.	.	.	.	.	.	.	.	.	.	.	.
175	63	95	99	.	.	.	.	.	.	.	.	.	.	.	.	.	.	.	.
200	70	97	.	.	.	.	.	.	.	.	.	.	.	.	.	.	.	.	.
225	76	99	.	.	.	.	.	.	.	.	.	.	.	.	.	.	.	.	.
250	80	99	.	.	.	.	.	.	.	.	.	.	.	.	.	.	.	.	.
300	88	.	.	.	.	.	.	.	.	.	.	.	.	.	.	.	.	.	.
400	96	.	.	.	.	.	.	.	.	.	.	.	.	.	.	.	.	.	.
500	99	.	.	.	.	.	.	.	.	.	.	.	.	.	.	.	.	.	.

Table 6.9. Sample size table for three-group independent samples ANOVA

Chart A. Independent samples ANOVA; pattern L/M, 3 groups at alpha = 0.01

Power	Hypothesized ES																		
	0.20	0.30	0.35	0.40	0.45	0.50	0.55	0.60	0.65	0.70	0.75	0.80	1.00	1.25	1.50	1.75	2.00	2.50	3.00
0.80	690	308	227	174	138	112	93	78	67	58	51	45	30	20	14	11	9	7	5
0.90	867	386	284	218	173	140	116	98	84	73	64	56	37	24	17	13	11	8	6

Chart B. Independent samples ANOVA; pattern H, 3 groups at alpha = 0.01

Power	Hypothesized ES																		
	0.20	0.30	0.35	0.40	0.45	0.50	0.55	0.60	0.65	0.70	0.75	0.80	1.00	1.25	1.50	1.75	2.00	2.50	3.00
0.80	518	231	170	131	104	85	70	59	51	44	39	34	23	15	11	9	7	6	5
0.90	650	290	214	164	130	106	88	74	63	55	48	43	28	19	14	11	9	6	5

Chart C. Independent samples ANOVA; pattern L/M, 3 groups at alpha = 0.05

Power	Hypothesized ES																		
	0.20	0.30	0.35	0.40	0.45	0.50	0.55	0.60	0.65	0.70	0.75	0.80	1.00	1.25	1.50	1.75	2.00	2.50	3.00
0.80	476	212	156	120	95	77	64	54	46	40	35	31	21	14	10	8	6	5	4
0.90	627	279	206	158	125	102	84	71	61	53	46	41	27	18	13	10	8	6	4

Chart D. Independent samples ANOVA; pattern H, 3 groups at alpha = 0.05

Power	Hypothesized ES																		
	0.20	0.30	0.35	0.40	0.45	0.50	0.55	0.60	0.65	0.70	0.75	0.80	1.00	1.25	1.50	1.75	2.00	2.50	3.00
0.80	357	160	118	90	72	58	49	41	35	31	27	24	16	11	8	6	5	4	3
0.90	470	210	155	119	94	77	64	54	46	40	35	31	20	14	10	8	6	5	4

Chart E. Independent samples ANOVA; pattern L/M, 3 groups at alpha = 0.10

Power	Hypothesized ES																		
	0.20	0.30	0.35	0.40	0.45	0.50	0.55	0.60	0.65	0.70	0.75	0.80	1.00	1.25	1.50	1.75	2.00	2.50	3.00
0.80	379	169	125	96	76	62	51	43	37	32	28	25	16	11	8	6	5	4	3
0.90	517	230	170	130	103	84	69	59	50	43	38	34	22	15	11	8	7	5	4

Chart F. Independent samples ANOVA; pattern H, 3 groups at alpha = 0.10

Power	Hypothesized ES																		
	0.20	0.30	0.35	0.40	0.45	0.50	0.55	0.60	0.65	0.70	0.75	0.80	1.00	1.25	1.50	1.75	2.00	2.50	3.00
0.80	285	127	94	72	57	47	39	33	28	24	21	19	13	9	6	5	4	3	3
0.90	388	173	127	98	78	63	52	44	38	33	29	25	17	11	8	6	5	4	3

Table 6.10. Sample size table for four-group independent samples ANOVA

Chart A. Independent samples ANOVA; pattern L, 4 groups at alpha = 0.01

Power	Hypothesized ES																		
	0.20	0.30	0.35	0.40	0.45	0.50	0.55	0.60	0.65	0.70	0.75	0.80	1.00	1.25	1.50	1.75	2.00	2.50	3.00
0.80	769	343	252	194	154	125	103	87	75	65	57	50	33	22	16	12	10	7	5
0.90	959	427	314	241	191	155	129	108	93	80	70	62	40	26	19	14		8	6

Chart B. Independent samples ANOVA; pattern M, 4 groups at alpha = 0.01

Power	Hypothesized ES																		
	0.20	0.30	0.35	0.40	0.45	0.50	0.55	0.60	0.65	0.70	0.75	0.80	1.00	1.25	1.50	1.75	2.00	2.50	3.00
0.80	693	309	227	175	138	112	93	79	67	58	51	45	30	20	14	11	9	6	5
0.90	863	385	283	217	172	140	116	98	83	72	63	56	36	24	17	13		8	6

Chart C. Independent samples ANOVA; pattern H, 4 groups at alpha = 0.01

Power	Hypothesized ES																		
	0.20	0.30	0.35	0.40	0.45	0.50	0.55	0.60	0.65	0.70	0.75	0.80	1.00	1.25	1.50	1.75	2.00	2.50	3.00
0.80	386	172	127	98	78	63	53	45	38	33	29	26	17	12	9	7	6	5	4
0.90	481	215	158	122	96	79	65	55	47	41	36	32	21	14	11	8		5	4

Chart D. Independent samples ANOVA; pattern L, 4 groups at alpha = 0.05

Power	Hypothesized ES																		
	0.20	0.30	0.35	0.40	0.45	0.50	0.55	0.60	0.65	0.70	0.75	0.80	1.00	1.25	1.50	1.75	2.00	2.50	3.00
0.80	540	241	177	135	108	88	73	61	53	46	40	35	23	15	11	9	7	5	4
0.90	705	314	231	177	140	114	95	80	68	59	52	45	30	20	14	11	9	6	5

Chart E. Independent samples ANOVA; pattern M, 4 groups at alpha = 0.05

Power	Hypothesized ES																		
	0.20	0.30	0.35	0.40	0.45	0.50	0.55	0.60	0.65	0.70	0.75	0.80	1.00	1.25	1.50	1.75	2.00	2.50	3.00
0.80	487	217	160	123	97	79	66	55	47	41	36	32	21	14	10	8	6	5	4
0.90	635	283	208	160	127	103	85	72	61	53	47	41	27	18	13	10	8	6	4

Chart F. Independent samples ANOVA; pattern H, 4 groups at alpha = 0.05

Power	Hypothesized ES																		
	0.20	0.30	0.35	0.40	0.45	0.50	0.55	0.60	0.65	0.70	0.75	0.80	1.00	1.25	1.50	1.75	2.00	2.50	3.00
0.80	271	121	89	69	55	45	37	31	27	24	21	18	12	8	6	5	4	3	3
0.90	353	158	116	89	71	58	48	41	35	30	27	23	16	11	8	6	5	4	3

Table 6.10. (*cont.*)

Chart G. Independent samples ANOVA; pattern L, 4 groups at alpha = 0.10																			
Power	Hypothesized ES																		
	0.20	0.30	0.35	0.40	0.45	0.50	0.55	0.60	0.65	0.70	0.75	0.80	1.00	1.25	1.50	1.75	2.00	2.50	3.00
0.80	435	194	143	110	87	71	59	49	42	37	32	28	19	12	9	7	6	4	3
0.90	586	261	192	148	117	95	79	66	57	49	43	38	25	16	12	9	7	5	4

Chart H. Independent samples ANOVA; pattern M, 4 groups at alpha = 0.10																			
Power	Hypothesized ES																		
	0.20	0.30	0.35	0.40	0.45	0.50	0.55	0.60	0.65	0.70	0.75	0.80	1.00	1.25	1.50	1.75	2.00	2.50	3.00
0.80	391	175	129	99	78	64	53	45	38	33	29	26	17	11	8	6	5	4	3
0.90	528	235	173	133	105	86	71	60	51	44	39	34	22	15	11	8	7	5	4

Chart I. Independent samples ANOVA; pattern H, 4 groups at alpha = 0.10																			
Power	Hypothesized ES																		
	0.20	0.30	0.35	0.40	0.45	0.50	0.55	0.60	0.65	0.70	0.75	0.80	1.00	1.25	1.50	1.75	2.00	2.50	3.00
0.80	218	98	72	55	44	36	30	25	22	19	17	15	10	7	5	4	4	3	3
0.90	294	131	97	74	59	48	40	34	29	25	22	20	13	9	7	5	4	3	3

Table 6.11. Sample size table for five-group independent samples ANOVA

Chart A. Independent samples ANOVA; pattern L, 5 groups at alpha = 0.01

Power	Hypothesized ES																		
	0.20	0.30	0.35	0.40	0.45	0.50	0.55	0.60	0.65	0.70	0.75	0.80	1.00	1.25	1.50	1.75	2.00	2.50	3.00
0.80	834	372	274	210	166	135	112	94	81	70	61	54	35	23	17	13	10	7	6
0.90	1035	461	339	260	206	167	138	117	100	86	75	66	43	28	20	15	12	9	7

Chart B. Independent samples ANOVA; pattern M, 5 groups at alpha = 0.01

Power	Hypothesized ES																		
	0.20	0.30	0.35	0.40	0.45	0.50	0.55	0.60	0.65	0.70	0.75	0.80	1.00	1.25	1.50	1.75	2.00	2.50	3.00
0.80	668	298	219	168	133	108	90	76	65	56	49	43	28	19	14	11	9	6	5
0.90	828	369	272	208	165	134	111	94	80	69	61	53	35	23	17	13	10	7	6

Chart C. Independent samples ANOVA; pattern H, 5 groups at alpha = 0.01

Power	Hypothesized ES																		
	0.20	0.30	0.35	0.40	0.45	0.50	0.55	0.60	0.65	0.70	0.75	0.80	1.00	1.25	1.50	1.75	2.00	2.50	3.00
0.80	349	156	115	89	70	57	48	40	35	30	27	24	16	11	8	6	5	4	4
0.90	432	193	142	109	87	71	59	50	43	37	32	29	19	13	10	8	6	5	4

Chart D. Independent samples ANOVA; pattern L, 5 groups at alpha = 0.05

Power	Hypothesized ES																		
	0.20	0.30	0.35	0.40	0.45	0.50	0.55	0.60	0.65	0.70	0.75	0.80	1.00	1.25	1.50	1.75	2.00	2.50	3.00
0.80	593	264	195	149	118	96	80	67	57	50	44	38	25	17	12	9	7	5	4
0.90	768	342	252	193	153	124	103	87	74	64	56	49	32	21	15	12	9	6	5

Chart E. Independent samples ANOVA; pattern M, 5 groups at alpha = 0.05

Power	Hypothesized ES																		
	0.20	0.30	0.35	0.40	0.45	0.50	0.55	0.60	0.65	0.70	0.75	0.80	1.00	1.25	1.50	1.75	2.00	2.50	3.00
0.80	475	212	156	120	95	77	64	54	46	40	35	31	20	14	10	8	6	5	4
0.90	615	274	202	155	123	100	83	70	60	52	45	40	26	17	12	10	8	5	4

Chart F. Independent samples ANOVA; pattern H, 5 groups at alpha = 0.05

Power	Hypothesized ES																		
	0.20	0.30	0.35	0.40	0.45	0.50	0.55	0.60	0.65	0.70	0.75	0.80	1.00	1.25	1.50	1.75	2.00	2.50	3.00
0.80	248	111	82	63	50	41	34	29	25	22	19	17	11	8	6	5	4	3	3
0.90	321	143	106	81	65	53	44	37	32	28	24	21	14	10	7	6	5	4	3

Table 6.11. (*cont.*)

Chart G. Independent samples ANOVA; pattern L, 5 groups at alpha = 0.10																			
Power	Hypothesized ES																		
	0.20	0.30	0.35	0.40	0.45	0.50	0.55	0.60	0.65	0.70	0.75	0.80	1.00	1.25	1.50	1.75	2.00	2.50	3.00
0.80	480	214	158	121	96	78	65	54	47	40	35	31	20	14	10	8	6	4	4
0.90	642	286	211	162	128	104	86	72	62	54	47	41	27	18	13	10	8	5	4

Chart H. Independent samples ANOVA; pattern M, 5 groups at alpha = 0.10																			
Power	Hypothesized ES																		
	0.20	0.30	0.35	0.40	0.45	0.50	0.55	0.60	0.65	0.70	0.75	0.80	1.00	1.25	1.50	1.75	2.00	2.50	3.00
0.80	384	171	126	97	77	63	52	44	38	33	29	25	17	11	8	6	5	4	3
0.90	514	229	169	129	103	83	69	58	50	43	38	33	22	14	10	8	6	5	4

Chart I. Independent samples ANOVA; pattern H, 5 groups at alpha = 0.10																			
Power	Hypothesized ES																		
	0.20	0.30	0.35	0.40	0.45	0.50	0.55	0.60	0.65	0.70	0.75	0.80	1.00	1.25	1.50	1.75	2.00	2.50	3.00
0.80	201	90	66	51	41	33	28	23	20	18	16	14	9	6	5	4	3	3	2
0.90	268	120	88	68	54	44	37	31	27	23	20	18	12	8	6	5	4	3	3

Table 6.12. Power table for independent samples ANOVA for Tukey HSD; 3 groups at alpha = 0.05

n	Hypothesized ES																		
	0.20	0.30	0.35	0.40	0.45	0.50	0.55	0.60	0.65	0.70	0.75	0.80	1.00	1.25	1.50	1.75	2.00	2.50	3.00
5	2	3	3	4	4	5	6	7	7	9	10	11	17	27	40	53	67	87	97
6	2	3	4	4	5	6	7	8	9	11	12	14	22	35	50	65	78	94	99
7	2	3	4	5	6	7	8	9	11	13	15	17	27	42	59	75	86	97	.
8	2	4	4	5	7	8	9	11	13	15	17	20	31	49	67	82	92	99	.
9	3	4	5	6	7	9	11	12	15	17	20	23	36	56	74	87	95	.	.
10	3	4	5	7	8	10	12	14	17	19	22	26	41	62	80	91	97	.	.
11	3	5	6	7	9	11	13	16	18	22	25	29	45	67	84	94	98	.	.
12	3	5	6	8	10	12	14	17	20	24	28	32	50	72	88	96	99	.	.
13	3	5	7	9	11	13	16	19	22	26	30	35	54	76	91	97	99	.	.
14	3	6	7	9	11	14	17	21	24	29	33	38	58	80	93	98	.	.	.
15	3	6	8	10	12	15	19	22	27	31	36	41	62	83	95	99	.	.	.
20	4	8	10	13	17	21	26	31	37	43	49	55	77	93	99	.	.	.	.
25	5	10	13	17	22	27	33	39	46	53	60	67	87	98	.	.	.	.	.
30	6	11	16	21	26	33	40	48	55	63	70	76	93	99	.	.	.	.	.
35	6	13	18	24	31	39	47	55	63	71	77	83	96	.	.	.	.	.	.
40	7	15	21	28	36	45	53	62	70	77	83	88	98	.	.	.	.	.	.
45	8	17	24	32	41	50	59	68	76	83	88	92	99	.	.	.	.	.	.
50	9	20	27	36	45	55	65	73	81	87	91	95	.	.	.	.	.	.	.
55	10	22	30	40	50	60	70	78	85	90	94	97	.	.	.	.	.	.	.
60	10	24	33	43	54	65	74	82	88	93	96	98	.	.	.	.	.	.	.
65	11	26	36	47	58	69	78	85	91	95	97	99	.	.	.	.	.	.	.
70	12	28	39	50	62	72	81	88	93	96	98	99	.	.	.	.	.	.	.
75	13	30	41	54	65	76	84	90	95	97	99	99	.	.	.	.	.	.	.
80	14	32	44	57	69	79	87	92	96	98	99	.	.	.	.	.	.	.	.
90	16	37	50	63	74	84	91	95	98	99	.	.	.	.	.	.	.	.	.
100	17	41	55	68	79	88	94	97	99	.	.	.	.	.	.	.	.	.	.
110	19	45	59	73	84	91	96	98	99	.	.	.	.	.	.	.	.	.	.
120	21	49	64	77	87	93	97	99	.	.	.	.	.	.	.	.	.	.	.
130	23	53	68	81	90	95	98	99	.	.	.	.	.	.	.	.	.	.	.
140	25	56	72	84	92	97	99	.	.	.	.	.	.	.	.	.	.	.	.
150	27	60	75	87	94	98	99	.	.	.	.	.	.	.	.	.	.	.	.
175	32	68	82	92	97	99	.	.	.	.	.	.	.	.	.	.	.	.	.
200	36	74	87	95	98	.	.	.	.	.	.	.	.	.	.	.	.	.	.
225	41	80	91	97	99	.	.	.	.	.	.	.	.	.	.	.	.	.	.
250	46	84	94	98	.	.	.	.	.	.	.	.	.	.	.	.	.	.	.
300	54	91	97	99	.	.	.	.	.	.	.	.	.	.	.	.	.	.	.
400	68	97	.	.	.	.	.	.	.	.	.	.	.	.	.	.	.	.	.
500	79	99	.	.	.	.	.	.	.	.	.	.	.	.	.	.	.	.	.

Table 6.13. Power table for independent samples ANOVA for Tukey HSD; 4 groups at alpha = 0.05

n	Hypothesized ES																		
	0.20	0.30	0.35	0.40	0.45	0.50	0.55	0.60	0.65	0.70	0.75	0.80	1.00	1.25	1.50	1.75	2.00	2.50	3.00
5	1	2	2	2	3	3	4	4	5	6	7	8	13	21	33	47	61	84	95
6	1	2	2	3	3	4	5	5	6	7	9	10	16	28	43	58	73	92	99
7	1	2	3	3	4	5	5	6	8	9	10	12	20	35	52	68	82	96	.
8	1	2	3	3	4	5	6	8	9	11	12	14	25	41	60	77	88	98	.
9	2	2	3	4	5	6	7	9	10	12	14	17	29	48	67	83	93	99	.
10	2	3	3	4	5	7	8	10	12	14	17	19	33	54	74	88	96	.	.
11	2	3	4	5	6	7	9	11	13	16	19	22	37	60	79	91	97	.	.
12	2	3	4	5	7	8	10	12	15	18	21	25	42	65	83	94	98	.	.
13	2	3	4	6	7	9	11	14	17	20	23	27	46	69	87	96	99	.	.
14	2	4	5	6	8	10	12	15	18	22	26	30	50	74	90	97	99	.	.
15	2	4	5	7	8	11	13	16	20	24	28	33	54	77	92	98	.	.	.
20	3	5	7	9	12	15	19	24	29	34	40	46	70	90	98	.	.	.	.
25	3	6	9	12	16	20	26	31	38	45	51	58	82	96	.	.	.	.	.
30	4	8	11	15	20	25	32	39	46	54	62	69	89	99	.	.	.	.	.
35	4	9	13	18	24	31	38	46	55	63	70	77	94	.	.	.	.	.	.
40	5	11	15	21	28	36	45	53	62	70	77	83	97	.	.	.	.	.	.
45	5	12	18	25	32	41	51	60	69	76	83	88	98	.	.	.	.	.	.
50	6	14	20	28	37	46	56	66	74	82	87	92	99	.	.	.	.	.	.
55	6	16	23	31	41	51	62	71	79	86	91	94	.	.	.	.	.	.	.
60	7	17	25	35	45	56	66	76	83	89	93	96	.	.	.	.	.	.	.
65	8	19	28	38	49	60	71	80	87	92	95	98	.	.	.	.	.	.	.
70	8	21	30	41	53	64	75	83	89	94	97	98	.	.	.	.	.	.	.
75	9	23	33	45	57	68	78	86	92	95	98	99	.	.	.	.	.	.	.
80	9	25	36	48	60	72	81	89	94	97	98	99	.	.	.	.	.	.	.
90	11	29	41	54	67	78	87	92	96	98	99	.	.	.	.	.	.	.	.
100	12	32	46	60	73	83	90	95	98	99	.	.	.	.	.	.	.	.	.
110	14	36	51	65	77	87	93	97	99	.	.	.	.	.	.	.	.	.	.
120	15	40	55	70	82	90	95	98	99	.	.	.	.	.	.	.	.	.	.
130	17	44	60	74	85	93	97	99	.	.	.	.	.	.	.	.	.	.	.
140	18	47	64	78	88	95	98	99	.	.	.	.	.	.	.	.	.	.	.
150	20	51	67	81	91	96	99	.	.	.	.	.	.	.	.	.	.	.	.
175	24	59	76	88	95	98	99	.	.	.	.	.	.	.	.	.	.	.	.
200	28	66	82	92	97	99	.	.	.	.	.	.	.	.	.	.	.	.	.
225	33	73	87	95	99	.	.	.	.	.	.	.	.	.	.	.	.	.	.
250	37	78	91	97	99	.	.	.	.	.	.	.	.	.	.	.	.	.	.
300	45	86	96	99	.	.	.	.	.	.	.	.	.	.	.	.	.	.	.
400	60	95	99	.	.	.	.	.	.	.	.	.	.	.	.	.	.	.	.
500	72	99	.	.	.	.	.	.	.	.	.	.	.	.	.	.	.	.	.

Table 6.14. Power table for independent samples ANOVA for Tukey HSD; 5 groups at alpha = 0.05

n	Hypothesized ES																		
	0.20	0.30	0.35	0.40	0.45	0.50	0.55	0.60	0.65	0.70	0.75	0.80	1.00	1.25	1.50	1.75	2.00	2.50	3.00
5	1	1	1	2	2	2	3	3	4	4	5	6	10	18	29	42	56	81	94
6	1	1	2	2	2	3	3	4	5	6	7	8	13	24	38	53	69	90	98
7	1	1	2	2	3	3	4	5	6	7	8	9	17	30	46	64	78	95	99
8	1	2	2	2	3	4	5	6	7	8	10	11	20	36	55	72	86	98	.
9	1	2	2	3	4	4	5	7	8	10	11	13	24	42	62	79	91	99	.
10	1	2	2	3	4	5	6	8	9	11	13	16	28	48	69	85	94	.	.
11	1	2	3	3	4	6	7	9	10	13	15	18	32	54	75	89	96	.	.
12	1	2	3	4	5	6	8	10	12	14	17	20	36	59	80	92	98	.	.
13	1	2	3	4	5	7	9	11	13	16	19	23	40	64	84	94	99	.	.
14	1	3	3	4	6	7	9	12	15	18	21	25	44	69	87	96	99	.	.
15	1	3	4	5	6	8	10	13	16	20	23	28	48	73	90	97	.	.	.
20	2	4	5	7	9	12	15	19	24	29	34	40	65	87	97	.	.	.	.
25	2	5	7	9	12	16	21	26	32	39	45	52	77	95	99	.	.	.	.
30	2	6	8	12	16	21	27	33	40	48	56	63	86	98	.	.	.	.	.
35	3	7	10	14	19	26	33	40	48	57	65	72	92	99	.	.	.	.	.
40	3	8	12	17	23	30	39	47	56	65	72	79	96	.	.	.	.	.	.
45	4	9	14	20	27	35	44	54	63	71	79	85	98	.	.	.	.	.	.
50	4	11	16	23	31	40	50	60	69	77	84	89	99	.	.	.	.	.	.
55	5	12	18	26	35	45	55	65	74	82	88	92	99	.	.	.	.	.	.
60	5	14	21	29	39	50	60	70	79	86	91	95	.	.	.	.	.	.	.
65	6	15	23	32	43	54	65	75	83	89	94	96	.	.	.	.	.	.	.
70	6	17	25	35	47	58	69	79	86	92	95	98	.	.	.	.	.	.	.
75	7	18	28	39	51	63	73	82	89	94	97	98	.	.	.	.	.	.	.
80	7	20	30	42	54	66	77	85	91	95	98	99	.	.	.	.	.	.	.
90	8	23	35	48	61	73	83	90	95	97	99	.	.	.	.	.	.	.	.
100	9	27	40	54	67	79	87	93	97	99	99	.	.	.	.	.	.	.	.
110	11	30	44	59	73	83	91	96	98	99	.	.	.	.	.	.	.	.	.
120	12	34	49	64	77	87	94	97	99	.	.	.	.	.	.	.	.	.	.
130	13	38	53	69	81	90	95	98	99	.	.	.	.	.	.	.	.	.	.
140	14	41	58	73	85	93	97	99	.	.	.	.	.	.	.	.	.	.	.
150	16	45	62	77	88	94	98	99	.	.	.	.	.	.	.	.	.	.	.
175	19	53	71	84	93	97	99	.	.	.	.	.	.	.	.	.	.	.	.
200	23	61	78	90	96	99	.	.	.	.	.	.	.	.	.	.	.	.	.
225	27	67	84	93	98	99	.	.	.	.	.	.	.	.	.	.	.	.	.
250	31	73	88	96	99	.	.	.	.	.	.	.	.	.	.	.	.	.	.
300	39	83	94	98	.	.	.	.	.	.	.	.	.	.	.	.	.	.	.
400	54	93	99	.	.	.	.	.	.	.	.	.	.	.	.	.	.	.	.
500	67	98	.	.	.	.	.	.	.	.	.	.	.	.	.	.	.	.	.

Table 6.15. Sample size table for between subjects ANOVA, MCP; $p = 0.05$

Chart A. 3 groups at alpha = 0.05; NK no intervening groups

Power	Hypothesized ES																		
	0.20	0.30	0.35	0.40	0.45	0.50	0.55	0.60	0.65	0.70	0.75	0.80	1.00	1.25	1.50	1.75	2.00	2.50	3.00
0.80	394	176	129	99	79	64	53	45	38	33	29	26	17	11	8	7	5	4	3
0.90	527	235	173	133	105	85	71	60	51	44	39	34	22	15	11	8	7	5	4

Chart B. 3 groups at alpha = 0.05; NK 1 intervening group, Tukey 3 groups

Power	Hypothesized ES																		
	0.20	0.30	0.35	0.40	0.45	0.50	0.55	0.60	0.65	0.70	0.75	0.80	1.00	1.25	1.50	1.75	2.00	2.50	3.00
0.80	509	227	167	128	102	83	69	58	50	43	38	33	22	15	11	8	7	5	4
0.90	659	294	216	166	131	107	88	74	64	55	48	43	28	18	13	10	8	6	5

Chart C. 4 groups at alpha = 0.05; NK no intervening groups

Power	Hypothesized ES																		
	0.20	0.30	0.35	0.40	0.45	0.50	0.55	0.60	0.65	0.70	0.75	0.80	1.00	1.25	1.50	1.75	2.00	2.50	3.00
0.80	394	175	126	99	79	64	53	45	38	33	29	26	17	11	8	6	5	4	3
0.90	526	235	173	132	105	85	71	59	51	44	38	34	22	15	10	8	6	5	4

Chart D. 4 groups at alpha = 0.05; NK 1 intervening group

Power	Hypothesized ES																		
	0.20	0.30	0.35	0.40	0.45	0.50	0.55	0.60	0.65	0.70	0.75	0.80	1.00	1.25	1.50	1.75	2.00	2.50	3.00
0.80	509	227	167	128	101	82	68	58	49	43	37	33	22	14	10	8	6	5	4
0.90	658	293	216	166	131	106	88	74	63	55	48	42	28	18	13	10	8	6	4

Chart E. 4 groups at alpha = 0.05; NK 2 intervening groups, Tukey 4 groups

Power	Hypothesized ES																		
	0.20	0.30	0.35	0.40	0.45	0.50	0.55	0.60	0.65	0.70	0.75	0.80	1.00	1.25	1.50	1.75	2.00	2.50	3.00
0.80	583	260	191	147	116	94	78	66	56	49	43	38	25	16	12	9	7	5	4
0.90	743	331	243	187	148	120	99	84	72	62	54	48	31	20	15	11	9	6	5

Chart F. 5 groups at alpha = 0.05; NK no intervening groups

Power	Hypothesized ES																		
	0.20	0.30	0.35	0.40	0.45	0.50	0.55	0.60	0.65	0.70	0.75	0.80	1.00	1.25	1.50	1.75	2.00	2.50	3.00
0.80	393	175	129	99	78	64	53	45	38	33	29	25	17	11	8	6	5	4	3
0.90	526	234	172	132	105	85	70	59	51	44	38	34	22	14	10	8	6	4	3

Table 6.15. (*cont.*)

Chart G. 5 groups at alpha = 0.05; NK 1 intervening group																			
Power	Hypothesized ES																		
	0.20	0.30	0.35	0.40	0.45	0.50	0.55	0.60	0.65	0.70	0.75	0.80	1.00	1.25	1.50	1.75	2.00	2.50	3.00
0.80	508	227	167	128	101	82	68	57	49	43	37	33	21	14	10	8	6	4	4
0.90	658	293	216	165	131	106	88	74	63	55	48	42	27	18	13	10	8	5	4

Chart H. 5 groups at alpha = 0.05; NK 2 intervening groups																			
Power	Hypothesized ES																		
	0.20	0.30	0.35	0.40	0.45	0.50	0.55	0.60	0.65	0.70	0.75	0.80	1.00	1.25	1.50	1.75	2.00	2.50	3.00
0.80	583	260	191	147	116	94	78	66	56	49	43	38	24	16	12	9	7	5	4
0.90	743	331	243	187	148	120	99	84	71	62	54	48	31	20	14	11	9	6	5

Chart I. 5 groups at alpha = 0.05; NK 3 intervening groups, Tukey 5 groups																			
Power	Hypothesized ES																		
	0.20	0.30	0.35	0.40	0.45	0.50	0.55	0.60	0.65	0.70	0.75	0.80	1.00	1.25	1.50	1.75	2.00	2.50	3.00
0.80	638	284	209	160	127	103	85	72	62	53	47	41	27	18	13	10	8	5	4
0.90	805	358	264	202	160	130	108	91	77	67	58	51	33	22	16	12	9	6	5

Table 6.16. Sample size table for between subjects ANOVA, MCP; $p = 0.01$

Chart A. 3 groups at alpha = 0.01; NK no intervening groups

Power	Hypothesized ES																	
	0.20	0.30	0.35	0.40	0.45	0.50	0.55	0.60	0.65	0.70	0.75	0.80	1.00	1.25	1.50	1.75	2.00	2.50	3.00
0.80	586	261	192	148	117	95	79	67	57	49	43	38	25	17	12	9	8	6	5
0.90	746	332	245	188	149	121	100	100	72	62	55	48	32	21	15	12	9	7	5

Chart B. 3 groups at alpha = 0.01; NK 1 intervening group, Tukey 3 groups

Power	Hypothesized ES																	
	0.20	0.30	0.35	0.40	0.45	0.50	0.55	0.60	0.65	0.70	0.75	0.80	1.00	1.25	1.50	1.75	2.00	2.50	3.00
0.80	707	315	232	178	141	115	95	80	69	60	52	46	30	20	15	11	9	7	5
0.90	882	393	289	222	176	143	118	100	85	74	65	57	37	25	18	14	11	8	6

Chart C. 4 groups at alpha = 0.01; NK no intervening groups

Power	Hypothesized ES																	
	0.20	0.30	0.35	0.40	0.45	0.50	0.55	0.60	0.65	0.70	0.75	0.80	1.00	1.25	1.50	1.75	2.00	2.50	3.00
0.80	585	261	192	147	117	95	79	66	57	49	43	38	25	16	12	9	7	5	4
0.90	745	332	244	187	148	120	100	84	72	62	54	48	31	20	15	11	9	6	5

Chart D. 4 groups at alpha = 0.01; NK 1 intervening group

Power	Hypothesized ES																	
	0.20	0.30	0.35	0.40	0.45	0.50	0.55	0.60	0.65	0.70	0.75	0.80	1.00	1.25	1.50	1.75	2.00	2.50	3.00
0.80	707	315	232	178	141	114	95	80	68	59	52	46	30	20	14	11	9	6	5
0.90	882	393	289	222	175	142	118	99	85	73	64	57	37	24	17	13	11	7	6

Chart E. 4 groups at alpha = 0.01; NK 2 intervening groups, Tukey 4 groups

Power	Hypothesized ES																	
	0.20	0.30	0.35	0.40	0.45	0.50	0.55	0.60	0.65	0.70	0.75	0.80	1.00	1.25	1.50	1.75	2.00	2.50	3.00
0.80	784	349	257	197	156	127	105	89	76	66	57	51	33	22	16	12	10	7	5
0.90	967	431	317	243	193	156	129	109	93	81	70	62	40	27	19	14	11	8	6

Chart F. 5 groups at alpha = 0.01; NK no intervening groups

Power	Hypothesized ES																	
	0.20	0.30	0.35	0.40	0.45	0.50	0.55	0.60	0.65	0.70	0.75	0.80	1.00	1.25	1.50	1.75	2.00	2.50	3.00
0.80	585	261	192	147	117	95	78	66	57	49	43	38	25	16	12	9	7	5	4
0.90	745	332	244	187	148	120	100	84	72	62	54	48	31	20	15	11	9	6	5

Table 6.16. (*cont.*)

Chart G. 5 groups at alpha = 0.01; NK 1 intervening group																			
Power	Hypothesized ES																		
	0.20	0.30	0.35	0.40	0.45	0.50	0.55	0.60	0.65	0.70	0.75	0.80	1.00	1.25	1.50	1.75	2.00	2.50	3.00
0.80	707	315	232	178	141	114	95	80	68	59	52	45	30	19	14	11	9	6	5
0.90	881	392	289	221	175	142	118	99	85	73	64	56	37	24	17	13	10	7	5

Chart H. 5 groups at alpha = 0.01; NK 2 intervening groups																			
Power	Hypothesized ES																		
	0.20	0.30	0.35	0.40	0.45	0.50	0.55	0.60	0.65	0.70	0.75	0.80	1.00	1.25	1.50	1.75	2.00	2.50	3.00
0.80	784	349	257	197	156	127	105	88	76	65	57	50	33	22	15	12	9	7	5
0.90	967	431	317	243	192	156	129	109	93	80	70	62	40	26	19	14	11	8	6

Chart I. 5 groups at alpha = 0.01; NK 3 intervening groups, Tukey 5 groups																			
Power	Hypothesized ES																		
	0.20	0.30	0.35	0.40	0.45	0.50	0.55	0.60	0.65	0.70	0.75	0.80	1.00	1.25	1.50	1.75	2.00	2.50	3.00
0.80	841	374	276	211	167	136	113	95	81	70	61	54	35	23	17	13	10	7	5
0.90	1030	459	338	259	205	166	138	116	99	86	75	66	43	28	20	15	12	8	6

Table 6.17. Sample size table for between subjects ANOVA, MCP; $p = 0.10$

Chart A. 3 groups at alpha = 0.10; NK no intervening groups

Power	Hypothesized ES																		
	0.20	0.30	0.35	0.40	0.45	0.50	0.55	0.60	0.65	0.70	0.75	0.80	1.00	1.25	1.50	1.75	2.00	2.50	3.00
0.80	310	138	102	78	62	51	42	35	30	26	23	20	13	9	7	5	4	3	3
0.90	429	191	141	108	86	70	58	49	42	36	32	28	18	12	9	7	5	4	3

Chart B. 3 groups at alpha = 0.10; NK 1 intervening group, Tukey 3 groups

Power	Hypothesized ES																		
	0.20	0.30	0.35	0.40	0.45	0.50	0.55	0.60	0.65	0.70	0.75	0.80	1.00	1.25	1.50	1.75	2.00	2.50	3.00
0.80	420	187	138	106	84	68	57	48	41	35	31	27	18	12	9	7	6	4	3
0.90	557	248	183	140	111	90	75	63	54	47	41	36	23	16	11	9	7	5	4

Chart C. 4 groups at alpha = 0.10; NK no intervening groups

Power	Hypothesized ES																		
	0.20	0.30	0.35	0.40	0.45	0.50	0.55	0.60	0.65	0.70	0.75	0.80	1.00	1.25	1.50	1.75	2.00	2.50	3.00
0.80	310	138	102	78	62	50	42	35	30	26	23	20	13	9	6	5	4	3	3
0.90	429	191	141	108	85	69	58	48	41	36	31	28	18	12	9	7	5	4	3

Chart D. 4 groups at alpha = 0.10; NK 1 intervening group

Power	Hypothesized ES																		
	0.20	0.30	0.35	0.40	0.45	0.50	0.55	0.60	0.65	0.70	0.75	0.80	1.00	1.25	1.50	1.75	2.00	2.50	3.00
0.80	420	187	138	106	84	68	56	48	41	35	31	27	18	12	9	7	5	4	3
0.90	557	248	183	140	111	90	75	63	54	46	41	36	23	15	11	8	7	5	4

Chart E. 4 groups at alpha = 0.10; NK 2 intervening groups, Tukey 4 groups

Power	Hypothesized ES																		
	0.20	0.30	0.35	0.40	0.45	0.50	0.55	0.60	0.65	0.70	0.75	0.80	1.00	1.25	1.50	1.75	2.00	2.50	3.00
0.80	492	219	161	124	98	80	66	56	48	41	36	32	21	14	10	8	6	4	4
0.90	639	285	210	161	127	103	86	72	62	53	47	41	27	18	13	10	8	5	4

Chart F. 5 groups at alpha = 0.10; NK no intervening groups

Power	Hypothesized ES																		
	0.20	0.30	0.35	0.40	0.45	0.50	0.55	0.60	0.65	0.70	0.75	0.80	1.00	1.25	1.50	1.75	2.00	2.50	3.00
0.80	310	138	102	78	62	50	42	35	30	26	23	20	13	9	6	5	4	3	3
0.90	429	191	141	108	85	69	57	48	41	36	31	28	18	12	8	6	5	4	3

Table 6.17. (*cont.*)

Chart G. 5 groups at alpha = 0.10; NK 1 intervening group																			
Power	Hypothesized ES																		
	0.20	0.30	0.35	0.40	0.45	0.50	0.55	0.60	0.65	0.70	0.75	0.80	1.00	1.25	1.50	1.75	2.00	2.50	3.00
0.80	420	187	138	106	84	68	56	47	41	35	31	27	18	12	8	6	5	4	3
0.90	557	248	182	140	111	90	74	63	54	46	40	36	23	15	11	8	7	5	4

Chart H. 5 groups at alpha = 0.10; NK 2 intervening groups																			
Power	Hypothesized ES																		
	0.20	0.30	0.35	0.40	0.45	0.50	0.55	0.60	0.65	0.70	0.75	0.80	1.00	1.25	1.50	1.75	2.00	2.50	3.00
0.80	492	219	161	124	98	80	66	56	47	41	36	32	21	14	10	8	6	4	3
0.90	639	285	209	161	127	103	85	72	61	53	46	41	27	17	12	9	7	5	4

Chart I. 5 groups at alpha = 0.10; NK 3 intervening groups, Tukey 5 groups																			
Power	Hypothesized ES																		
	0.20	0.30	0.35	0.40	0.45	0.50	0.55	0.60	0.65	0.70	0.75	0.80	1.00	1.25	1.50	1.75	2.00	2.50	3.00
0.80	546	243	179	137	109	88	73	62	53	46	40	35	23	15	11	8	7	5	4
0.90	701	312	230	176	139	113	94	79	67	58	51	45	29	19	14	10	8	6	4

7 One-way between subjects analysis of covariance

Purpose of the statistic

The one-way analysis of covariance (ANCOVA) is used to ascertain how likely the difference(s) in means among two or more groups would be to occur by chance alone *after* the effects of one or more pre-existing variables have been taken into account (i.e., statistically controlled). As such, it is an extension of both the independent samples *t*-test (since it can be employed with only two groups) *and* the between subjects ANOVA model. This statistical control of a pre-existing variable (called a covariate) can be visualized as accomplishing three distinct functions:[1]

(1) In the presence of a pretest or baseline measure *identical* to the outcome variable, the results of a one-way ANCOVA can be interpreted as the extent to which the groups involved have *changed* over time. In this sense it is interpreted identically to the time (e.g., baseline vs. end-of-treatment) × treatment interaction in a mixed within group (repeated measures) design except that it does not provide for a main effect test of the hypothesis regarding an overall baseline to end-of-treatment change (see Chapter 9). As was discussed in Chapter 2, however, it is usually a slightly more powerful test of treatment changes than this two-factor model. If more than one post-baseline interval is used, as is often the case, then a factorial ANCOVA must be employed, although the effect of interest may now be the between group main effect (which combines means across these post-intervention assessments) rather than the interaction (see Chapter 10).

(2) If the covariate is not identical to the outcome variable, it still serves to increase the study's overall power by decreasing the error term. The results, however, are strictly speaking not interpretable as group changes due to the experimental intervention(s). However, when subjects are randomly assigned to groups, then these groups can be assumed to be equal with respect to baseline measures even though none exist – hence this issue becomes a moot point.

(3) For both of the above scenarios, the existence of one or more covariates helps to equate statistically the groups' follow-up means for minor pre-existing differences. As such the resulting F-ratio really tests differences between *adjusted* (i.e., for these pre-existing covariate differences) outcome (dependent variable) means. Because these adjustments should be quite minor in the presence of randomization or appropriate control group selection in non-randomized but controlled experiments (both of which should theoretically result in no pre-existing differences), we will assume that no adjustment occurs in the discussions that follow. In those rare circumstances in which such an adjustment *is* hypothesized, then this adjustment should simply be added (if it increases differences between groups) or subtracted (if it decreases them) from the hypothesized ES.

The ANCOVA, then, while it is a direct extension of the t-test and one-way between subjects ANOVA, is a more sensitive, powerful statistic than either of these options. *It is, in many ways, the least expensive and least burdensome existing mechanism for increasing a trial's statistical power since all that its use requires is the existence of a pre-existing correlate of the dependent variable. If such a correlate can be identified, then a one-way ANCOVA is always preferable to the use of a one-way between subjects analysis of variance design.* The only caveat to this generalization is the occasional difficulty involved in identifying a covariate that correlates substantially (i.e., 0.40 or more) with the dependent variable. Such variables usually exist, but they are identifiable only through the previous literature or extensive pilot work. They can almost never be identified via post hoc "fishing" expeditions using data collected for other trial purposes, nor should they be because of the likelihood of producing spurious, non-replicable results.

As with ANOVA, the groups themselves may be experimental conditions to which subjects have been randomly assigned or they may be defined by a naturally occurring phenomenon. Also as with ANOVA (other than the standard assumptions of the procedure itself), the only stipulations for its use are that the hypotheses being tested involve group means (which implies that the dependent variable is continuous in nature) and the groups are not made up of the same or matched subjects. (As discussed in Chapters 9 and 10, ANCOVA may still be appropriate for such designs, but the tables presented in this chapter are not useful for ascertaining their power/sample size requirements.)

The procedure does assume that the pre-existing variable (for convenience we will assume one covariate although multiple variables can be employed)[2] is either continuous or dichotomous in nature[3] and that it is linearly related to the dependent variable.

A one-way between subjects ANCOVA, then, is used when:

(1) a single, independent variable is employed (i.e., membership in two or more groups: note that a one-way ANOVA assumes three or more groups, but ANCOVA is appropriate for two groups because the independent samples *t*-test cannot incorporate a pre-existing variable, called a covariate, into an analysis),
(2) the dependent variable is measured in such a way that it can be described by a mean (i.e., it is continuous in nature and not categorical, although a strongly correlated dichotomous variable can be employed as a covariate if it is not expected to interact with the treatment),
(3) an appropriate[4] pre-existing variable (i.e., the covariate) exists which is linearly related to the dependent variable,
(4) this linear correlation is approximately the same for each of the experimental groups (which is called homogeneity of the regression slopes),
(5) this pre-existing variable does not interact with group membership, and
(6) the subjects within the different treatment groups are statistically independent of one another (i.e., they are different individuals and are not specifically matched in some way).

A one-way between subjects ANCOVA is *not* used when:

(1) a one-way ANOVA (in the absence of a covariate) would not be appropriate, and/or
(2) the investigator is interested in ascertaining whether the treatments are differentially effective for individuals with certain levels of the covariate (in which case, as described in Chapter 9, the various levels of the covariate would be entered as a second independent variable in a factorial ANOVA and these differential treatment effects would be tested by the interaction term).

Like a one-way ANOVA, a one-way ANCOVA is a two-step process when three or more groups are employed. The first step involves the computation of an overall *F*-ratio to ascertain whether indeed the treatments do differ from one another after the covariate has been statistically controlled. The second step, which occurs only in the presence of a statistically significant *F*-ratio, involves ascertaining which of the individual treatments do differ significantly from one another via the use of a multiple comparison procedure. (This latter step is not necessary for a two-group ANCOVA since there is only one mean difference involved, hence a statistically significant overall *F* indicates that this one mean difference is statistically significant.) This in turn dictates that a power analysis for such a design is also a two-step process.

The remainder of this chapter is therefore divided into two parts. The first presents power and sample size tables for the overall *F*-ratio. The second presents tables that allow the investigator to estimate the power (or sample size requirements) for individual contrasts using the Tukey HSD and Newman–Keuls multiple comparison procedures. In effect, the use of these tables is identical to the process described in Chapter 6 with one exception: the necessity of estimating the correlation that will accrue between the covariate and the dependent variable. Once this *r* is hypothesized, the estimation of power involves nothing more than accessing the appropriate table(s) indicated in the templates at the end of this chapter.

Hypothesizing the correlation between a covariate and a dependent variable. The most direct way of hypothesizing the size of the relationship between the variable chosen as the covariate and a study's dependent variable is from the data collected in the pilot study (which should always precede a full blown experiment). This can, of course, be done by simply running a Pearson *r* between the two variables.

Other sources of this information include previous research employing the variables of interest, although unfortunately journal articles do not often directly report correlations such as these since their relationship to the power analytic process is not well understood. The authors of these former studies can be contacted directly, however, although many will not know this value since they did not report it in their original research report. Occasionally, however, an investigator will be kind enough to look up or recompute the statistic.

In lieu of all of the above, it is truly difficult to estimate an exact correlation between a pre-existing variable (the covariate) and a variable that is expected to be influenced by the study itself (the dependent variable), although minor variations in these estimates do not greatly influence the results of a power analysis. With this said, we recommend the following guidelines:

(1) If the covariate is conceptualized as baseline values of the study's actual dependent variable (and if this variable is a well established empirical outcome that has been employed in similar studies with similar types of subjects), then it is probably safe to assume that the

correlation between the two variables may be as high as 0.60.

(2) If the covariate is not identical to the dependent variable, but both constitute relatively stable attributes (e.g., not something such as blood pressure which varies considerably across time and situation) with reliabilities in excess of 0.70, then it may be safe to assume $r = 0.60$ if this type of relationship has been observed consistently in previous research. *If all of these conditions do not hold, then an r of 0.40 is a safer assumption as long as there is previous evidence for such a correlation. If no previous evidence exists, such a variable should not be selected as a covariate at all in the absence of pilot data.*

In either of the above situations, however, it is always wise to model both of these values in the power analytic process.

Chapter 10 provides a table by which a wider range of covariate–dependent variable correlation coefficients can be modeled. As with Chapter 6, sample size tables are provided for different alpha levels (0.01 and 0.10) for the overall *F*-ratio as well as MCPs (Tukey and Newman–Keuls).

Part I. Power of the overall *F*-ratio

Tables 7.1 through 7.24 present power and sample size tables for a wide range of parameters for *F*-ratios involving two through five groups. These tables are used in the same way as those presented in Chapter 6 except that for an ANCOVA it is also necessary to estimate the most likely correlation between the covariate(s) and the dependent (outcome) variable. As discussed above, this can be relatively difficult; hence we recommend that the investigator model his/her power estimates based upon *r* values of both 0.40 and 0.60 in the absence of reliable pilot data (for other values, see Chapter 10).

Example. To illustrate how these tables are employed, let us posit a trial in which the investigator is interested in learning whether traditional Chinese acupuncture is capable of reducing the pain associated with osteoarthritis of the knee. Because an education/self-help course offered by the Arthritis Foundation has been shown also to produce salutary outcomes among this type of patient, a decision is made to employ this intervention as a second active non-pharmacological comparison group. Also, since it is quite possible that the novelty of receiving acupuncture could result in a placebo effect in the absence of a true analgesic mechanism, a sham procedure is employed to control for this possibility. This control mechanism mimics the true acupuncture process but does not employ needle insertion in active points. Pilot work has indicated that patients cannot distinguish it from true acupuncture.

The basic protocol for this trial, then, involves (a) obtaining a baseline assessment of the eligible participants' pain prior to the initiation of the study, (b) the random assignment of these subjects to one of three groups (true acupuncture, education/self-help groups, and sham acupuncture), (c) the implementation of the interventions, and (d) the re-assessment of all subjects at the end of the study. The functional hypotheses (which can also be written as specific aims, objectives, or research questions as long as the explicit contrasts are specified) for this study are:

(1) Patients with osteoarthritis of the knee who receive acupuncture will experience less pain than patients who receive sham acupuncture after controlling for pre-existing pain levels.
(2) Patients with osteoarthritis of the knee who receive acupuncture will experience less pain than patients who receive an education self-help intervention after controlling for pre-existing pain levels.
(3) Patients with osteoarthritis of the knee who receive an education self-help intervention will experience less pain than patients who receive sham acupuncture after controlling for pre-existing pain levels.

Prior to submitting this protocol for funding and IRB approval, the most pressing issues facing the investigator are (a) how many subjects will be required to provide the trial with a reasonable chance of obtaining statistical significance among the three groups and (b) how much power will be available for testing the hypothesis given a fixed sample size. In the present case, let us begin by assuming this latter scenario.

To answer the question of how much power would be available for a fixed sample size, then, the following parameters must be specified:

(1) the ES for the largest mean difference among the three groups,
(2) the projected *N*/group,
(3) whether the spread of means is likely to reflect a low/medium, or high dispersion pattern, and
(4) the most likely covariate–dependent variable relationship.

Templates are provided at the end of this chapter to facilitate the computation of power for this and any one-way ANCOVA design employing from two to five groups. Let us begin this process, then, by employing preliminary Template 7.1.

The first steps involve specifying the hypothesized order (step 1) and magnitude (step 2) of the differences among the three group means, both of which are identical to the process employed in a one-way ANOVA. Step 1 is normally already accomplished by the time the study's

Chart 7.1. *Converting a negative dependent variable scale for entry on the ES line*

	Sham acupuncture Mean (SD)	Education control Mean (SD)	True acupuncture Mean (SD)
Baseline	6.00 (2.5)	6.04 (2.4)	6.10 (2.6)
End-of-treatment	6.50 (2.4)	6.29 (2.5)	5.10 (2.6)
Raw change	−0.50	−0.25	1.00
Lowest treatment effect set to zero	−0.50 + 0.50 = 0	−0.25 + (0.50) = 0.25	1.00 + 0.50 = 1.50
Conversion to ES	0/2.5 = 0	0.25/2.5 = 0.1	1.5/2.5 = 0.6

hypotheses are formulated. In this case, the investigator believes that true Chinese acupuncture will be the most effective treatment and has included the education self-help comparison only because it is known to have a statistically significant but small effect upon arthritis pain; hence the order of treatment effects requested by step 1 of the preliminary template would be as follows:

Step 1. Write in the names/codes of the groups in the chart to the right in ascending order based upon their expected means (i.e., the name of the group expected to have the lowest mean or the weakest effect will be written next to ①, followed by the next strongest treatment and so forth).	Group definitions ① Sham acupuncture ② Education self-help ③ True acupuncture ④

Step 2 involves specifying the individual ES values for each of the groups listed above (with the exception of the "weakest" condition, which is arbitrarily set at zero). Chart 7.1 illustrates one of a number of ways to approach this task. (If the Pearson *r* is not readily available (which is often the case when summary statistics from previously published research are used), for example, one can always use the standard deviations of the differences between baseline and end-of-treatment (if presented) to calculate both *r* and the EOT standard deviations via Formula 7.1 in the Technical appendix.) For present purposes, however, let us assume that these results accrue from the investigator's own pilot study and a Pearson *r* is already available.

The first step, then, in standardizing the EOT means for entry onto the ES line is to calculate raw changes from the baseline to EOT. This is done by (a) subtracting baseline from EOT values, (b) performing whatever addition or subtraction operation across the groups that will produce a zero value for the poorest performing group (in this case, since lower scores on the pain measure are optimal, this will involve adding the absolute value of the negative gain to each of the raw change scores), and (c) dividing by the pooled standard deviation $(2.4+2.5+2.6)/3=2.5$). This process (Chart 7.1) results in an arbitrary zero score for the sham group, an ES of 0.1 for the sham vs. education control contrast, and an ES of 0.6 for the sham vs. true acupuncture contrast. (The same power analytic results will accrue if the direction of the scale is not corrected; we simply believe that it is conceptually easier to do so.)

Step 2 of the preliminary template would then consist of simply plotting these values:

Step 2. Plot these groups, using the circled numbers rather than their codes, on the ES line below. In the case of a tie between two groups with respect to their actual ES value, place one to the right of the other based upon the best theoretical evidence for which might be slightly superior.

ES line

①	②					③									
0	0.1	0.2	0.3	0.4	0.5	0.6	0.7	0.8	0.9	1.0	1.1	1.2	1.3	1.4	1.5

There are now only three parameters that need to be specified prior to calculating the power of the overall *F*-ratio: (a) the *N*/group that the investigator expects to be available for the trial, (b) the estimated pattern of means (which is obtained by simply comparing the ES line above to the two options for three-group designs in step 4), and (c) the most likely covariate–dependent variable relationship to accrue (step 5).

Let us assume that our investigator knew that he/she would have approximately 150 potential subjects available for the study, without resorting to extensive (and expensive) recruitment efforts, of which no more than 20% would be likely to withdraw from the trial prior to its completion. This would mean that effectively the *N*/group would be 40 (assuming equal attrition among the groups, which is not always the case). The pattern of means resulting from the hypothesized ES line above most closely approximates the high dispersion pattern in step 4(a). Finally, step 5 requires the estimation of the most likely covariate–dependent variable correlation coefficient. In this example, although baseline pain measures are to be employed whose measurement is identical to the end-of-treatment dependent variable assessment, let us assume that previous research has indicated that pain related to osteoarthritis of the knee can be relatively variable, hence the investigator decides to opt for the more conservative estimate of 0.40 for the covariate–dependent variable correlation, at which point the final instruction in step 5 sends the investigator to Template 7.2.

Based upon this information, the three power analytic steps related to the overall *F*-ratio dictated by Template 7.2 can be accomplished quite easily. The largest ES based upon the ES line is 0.60 for group 3 (step 6), and step 7 indicates that (for three-group studies hypothesizing a high mean pattern and a covariate–dependent variable correlation of 0.40) the best estimate for power can be found in Table 7.5.

By locating the intersection between the ES column of 0.60 and the *N*/group row of 40 (following the instructions in step 8), the investigator learns that the power for the overall ANCOVA *F*-ratio based upon the parameters specified above would be 0.86. In the past, most funding agencies would have accepted this as adequate justification for the proposed sample size, but a properly written description of this power analysis indicates a basic weakness with this particular end point:

> A power analysis indicated that the hypothesized ES between the true and sham acupuncture arms of 0.60 (assuming a high dispersion pattern for the three means and an average correlation between the covariate and the dependent variable of 0.40) would provide an 86% chance of obtaining statistical significance at the 0.05 level for the overall *F*-ratio.

If this statement were compared to the trial's original actual hypotheses and/or specific aims as presented above, however, a discrepancy would most likely be apparent because *this* power level tests the following hypothesis:

> There is difference in the amount of pain associated with osteoarthritis of the knee between patients who receive true Chinese acupuncture, sham acupuncture, and an education self-help intervention after controlling for pre-existing pain levels.

Note that this hypothesis does not specify which groups would be superior to which other groups because the overall *F*-ratio does not test for directionality. What the investigator (and his/her funding agency) would be most interested in ascertaining would be the individual contrasts among the three groups, especially the contrasts between (a) true and sham acupuncture (group 3 vs. group 1), which would indicate whether or not true acupuncture actually had an analgesic effect over and above a simple placebo effect, and (b) true acupuncture vs. the education self-help intervention (group 3 vs. group 2), which would indicate whether or not the study's primary intervention (acupuncture) possessed an analgesic effect over-and-above the only other non-pharmacological treatment known to be effective in reducing pain from osteoarthritis of the knee. From a scientific point of view, even the group 2 vs.

group 1 contrast might be interesting as a preliminary assessment of the extent to which this non-pharmacological treatment could be explained solely in terms of a placebo effect. Let us assume, therefore, that the investigator is indeed interested in testing all three of the hypotheses logically generated by this study, although it is a rare trial that will produce sufficient power for all of the pairwise contrasts available, even in a three-group study such as this.

Part II. Power of individual pairwise contrasts

To assess how much power would be available for these three pairwise contrasts, then, the investigator would need to complete steps 9 and 10 of Template 7.2. Step 9 simply asks him/her to compute the individual pairwise ES values based upon the already completed ES line, which would entail the following operations: group 2 ES − group 1 ES = 0.1 − 0 = 0.1, group 3 ES − group 1 ES = 0.6 − 0 = 0.6, and group 3 ES − group 2 ES = 0.6 − 0.1 = 0.5.

Step 9. Using the hypothesized values from step 2 in the preliminary template (7.1), fill out the hypothesized ES values by performing the indicated subtractions listed under the "Contrasts" column in the chart below. Note that for two-group designs, the power of the single pairwise contrast is identical to the power of the overall F (i.e., if $r = 0.40$, Table 7.1 is used and for $r = 0.60$, Table 7.2 is used).

ES ② − ① = 0.1	ES ④ − ① =	ES ⑤ − ① =
ES ③ − ① = 0.6	ES ④ − ② =	ES ⑤ − ② =
ES ③ − ② = 0.5	ES ④ − ③ =	ES ⑤ − ③ =
		ES ⑤ − ④ =

Step 10 indicates the appropriate power table for the Tukey HSD procedure (Table 7.25). Assuming that the investigator is interested in computing all three pairwise power levels, these values will be found at the intersection of the appropriate ES columns (i.e., 0.1, 0.5, and 0.6) and the *N*/group row of 40 in Table 7.25. Since our power tables do not provide for powers of ES below 0.20 because there is seldom enough power for effects this small in experimental research, we suggest that the reader simply estimate power to be *less than* the closest tabled value. Following the instructions in step 10, then, produces the following results:

Tukey HSD	**Newman–Keuls**
Power ② − ① < 0.08	Power ② − ① < 0.16
Power ③ − ① = 0.71	Power ③ − ① = 0.71
Power ③ − ② = 0.53	Power ③ − ② = 0.67

What has resulted here, then, is a situation in which the power of the overall *F*-ratio is quite adequate but each of the individual contrasts is underpowered. These results might be appended to the write-up of the power analysis conducted for the overall *F*-ratio as follows:

> A power analysis indicated that the hypothesized ES between the true and sham acupuncture arms of 0.60 (assuming a high dispersion pattern for the three means and an average correlation between the covariate and the dependent variable of 0.40) would provide an 86% chance of obtaining statistical significance at the 0.05 level for the overall *F*-ratio. The powers for the individual contrasts assuming the analytic use of the Tukey HSD multiple comparison procedure were as follows: (a) true acupuncture vs. education self-help (power = 0.53), (b) true vs. sham acupuncture (power = 0.71), and (c) education self-help vs. sham acupuncture (power < 0.08) (Bausell & Li, 2002).

As mentioned elsewhere, it is not unusual for the power of the overall *F*-ratio to be quite adequate, with the corresponding values for the more important individual contrasts to be considerably less than adequate. In such a situation the investigator would need to make a decision regarding whether to proceed with the experiment or to adopt another strategy. In the present case, from a practical perspective the manner in which this trial is designed will preclude any true test of the group 2 vs. group 1 contrast short of increasing the sample size by 500 subjects or so. One benefit of a power analysis such as this, however, is *to force the investigator to come to grips with this fact and potentially make some hard decisions.* Some of his/her options include:

(1) Dropping either the education self-help or the sham acupuncture group. This would allow each of the remaining groups to share the 40 extra subjects and would result in an increase of power to 0.85 if the sham control is dropped or to 0.95 if the education self-help group is dropped (see Table 7.1).
(2) Increasing the sample size.
(3) Implementing a strategy to decrease the attrition rate (since the *N*/group of 40 is based upon the sample size following attrition).
(4) Increasing the ES for the primary intervention (e.g., increasing the dose by adding more acupuncture sessions or performing acupuncture for a longer period of time) or weakening the control (e.g., by employing a treatment-as-usual or a wait list control).
(5) Employing a dependent variable or an additional covariate that would raise the estimated covariate–dependent variable correlation. If, for example, a means were found to raise this value to 0.60, the following power results would be obtained from Table 7.26.

Tukey HSD

Power ② − ① <0.11
Power ③ − ① = 0.83
Power ③ − ② = 0.66

All or none of these options could be viable, depending upon the science being conducted. The bottom line is always that scientific considerations must take precedence over statistical ones, but a power analysis is capable of telling an investigator how likely his/her scientific objectives – whatever they may be – are to be realized and this is a valuable service indeed.

A power analysis also can provide the investigator with some valuable information for the process of making these decisions. One helpful strategy in this regard is to determine the maximum detectable ES that can be detected with the available sample size.

Conducting a maximum detectable ES analysis. When an investigator's available sample size is fixed it is often helpful for him/her to know exactly what size effect can be detected at various power levels given the proposed design. In our example above, suppose that our investigator wondered what the maximum detectable ES size would be for 40 subjects per group using the Tukey HSD procedure. If a power level of 0.80 or 0.90 were deemed essential (assuming as always an alpha level of 0.05), he/she could employ Table 7.31 Chart B and locate the closest approximation to an *N*/group of 40. For a power level of 0.80, this would fall between an ES of 0.65 and 0.70, which might provide some valuable guidance if a decision were made to strengthen the intervention or weaken the control.

If a lower power level were deemed sufficient, the tables listed in step 10 of Template 7.2 could be employed in reverse. In other words, the power closest to the desired value could be located in the *N*/group = 40 row and the maximum detectable ES could be read at the top of its intersecting column. Let us assume in the present case that a power of 0.60 was determined to be adequate. Template 7.2 has already indicated that Table 7.25 is appropriate for a Tukey HSD procedure involving three groups. Reading across the *N*/group row of 40, 0.62 is the closest value to 0.60 and it is located in the 0.55 ES column, which means that: using a Tukey HSD multiple comparison procedure, approximately 40 subjects per group will produce a 60% chance of detecting a pairwise ES of 0.55.

Sample size analysis. Template 7.3 is provided for those instances in which the investigator wishes to determine how many subjects per group will be necessary to achieve a given level of power. This template, like the power template (7.2) whose use we have just illustrated, is designed to be employed following the preliminary steps suggested in Template 7.1.

Using the same hypothetical study, let us assume that our investigator had wished to ascertain the necessary *N*/group he/she would have needed to achieve an 80% chance of obtaining statistical significance. All of the preliminary steps involved in specifying the necessary parameters employed to determine power would be used in this scenario with the exception of step 3, where the desired power level would be specified rather than the *N*/group. The instructions in Template 7.3 would therefore yield the following results:

Step 6. As in step 6 of the power template, the largest ES corresponds to group 3, which is 0.6.

Step 7. Since the desired power level is 0.80, three groups are involved, the hypothesized *r* is 0.40, and a high dispersion pattern of means is predicted, the investigator is directed to find the intersection of the 0.80 row and the ES = 0.60 column of Table 7.19 Chart C which yields the minimum *N*/group value of 35. This analysis could be described as follows:

> A sample size analysis indicated that the hypothesized ES between the true and sham acupuncture arms of 0.60 (assuming a high dispersion pattern for the three means and an average correlation between the covariate and the dependent variable of 0.40) would require 35 subjects per group (total *N* = 105) to provide an 80% chance of obtaining statistical significance at the 0.05 level for the overall ANCOVA *F*-ratio (Bausell & Li, 2002).

To determine the minimum *N*/group required to obtain statistical significance for the pairwise contrasts of interest, the investigator would proceed to the second half of Template 7.3. Let us assume that the group 1 vs. group 2 ES of 0.1 was not deemed to be of sufficient scientific interest to enter into this analysis, but that all of the effects involving the true acupuncture group were deemed crucial.

Step 9 has already been performed in the present example producing ES values of 0.5 for the group 3 vs. group 2 contrast and 0.6 for group 3 vs. group 1.

Step 10 instructs the reader to find the required *N*/group for the Tukey procedure in Table 7.31 Chart B at the intersections of the 0.50 and 0.60 columns and the power = 0.80 row. Following these instructions yields values of *N*/group of 70 and 49 respectively, which in effect means that the *N*/group for this experiment must be 70 in order to produce at least an 80%

chance of achieving statistical significance for both contrasts. (Naturally the available power using an *N*/group of 70 for the true vs. sham acupuncture comparison would be considerably greater, >0.90 rather than 0.80.) For the Newman–Keuls procedure, the *N*/group for the 0.50 ES is found to be 54 in Table 7.31 Chart A (and 49 in Table 7.31 Chart B for the true vs. sham acupuncture ES).

The results of this analysis might be appended to the earlier results for the overall *F*-ratio as follows:

> A sample size analysis indicated that the hypothesized ES between the true and sham acupuncture arms of 0.60 (assuming a high dispersion pattern for the three means and an average correlation between the covariate and the dependent variable of 0.40) would require 35 subjects per group (total $N = 105$) to provide an 80% chance of obtaining statistical significance at the 0.05 level for the overall ANCOVA *F*-ratio. However, using the Tukey HSD multiple comparison procedure, an *N* of 70 per group would be required to assure an 80% chance of demonstrating that true acupuncture was superior to the education control after statistically controlling for patients' pre-existing levels of pain.

This, then, illustrates another potential benefit of conducting a power analysis at an experiment's design stage: namely that not only does such an analysis force an investigator to refine and prioritize his/her objectives/hypotheses, it also permits the optimal marshalling of study resources for achieving those objectives. Sometimes this can be a painful process when it requires the acknowledgement that a cherished objective for conducting the study in the first place cannot be achieved with the resources available, but in science knowledge is ultimately advantageous in comparison to the alternative, which is ignorance.

Summary

The analysis of covariance (ANCOVA) is a powerful analytic tool for increasing the sensitivity and cost-effectiveness of conducting experimental research. The higher the covariate–dependent variable correlation, the more effective this procedure becomes. Assuming an ES of 0.50, for example, the existence of a covariate possessing a 0.40 correlation with the dependent variable would allow an investigator to reduce his/her total sample size by 48 subjects in a four-group design (assuming that a 0.80 level of power was desired for the significance of the overall *F*-ratio and a medium dispersion pattern were predicted), while an *r* of 0.60 would allow for a total 112 subject reduction (or a 35% reduction in the total sample size).

Other designs employing analysis of covariance are discussed elsewhere in this text. Computing the power for ANCOVA interactions, for example, is discussed in Chapter 9, while ANCOVA main effects for higher level designs are discussed in Chapter 10. Analysis of covariance is, in many ways, the least expensive and least burdensome existing mechanism for increasing a trial's statistical power since all that its use requires is the existence of a pre-existing correlate of the dependent variable. If such a correlate can be identified, and if it makes theoretical sense as a control variable, then a one-way ANCOVA is always preferable to the use of a one-way between subjects analysis of variance design.

Endnotes

1 Actually there is a fourth function that will not be considered here. The ANCOVA model is often employed as an analytic approach in quasi-experimental designs (i.e., studies that employ non-randomized comparison groups). We will not discuss this option here since the covariate is used to adjust the dependent variable based upon *expected* pre-experimental differences among the groups. In this chapter, we assume that the experimental groups will be equivalent on the covariate at the initiation of the trial.

2 If two or more covariates are employed, it is their multiple correlation [R] with the dependent variable that is of interest. This will add some imprecision to the use of the tables in this chapter since one degree of freedom is subtracted from the error term for each covariate employed. The effect will be noticeable only for extremely small values of N/group, hence the use of multiple covariates in such studies should be employed with care.

3 Although rarely done in experimental research, a categorical variable may be dummy coded into multiple variables and the resulting constructions may be employed as multiple covariates.

4 An investigator must be careful both in non-experimental research and in experimental factorial studies in which a blocking variable is employed, that statistically controlling for pre-existing differences among groups does not run counter to the purposes of the research. As an example, if severity of an illness were used as a blocking variable in a factorial study investigating an intervention's effectiveness, a researcher should not use a covariate such as health status as a covariate, since this would decrease the effect due to severity.

Template 7.1. Preliminary one-way between subjects ANCOVA power/sample size template

This preliminary template is applicable to all one-way between subjects ANCOVA designs employing between two and five groups.

Step 1. Write in the names/codes of the groups in the chart to the right in ascending order based upon their expected means (i.e., the name of the group expected to have the lowest mean or the weakest effect will be written next to ①, followed by the next strongest treatment and so forth).

Group definitions
①
②
③
④
⑤

Step 2. Plot these groups, using the circled numbers rather than their codes, on the ES line below. In the case of a tie between two groups with respect to their actual ES value, place one to the right of the other based upon the best theoretical evidence for which might be slightly superior.

ES line

①

0 0.1 0.2 0.3 0.4 0.5 0.6 0.7 0.8 0.9 1.0 1.1 1.2 1.3 1.4 1.5

Step 3. For the estimation of power, specify the *N*/group available. For required sample size, enter the desired power level.

N/group = or, desired power =

Step 4. Compare the low, medium, and high ES patterns below with the graphed ES line from step 2 above. Choose the pattern which most closely matches the hypothesized pattern of means (step 2). Note that this step does not need to be performed for two-group designs, since there is only one possible pattern of means.

(a) Three-group designs
ES pattern for low/medium *F* power ① ② ③
ES pattern for high *F* power ①② ③
or ① ②③

(b) Four-group designs
ES pattern for low *F* power ① ②③ ④
ES pattern for medium *F* power ① ② ③ ④
ES pattern for high *F* power ①② ③④

(c) Five-group designs
ES pattern for low *F* power ① ②③④ ⑤
ES pattern for medium *F* power ① ② ③ ④ ⑤
ES pattern for high *F* power ①②③ ④⑤
or ①② ③④⑤

Step 5. Select the most likely correlation between the covariate and the dependent variable. (For values other than 0.40 or 0.60, see Chapter 10.)

$r = 0.40$ ___ $r = 0.60$ ___

To compute power, turn to Template 7.2. To determine *N*/group, turn to Template 7.3.

Template 7.2. One-way between subjects ANCOVA power template

Power of the overall *F*-ratio

Steps 1–5, see preliminary Template 7.1.

Step 6. Use the largest hypothesized ES produced in step 2 (Template 7.1). For two-group studies, this will be the ES for group ②, for three-group studies it will be the ES for group ③, and so forth.

Largest ES from preliminary step 2:

Step 7. Use the chart below to find the appropriate power table.

(a) For two groups, if $r = 0.40$, use Table 7.1, if $r = 0.60$, use Table 7.2.

(b) For three groups, choose the appropriate model below:

$r = 0.40$	Table	$r = 0.60$	Table
L/M pattern	7.3	L/M pattern	7.4
H pattern	7.5	H pattern	7.6

(c) For four groups, choose the appropriate model below:

$r = 0.40$	Table	$r = 0.60$	Table
L pattern	7.7	L pattern	7.8
M pattern	7.9	M pattern	7.10
H pattern	7.11	H pattern	7.12

(d) For five groups, choose the appropriate model below:

$r = 0.40$	Table	$r = 0.60$	Table
L pattern	7.13	L pattern	7.14
M pattern	7.15	M pattern	7.16
H pattern	7.17	H pattern	7.18

Step 8. Turn to the table identified in step 7 and find the power at the intersection between the ES column from step 6 and the *N*/group specified in step 3 of the preliminary template. Interpolate as desired.

Power =

Power of the pairwise contrasts

Step 9. Using the hypothesized values from step 2 in the preliminary template (7.1), fill out the hypothesized ES values by performing the indicated subtractions listed under the "Contrasts" column in the chart below. Note that for two-group designs, the power of the single pairwise contrast is identical to the power of the overall *F* (i.e., if $r = 0.40$, Table 7.1 is used and for $r = 0.60$, Table 7.2 is used).

ES ② − ① =	ES ④ − ① =	ES ⑤ − ① =
ES ③ − ① =	ES ④ − ② =	ES ⑤ − ② =
ES ③ − ② =	ES ④ − ③ =	ES ⑤ − ③ =
		ES ⑤ − ④ =

Step 10. For power of the above pairwise contrasts (step 9) using the **Tukey HSD**, use the tables indicated in the "Power tables for the Tukey HSD procedure" chart (only sample size tables are available for the **Newman–Keuls** procedure, see Template 7.3). Locate the

power in the indicated table at the intersection of the ES column (step 9) and the *N*/group row (preliminary step 3). Interpolate as desired.

Power tables for the Tukey HSD procedure

Three-group studies	If $r=0.40$, use Table 7.25	If $r=0.60$, use Table 7.26
Four-group studies	If $r=0.40$, use Table 7.27	If $r=0.60$, use Table 7.28
Five-group studies	If $r=0.40$, use Table 7.29	If $r=0.60$, use Table 7.30

Power ② – ① =	Power ④ – ① =	Power ⑤ – ① =
Power ③ – ① =	Power ④ – ② =	Power ⑤ – ② =
Power ③ – ② =	Power ④ – ③ =	Power ⑤ – ③ =
		Power ⑤ – ④ =

Template 7.3. One-way between subjects ANCOVA sample size template

---

Required sample size for a statistically significant overall *F*-ratio

Steps 1–5, see preliminary Template 7.1.

Step 6. Use the largest hypothesized ES produced in preliminary step 2. For two-group studies, this will be the ES for group ②, for three-group studies it will be the ES for group ③, and so forth.

Largest ES from preliminary step 2:

Step 7. For desired power values of 0.80 and 0.90, use the chart below to find the appropriate sample size table. Note that sample size tables are also provided for $p=0.01$ (Table 7.21 for $r=0.40$; Table 7.22 for $r=0.60$) and $p=0.10$ (Tables 7.23 and 7.24).

(a) For two groups, use Table 7.19 Chart A for $r=0.40$ and Table 7.20 Chart A for $r=0.60$.

(b) For three groups, choose the appropriate model below:

$r=0.40$	Table	$r=0.60$	Table
L/M pattern	7.19 Chart B	L/M pattern	7.20 Chart B
H pattern	7.19 Chart C	H pattern	7.20 Chart C

(c) For four groups, choose the appropriate model below:

$r=0.40$	Table	$r=0.60$	Table
L pattern	7.19 Chart D	L pattern	7.20 Chart D
M pattern	7.19 Chart E	M pattern	7.20 Chart E
H pattern	7.19 Chart F	H pattern	7.20 Chart F

(d) For five groups, choose the appropriate model below:

$r=0.40$	Table	$r=0.60$	Table
L pattern	7.19 Chart G	L pattern	7.20 Chart G
M pattern	7.19 Chart H	M pattern	7.20 Chart H
H pattern	7.19 Chart I	H pattern	7.20 Chart I

Step 8. For desired powers of 0.80 and 0.90, turn to the table identified in step 7 and find the *N*/group at the intersection between the ES column from step 6 and the *N*/group specified in step 3 of the preliminary template. For powers other than 0.80 and 0.90, find

the appropriate table from step 7 in Template 7.2. Locate the nearest power value in the ES row (step 6) and read the required *N*/group associated with that row. Interpolate as desired.

N/group =

Sample size for the pairwise contrasts

Step 9. Using the hypothesized values from step 2 in the preliminary template, fill out the hypothesized ES values by performing the indicated subtractions listed under the "Contrasts" column in the chart below. Note that for two-group designs, the *N*/group of the single pairwise contrast is identical to that for the overall *F* as given in Tables 7.1 ($r=0.40$) and 7.2 ($r=0.60$). For two-group desired powers of 0.80 and 0.90, see Table 7.19 Chart A ($r=0.40$) and 7.19 Chart B ($r=0.60$).

ES ② − ① =	ES ④ − ① =	ES ⑤ − ① =
ES ③ − ① =	ES ④ − ② =	ES ⑤ − ② =
ES ③ − ② =	ES ④ − ③ =	ES ⑤ − ③ =
		ES ⑤ − ④ =

Step 10. For the required sample size for powers of 0.80 or 0.90 of the pairwise contrasts listed in step 9 for the **Tukey HSD** procedure, use the tables indicated in the "Tukey HSD" chart below. For powers of 0.80 or 0.90 for the **Newman–Keuls** procedure, use the tables indicated in the "Newman–Keuls" chart. (Note that the table specified for the Tukey procedure are based *only* upon the number of groups and *not* the particular contrast involved while the tables used for Newman–Keuls are based only upon the contrasts involved and *not* the number of groups.) Locate the *N*/group in the indicated table at the intersection of the ES column (step 9) and the desired power row (preliminary step 3). For powers other than 0.80 and 0.90 (for the Tukey HSD procedure), use the appropriate tables indicated in step 10 of Template 7.2 by finding the nearest power value in the ES column (step 9) and reading the *N*/group associated with that row. Interpolate as desired. (Note that sample size requirements for $p=0.10$ and $p=0.01$ are provided in Tables 7.33 to 7.36 and are mirror images of the $p=0.05$ tables/charts presented in the template below.)

Sample size tables for the Tukey HSD procedure

Three-group studies
If $r=0.40$, use Table 7.31 Chart B — If $r=0.60$, use Table 7.32 Chart B

Four-group studies
If $r=0.40$, use Table 7.31 Chart E — If $r=0.60$, use Table7.32 Chart E

Five-group studies
If $r=0.40$, use Table 7.31 Chart I — If $r=0.60$, use Table 7.32 Chart I

Sample size tables for the Newman–Keuls procedure

Determining the sample size requirements for the Newman–Keuls procedure is slightly more complicated since different charts within Tables 7.31 ($r=0.40$, $p=0.05$), 7.32 ($r=0.60$, $p=0.05$), 7.33 ($r=0.40$, $p=0.01$), 7.34 ($r=0.60$, $p=0.01$), 7.35 ($r=0.40$, $p=0.10$), and 7.36 ($r=0.60$, $p=0.10$) are used to determine sample sizes for both the number of groups and the number of intervening variables:

Three groups

Contrast	Chart
② − ①	A
③ − ①	B
③ − ②	A

Four groups

Contrast	Chart	Contrast	Chart
② − ①	C	④ − ①	E
③ − ①	D	④ − ②	D
③ − ②	C	④ − ③	C

Five groups

Contrast	Chart	Contrast	Chart	Contrast	Chart
② − ①	F	④ − ①	H	⑤ − ①	I
③ − ①	G	③ − ①	G	⑤ − ②	H
③ − ②	F	④ − ③	F	⑤ − ③	G
				⑤ − ④	F

N/group ② − ① =	*N*/group ④ − ① =	*N*/group ⑤ − ① =
N/group ③ − ① =	*N*/group ④ − ② =	*N*/group ⑤ − ② =
N/group ③ − ② =	*N*/group ④ − ③ =	*N*/group ⑤ − ③ =
		N/group ⑤ − ④ =

Table 7.1. Power table for ANCOVA F-ratio and pairwise contrast; $r = 0.40$, 2 groups at alpha = 0.05

n	Hypothesized ES																		
	0.20	0.30	0.35	0.40	0.45	0.50	0.55	0.60	0.65	0.70	0.75	0.80	1.00	1.25	1.50	1.75	2.00	2.50	3.00
5	6	7	8	9	10	12	13	15	16	18	20	22	32	46	61	74	84	96	99
6	6	8	9	10	12	13	15	17	20	22	25	27	40	56	72	84	92	99	.
7	6	8	10	11	13	15	18	20	23	26	29	32	47	65	80	91	96	.	.
8	6	9	10	12	15	17	20	23	26	30	33	37	53	72	86	94	98	.	.
9	6	9	11	14	16	19	22	26	29	33	37	42	59	78	90	97	99	.	.
10	7	10	12	15	18	21	25	28	33	37	41	46	64	83	94	98	.	.	.
11	7	11	13	16	19	23	27	31	36	40	45	50	69	86	96	99	.	.	.
12	7	11	14	17	21	25	29	34	39	44	49	54	73	90	97	99	.	.	.
13	7	12	15	18	22	27	31	37	42	47	52	58	77	92	98	.	.	.	.
14	8	13	16	20	24	29	34	39	45	50	56	61	80	94	99	.	.	.	.
15	8	13	17	21	26	31	36	42	47	53	59	65	83	95	99	.	.	.	.
20	9	17	22	27	33	40	46	53	60	66	72	78	92	99	.	.	.	.	.
25	11	20	26	33	41	48	56	63	70	76	82	86	97	.	.	.	.	.	.
30	13	24	31	39	47	56	64	71	78	84	88	92	99	.	.	.	.	.	.
35	14	27	36	45	54	63	71	78	84	89	93	95	99	.	.	.	.	.	.
40	16	31	40	50	59	68	77	83	89	93	95	97	.	.	.	.	.	.	.
45	17	34	44	55	65	74	81	87	92	95	97	98	.	.	.	.	.	.	.
50	19	37	48	59	69	78	85	91	94	97	98	99	.	.	.	.	.	.	.
55	20	41	52	63	73	82	88	93	96	98	99	.	.	.	.	.	.	.	.
60	22	44	56	67	77	85	91	95	97	99	99	.	.	.	.	.	.	.	.
65	23	47	59	71	80	88	93	96	98	99	.	.	.	.	.	.	.	.	.
70	25	50	62	74	83	90	95	97	99	99	.	.	.	.	.	.	.	.	.
75	27	52	65	77	86	92	96	98	99	.	.	.	.	.	.	.	.	.	.
80	28	55	68	79	88	93	97	99	99	.	.	.	.	.	.	.	.	.	.
90	31	60	73	84	91	96	98	99	.	.	.	.	.	.	.	.	.	.	.
100	34	65	78	87	94	97	99	.	.	.	.	.	.	.	.	.	.	.	.
110	37	69	81	90	96	98	99	.	.	.	.	.	.	.	.	.	.	.	.
120	40	73	85	93	97	99	.	.	.	.	.	.	.	.	.	.	.	.	.
130	43	76	87	94	98	99	.	.	.	.	.	.	.	.	.	.	.	.	.
140	45	79	90	96	98	.	.	.	.	.	.	.	.	.	.	.	.	.	.
150	48	82	91	97	99	.	.	.	.	.	.	.	.	.	.	.	.	.	.
175	54	87	95	98	.	.	.	.	.	.	.	.	.	.	.	.	.	.	.
200	60	91	97	99	.	.	.	.	.	.	.	.	.	.	.	.	.	.	.
225	65	94	98	.	.	.	.	.	.	.	.	.	.	.	.	.	.	.	.
250	69	96	99	.	.	.	.	.	.	.	.	.	.	.	.	.	.	.	.
300	77	98	.	.	.	.	.	.	.	.	.	.	.	.	.	.	.	.	.
350	83	99	.	.	.	.	.	.	.	.	.	.	.	.	.	.	.	.	.
400	88	.	.	.	.	.	.	.	.	.	.	.	.	.	.	.	.	.	.
450	91	.	.	.	.	.	.	.	.	.	.	.	.	.	.	.	.	.	.
500	94	.	.	.	.	.	.	.	.	.	.	.	.	.	.	.	.	.	.
600	97	.	.	.	.	.	.	.	.	.	.	.	.	.	.	.	.	.	.
700	98	.	.	.	.	.	.	.	.	.	.	.	.	.	.	.	.	.	.
800	99	.	.	.	.	.	.	.	.	.	.	.	.	.	.	.	.	.	.
900	.	.	.	.	.	.	.	.	.	.	.	.	.	.	.	.	.	.	.
1000	.	.	.	.	.	.	.	.	.	.	.	.	.	.	.	.	.	.	.

Table 7.2. Power table for ANCOVA F-ratio and pairwise contrast; $r = 0.60$, 2 groups at alpha = 0.05

n	Hypothesized ES																		
	0.20	0.30	0.35	0.40	0.45	0.50	0.55	0.60	0.65	0.70	0.75	0.80	1.00	1.25	1.50	1.75	2.00	2.50	3.00
5	6	8	9	10	12	14	16	18	20	23	25	28	40	57	73	85	92	99	.
6	6	9	10	12	14	16	19	21	24	28	31	34	50	68	83	92	97	.	.
7	7	9	11	13	16	19	22	25	29	33	36	41	58	77	89	96	99	.	.
8	7	10	12	15	18	21	25	29	33	37	42	46	65	83	94	98	.	.	.
9	7	11	14	17	20	24	28	32	37	42	47	52	71	88	96	99	.	.	.
10	8	12	15	18	22	26	31	36	41	46	52	57	76	91	98	.	.	.	.
11	8	13	16	20	24	29	34	39	45	50	56	61	80	94	99	.	.	.	.
12	8	14	17	21	26	31	37	43	48	54	60	66	84	96	99	.	.	.	.
13	9	15	19	23	28	34	40	46	52	58	64	69	87	97	.	.	.	.	.
14	9	15	20	25	30	36	42	49	55	61	67	73	89	98	.	.	.	.	.
15	9	16	21	26	32	39	45	52	58	65	71	76	91	99	.	.	.	.	.
20	11	21	27	34	42	50	57	65	72	78	83	88	97	.	.	.	.	.	.
25	13	26	33	42	51	59	67	75	81	86	91	94	99	.	.	.	.	.	.
30	15	30	39	49	58	67	75	82	88	92	95	97	.	.	.	.	.	.	.
35	17	35	45	55	65	74	82	88	92	95	97	99	.	.	.	.	.	.	.
40	19	39	50	61	71	80	87	92	95	97	99	99	.	.	.	.	.	.	.
45	22	43	55	66	76	84	90	94	97	99	99	.	.	.	.	.	.	.	.
50	24	47	59	71	80	88	93	96	98	99	.	.	.	.	.	.	.	.	.
55	26	51	63	75	84	91	95	98	99	.	.	.	.	.	.	.	.	.	.
60	28	54	67	79	87	93	96	98	99	.	.	.	.	.	.	.	.	.	.
65	30	58	71	82	90	95	98	99	.	.	.	.	.	.	.	.	.	.	.
70	32	61	74	84	92	96	98	99	.	.	.	.	.	.	.	.	.	.	.
75	34	64	77	87	93	97	99	.	.	.	.	.	.	.	.	.	.	.	.
80	36	67	80	89	95	98	99	.	.	.	.	.	.	.	.	.	.	.	.
90	39	72	84	92	97	99	.	.	.	.	.	.	.	.	.	.	.	.	.
100	43	76	88	94	98	99	.	.	.	.	.	.	.	.	.	.	.	.	.
110	46	80	90	96	99	.	.	.	.	.	.	.	.	.	.	.	.	.	.
120	50	83	93	97	99	.	.	.	.	.	.	.	.	.	.	.	.	.	.
130	53	86	94	98	.	.	.	.	.	.	.	.	.	.	.	.	.	.	.
140	56	89	96	99	.	.	.	.	.	.	.	.	.	.	.	.	.	.	.
150	59	91	97	99	.	.	.	.	.	.	.	.	.	.	.	.	.	.	.
175	66	94	98	.	.	.	.	.	.	.	.	.	.	.	.	.	.	.	.
200	71	97	99	.	.	.	.	.	.	.	.	.	.	.	.	.	.	.	.
225	76	98	.	.	.	.	.	.	.	.	.	.	.	.	.	.	.	.	.
250	81	99	.	.	.	.	.	.	.	.	.	.	.	.	.	.	.	.	.
300	87	.	.	.	.	.	.	.	.	.	.	.	.	.	.	.	.	.	.
350	92	.	.	.	.	.	.	.	.	.	.	.	.	.	.	.	.	.	.
400	95	.	.	.	.	.	.	.	.	.	.	.	.	.	.	.	.	.	.
450	97	.	.	.	.	.	.	.	.	.	.	.	.	.	.	.	.	.	.
500	98	.	.	.	.	.	.	.	.	.	.	.	.	.	.	.	.	.	.
600	99	.	.	.	.	.	.	.	.	.	.	.	.	.	.	.	.	.	.
700	.	.	.	.	.	.	.	.	.	.	.	.	.	.	.	.	.	.	.
800	.	.	.	.	.	.	.	.	.	.	.	.	.	.	.	.	.	.	.
900	.	.	.	.	.	.	.	.	.	.	.	.	.	.	.	.	.	.	.
1000	.	.	.	.	.	.	.	.	.	.	.	.	.	.	.	.	.	.	.

Table 7.3. Power table for ANCOVA F-ratio; $r = 0.40$, pattern L/M, 3 groups at alpha = 0.05

n	Hypothesized ES																		
	0.20	0.30	0.35	0.40	0.45	0.50	0.55	0.60	0.65	0.70	0.75	0.80	1.00	1.25	1.50	1.75	2.00	2.50	3.00
5	6	7	7	8	9	10	11	12	13	14	16	18	25	38	51	65	77	93	99
6	6	7	8	9	10	11	12	14	15	17	19	21	31	46	62	76	87	97	.
7	6	7	8	9	11	12	14	16	18	20	22	25	37	55	71	84	93	99	.
8	6	8	9	10	12	13	15	18	20	23	26	29	43	62	78	90	96	.	.
9	6	8	9	11	13	15	17	20	23	26	29	33	48	68	84	94	98	.	.
10	6	8	10	12	14	16	19	22	25	29	32	36	54	74	88	96	99	.	.
11	6	9	11	13	15	18	21	24	28	32	36	40	58	79	92	98	99	.	.
12	7	9	11	13	16	19	22	26	30	34	39	44	63	83	94	99	.	.	.
13	7	10	12	14	17	21	24	28	33	37	42	47	67	86	96	99	.	.	.
14	7	10	13	15	18	22	26	30	35	40	45	51	71	89	97	99	.	.	.
15	7	11	13	16	20	24	28	33	38	43	48	54	74	91	98	.	.	.	.
20	8	13	17	21	26	31	37	43	49	56	62	68	87	97	.	.	.	.	.
25	9	16	20	26	32	38	45	52	59	66	73	78	93	99	.	.	.	.	.
30	10	18	24	30	38	45	53	61	68	75	81	86	97	.	.	.	.	.	.
35	11	21	28	35	43	52	60	68	75	82	87	91	99	.	.	.	.	.	.
40	12	24	31	40	49	58	67	74	81	87	91	94	99	.	.	.	.	.	.
45	13	26	35	44	54	63	72	80	86	91	94	97	.	.	.	.	.	.	.
50	15	29	38	48	59	68	77	84	90	94	96	98	.	.	.	.	.	.	.
55	16	32	42	52	63	73	81	87	92	96	98	99	.	.	.	.	.	.	.
60	17	34	45	56	67	77	84	90	94	97	98	99	.	.	.	.	.	.	.
65	18	37	48	60	71	80	87	93	96	98	99	.	.	.	.	.	.	.	.
70	19	40	52	63	74	83	90	94	97	99	99	.	.	.	.	.	.	.	.
75	20	42	55	67	77	86	92	96	98	99	.	.	.	.	.	.	.	.	.
80	22	45	57	70	80	88	93	97	99	99	.	.	.	.	.	.	.	.	.
90	24	49	63	75	85	92	96	98	99	.	.	.	.	.	.	.	.	.	.
100	26	54	68	80	89	94	97	99	.	.	.	.	.	.	.	.	.	.	.
110	29	58	72	84	91	96	98	99	.	.	.	.	.	.	.	.	.	.	.
120	31	62	76	87	94	97	99	.	.	.	.	.	.	.	.	.	.	.	.
130	33	66	80	89	95	98	99	.	.	.	.	.	.	.	.	.	.	.	.
140	36	69	83	92	97	99	.	.	.	.	.	.	.	.	.	.	.	.	.
150	38	72	85	93	98	99	.	.	.	.	.	.	.	.	.	.	.	.	.
175	44	79	90	96	99	.	.	.	.	.	.	.	.	.	.	.	.	.	.
200	49	85	94	98	.	.	.	.	.	.	.	.	.	.	.	.	.	.	.
225	54	89	96	99	.	.	.	.	.	.	.	.	.	.	.	.	.	.	.
250	59	92	98	.	.	.	.	.	.	.	.	.	.	.	.	.	.	.	.
300	67	96	99	.	.	.	.	.	.	.	.	.	.	.	.	.	.	.	.
350	74	98	.	.	.	.	.	.	.	.	.	.	.	.	.	.	.	.	.
400	80	99	.	.	.	.	.	.	.	.	.	.	.	.	.	.	.	.	.
450	85	.	.	.	.	.	.	.	.	.	.	.	.	.	.	.	.	.	.
500	88	.	.	.	.	.	.	.	.	.	.	.	.	.	.	.	.	.	.
600	93	.	.	.	.	.	.	.	.	.	.	.	.	.	.	.	.	.	.
700	96	.	.	.	.	.	.	.	.	.	.	.	.	.	.	.	.	.	.
800	98	.	.	.	.	.	.	.	.	.	.	.	.	.	.	.	.	.	.
900	99	.	.	.	.	.	.	.	.	.	.	.	.	.	.	.	.	.	.
1000	.	.	.	.	.	.	.	.	.	.	.	.	.	.	.	.	.	.	.

Table 7.4. Power table for ANCOVA F-ratio; $r = 0.60$, pattern L/M, 3 groups at alpha $= 0.05$

n	Hypothesized ES																		
	0.20	0.30	0.35	0.40	0.45	0.50	0.55	0.60	0.65	0.70	0.75	0.80	1.00	1.25	1.50	1.75	2.00	2.50	3.00
5	6	7	8	9	10	11	13	14	16	18	20	22	32	48	63	77	88	98	.
6	6	7	9	10	11	13	15	17	19	21	24	27	40	58	75	87	94	99	.
7	6	8	9	11	13	15	17	19	22	25	28	32	47	67	83	93	98	.	.
8	6	9	10	12	14	16	19	22	25	29	33	37	54	74	89	96	99	.	.
9	6	9	11	13	16	18	21	25	29	33	37	42	60	80	93	98	.	.	.
10	7	10	12	14	17	20	24	28	32	37	41	46	66	85	95	99	.	.	.
11	7	10	13	15	19	22	26	31	35	40	45	51	71	89	97	99	.	.	.
12	7	11	14	17	20	24	29	33	39	44	49	55	75	92	98	.	.	.	.
13	7	12	14	18	22	26	31	36	42	47	53	59	79	94	99	.	.	.	.
14	8	12	15	19	23	28	33	39	45	51	57	63	82	96	99	.	.	.	.
15	8	13	16	20	25	30	36	42	48	54	60	66	85	97	.	.	.	.	.
20	9	16	21	26	33	40	47	54	61	68	74	80	94	99	.	.	.	.	.
25	11	20	26	33	40	48	57	65	72	79	84	89	98	.	.	.	.	.	.
30	12	23	31	39	48	57	65	73	80	86	91	94	99	.	.	.	.	.	.
35	14	27	35	45	54	64	73	80	86	91	94	97	.	.	.	.	.	.	.
40	15	30	40	50	61	70	79	86	91	94	97	98	.	.	.	.	.	.	.
45	17	34	44	55	66	76	84	90	94	97	98	99	.	.	.	.	.	.	.
50	18	37	49	60	71	80	88	93	96	98	99	.	.	.	.	.	.	.	.
55	20	40	53	65	75	84	91	95	97	99	.	.	.	.	.	.	.	.	.
60	21	44	57	69	79	87	93	96	98	99	.	.	.	.	.	.	.	.	.
65	23	47	60	73	83	90	95	98	99	.	.	.	.	.	.	.	.	.	.
70	24	50	64	76	85	92	96	98	99	.	.	.	.	.	.	.	.	.	.
75	26	53	67	79	88	94	97	99	.	.	.	.	.	.	.	.	.	.	.
80	28	56	70	82	90	95	98	99	.	.	.	.	.	.	.	.	.	.	.
90	31	61	75	86	93	97	99	.	.	.	.	.	.	.	.	.	.	.	.
100	34	66	80	90	95	98	99	.	.	.	.	.	.	.	.	.	.	.	.
110	37	71	84	92	97	99	.	.	.	.	.	.	.	.	.	.	.	.	.
120	40	75	87	94	98	99	.	.	.	.	.	.	.	.	.	.	.	.	.
130	43	78	90	96	99	.	.	.	.	.	.	.	.	.	.	.	.	.	.
140	46	81	92	97	99	.	.	.	.	.	.	.	.	.	.	.	.	.	.
150	48	84	93	98	99	.	.	.	.	.	.	.	.	.	.	.	.	.	.
175	55	89	96	99	.	.	.	.	.	.	.	.	.	.	.	.	.	.	.
200	61	93	98	.	.	.	.	.	.	.	.	.	.	.	.	.	.	.	.
225	66	96	99	.	.	.	.	.	.	.	.	.	.	.	.	.	.	.	.
250	71	97	.	.	.	.	.	.	.	.	.	.	.	.	.	.	.	.	.
300	79	99	.	.	.	.	.	.	.	.	.	.	.	.	.	.	.	.	.
350	85	.	.	.	.	.	.	.	.	.	.	.	.	.	.	.	.	.	.
400	90	.	.	.	.	.	.	.	.	.	.	.	.	.	.	.	.	.	.
450	93	.	.	.	.	.	.	.	.	.	.	.	.	.	.	.	.	.	.
500	95	.	.	.	.	.	.	.	.	.	.	.	.	.	.	.	.	.	.
600	98	.	.	.	.	.	.	.	.	.	.	.	.	.	.	.	.	.	.
700	99	.	.	.	.	.	.	.	.	.	.	.	.	.	.	.	.	.	.
800	.	.	.	.	.	.	.	.	.	.	.	.	.	.	.	.	.	.	.
900	.	.	.	.	.	.	.	.	.	.	.	.	.	.	.	.	.	.	.
1000	.	.	.	.	.	.	.	.	.	.	.	.	.	.	.	.	.	.	.

Table 7.5. Power table for ANCOVA F-ratio; $r = 0.40$, pattern H, 3 groups at alpha = 0.05

n	Hypothesized ES																		
	0.20	0.30	0.35	0.40	0.45	0.50	0.55	0.60	0.65	0.70	0.75	0.80	1.00	1.25	1.50	1.75	2.00	2.50	3.00
5	6	7	8	9	10	11	13	14	16	18	20	22	33	48	64	78	88	98	.
6	6	8	9	10	11	13	15	17	19	22	24	27	41	59	75	88	95	.	.
7	6	8	9	11	13	15	17	20	22	26	29	32	48	68	84	93	98	.	.
8	6	9	10	12	14	17	19	22	26	29	33	37	55	75	89	96	99	.	.
9	7	9	11	13	16	19	22	25	29	33	38	42	61	81	93	98	.	.	.
10	7	10	12	14	17	21	24	28	33	37	42	47	67	86	96	99	.	.	.
11	7	10	13	16	19	22	27	31	36	41	46	51	72	89	97	.	.	.	.
12	7	11	14	17	20	24	29	34	39	45	50	56	76	92	98	.	.	.	.
13	7	12	15	18	22	26	31	37	42	48	54	60	80	94	99	.	.	.	.
14	8	12	16	19	24	28	34	39	45	51	57	63	83	96	99	.	.	.	.
15	8	13	16	21	25	30	36	42	48	55	61	67	86	97	.	.	.	.	.
20	9	16	21	27	33	40	47	55	62	69	75	81	95	99	.	.	.	.	.
25	11	20	26	33	41	49	57	65	73	79	85	89	98	.	.	.	.	.	.
30	12	23	31	39	48	57	66	74	81	87	91	94	99	.	.	.	.	.	.
35	14	27	36	45	55	65	73	81	87	92	95	97	.	.	.	.	.	.	.
40	15	31	40	51	61	71	79	86	91	95	97	98	.	.	.	.	.	.	.
45	17	34	45	56	67	76	84	90	94	97	98	99	.	.	.	.	.	.	.
50	18	38	49	61	72	81	88	93	96	98	99	.	.	.	.	.	.	.	.
55	20	41	53	65	76	85	91	95	98	99	.	.	.	.	.	.	.	.	.
60	22	44	57	70	80	88	93	97	98	99	.	.	.	.	.	.	.	.	.
65	23	48	61	73	83	90	95	98	99	.	.	.	.	.	.	.	.	.	.
70	25	51	64	77	86	92	96	98	99	.	.	.	.	.	.	.	.	.	.
75	26	54	68	80	88	94	97	99	.	.	.	.	.	.	.	.	.	.	.
80	28	57	71	82	90	95	98	99	.	.	.	.	.	.	.	.	.	.	.
90	31	62	76	87	94	97	99	.	.	.	.	.	.	.	.	.	.	.	.
100	34	67	81	90	96	98	.	.	.	.	.	.	.	.	.	.	.	.	.
110	37	71	84	93	97	99	.	.	.	.	.	.	.	.	.	.	.	.	.
120	40	75	88	95	98	99	.	.	.	.	.	.	.	.	.	.	.	.	.
130	43	79	90	96	99	.	.	.	.	.	.	.	.	.	.	.	.	.	.
140	46	82	92	97	99	.	.	.	.	.	.	.	.	.	.	.	.	.	.
150	49	85	94	98	.	.	.	.	.	.	.	.	.	.	.	.	.	.	.
175	56	90	97	99	.	.	.	.	.	.	.	.	.	.	.	.	.	.	.
200	62	93	98	.	.	.	.	.	.	.	.	.	.	.	.	.	.	.	.
225	67	96	99	.	.	.	.	.	.	.	.	.	.	.	.	.	.	.	.
250	72	97	.	.	.	.	.	.	.	.	.	.	.	.	.	.	.	.	.
300	80	99	.	.	.	.	.	.	.	.	.	.	.	.	.	.	.	.	.
350	86	.	.	.	.	.	.	.	.	.	.	.	.	.	.	.	.	.	.
400	90	.	.	.	.	.	.	.	.	.	.	.	.	.	.	.	.	.	.
450	93	.	.	.	.	.	.	.	.	.	.	.	.	.	.	.	.	.	.
500	96	.	.	.	.	.	.	.	.	.	.	.	.	.	.	.	.	.	.
600	98	.	.	.	.	.	.	.	.	.	.	.	.	.	.	.	.	.	.
700	99	.	.	.	.	.	.	.	.	.	.	.	.	.	.	.	.	.	.
800	.	.	.	.	.	.	.	.	.	.	.	.	.	.	.	.	.	.	.
900	.	.	.	.	.	.	.	.	.	.	.	.	.	.	.	.	.	.	.
1000	.	.	.	.	.	.	.	.	.	.	.	.	.	.	.	.	.	.	.

Table 7.6. Power table for ANCOVA F-ratio; $r = 0.60$, pattern H, 3 groups at alpha = 0.05

n	Hypothesized ES																		
	0.20	0.30	0.35	0.40	0.45	0.50	0.55	0.60	0.65	0.70	0.75	0.80	1.00	1.25	1.50	1.75	2.00	2.50	3.00
5	6	8	9	10	12	13	15	17	20	22	25	28	41	60	76	88	95	.	.
6	6	8	10	12	13	16	18	21	24	27	31	35	51	71	86	95	99	.	.
7	7	9	11	13	15	18	21	25	28	33	37	41	60	80	92	98	.	.	.
8	7	10	12	15	17	21	24	29	33	38	42	47	67	86	96	99	.	.	.
9	7	11	13	16	20	23	28	32	37	42	48	53	74	91	98	.	.	.	.
10	8	12	14	18	22	26	31	36	41	47	53	59	79	94	99	.	.	.	.
11	8	12	16	19	24	29	34	40	46	52	58	64	83	96	99	.	.	.	.
12	8	13	17	21	26	31	37	43	49	56	62	68	87	97	.	.	.	.	.
13	9	14	18	23	28	34	40	47	53	60	66	72	90	98	.	.	.	.	.
14	9	15	19	24	30	36	43	50	57	64	70	76	92	99	.	.	.	.	.
15	9	16	21	26	32	39	46	53	60	67	73	79	94	99	.	.	.	.	.
20	11	21	27	34	42	51	59	67	74	81	86	90	98	.	.	.	.	.	.
25	13	25	33	42	52	61	70	78	84	89	93	96	.	.	.	.	.	.	.
30	15	30	39	50	60	70	78	85	91	94	97	98	.	.	.	.	.	.	.
35	17	35	45	57	67	77	85	91	95	97	99	99	.	.	.	.	.	.	.
40	19	39	51	63	74	83	89	94	97	99	99	.	.	.	.	.	.	.	.
45	21	44	56	69	79	87	93	96	98	99	.	.	.	.	.	.	.	.	.
50	23	48	61	73	83	91	95	98	99	.	.	.	.	.	.	.	.	.	.
55	25	52	66	78	87	93	97	99	.	.	.	.	.	.	.	.	.	.	.
60	27	56	70	81	90	95	98	99	.	.	.	.	.	.	.	.	.	.	.
65	29	59	73	85	92	96	99	.	.	.	.	.	.	.	.	.	.	.	.
70	32	63	77	87	94	98	99	.	.	.	.	.	.	.	.	.	.	.	.
75	34	66	80	90	95	98	99	.	.	.	.	.	.	.	.	.	.	.	.
80	36	69	82	91	96	99	.	.	.	.	.	.	.	.	.	.	.	.	.
90	40	74	87	94	98	99	.	.	.	.	.	.	.	.	.	.	.	.	.
100	44	79	90	96	99	.	.	.	.	.	.	.	.	.	.	.	.	.	.
110	47	83	93	98	99	.	.	.	.	.	.	.	.	.	.	.	.	.	.
120	51	86	95	99	.	.	.	.	.	.	.	.	.	.	.	.	.	.	.
130	54	89	96	99	.	.	.	.	.	.	.	.	.	.	.	.	.	.	.
140	58	91	97	99	.	.	.	.	.	.	.	.	.	.	.	.	.	.	.
150	61	93	98	.	.	.	.	.	.	.	.	.	.	.	.	.	.	.	.
175	68	96	99	.	.	.	.	.	.	.	.	.	.	.	.	.	.	.	.
200	74	98	.	.	.	.	.	.	.	.	.	.	.	.	.	.	.	.	.
225	79	99	.	.	.	.	.	.	.	.	.	.	.	.	.	.	.	.	.
250	84	99	.	.	.	.	.	.	.	.	.	.	.	.	.	.	.	.	.
300	90	.	.	.	.	.	.	.	.	.	.	.	.	.	.	.	.	.	.
350	94	.	.	.	.	.	.	.	.	.	.	.	.	.	.	.	.	.	.
400	96	.	.	.	.	.	.	.	.	.	.	.	.	.	.	.	.	.	.
450	98	.	.	.	.	.	.	.	.	.	.	.	.	.	.	.	.	.	.
500	99	.	.	.	.	.	.	.	.	.	.	.	.	.	.	.	.	.	.
600	.	.	.	.	.	.	.	.	.	.	.	.	.	.	.	.	.	.	.
700	.	.	.	.	.	.	.	.	.	.	.	.	.	.	.	.	.	.	.
800	.	.	.	.	.	.	.	.	.	.	.	.	.	.	.	.	.	.	.
900	.	.	.	.	.	.	.	.	.	.	.	.	.	.	.	.	.	.	.
1000	.	.	.	.	.	.	.	.	.	.	.	.	.	.	.	.	.	.	.

Table 7.7. Power table for ANCOVA F-ratio; $r = 0.40$, pattern L, 4 groups at alpha $= 0.05$

n	Hypothesized ES																		
	0.20	0.30	0.35	0.40	0.45	0.50	0.55	0.60	0.65	0.70	0.75	0.80	1.00	1.25	1.50	1.75	2.00	2.50	3.00
5	5	6	7	7	8	9	9	10	11	13	14	15	22	33	46	60	72	90	98
6	5	6	7	8	9	10	11	12	13	15	17	18	27	41	56	71	83	96	99
7	6	7	7	8	9	11	12	14	15	17	19	22	32	49	65	80	90	98	.
8	6	7	8	9	10	12	13	15	17	20	22	25	37	56	73	86	94	99	.
9	6	7	8	10	11	13	15	17	19	22	25	28	42	62	79	91	97	.	.
10	6	8	9	10	12	14	16	19	22	25	28	31	47	68	84	94	98	.	.
11	6	8	9	11	13	15	18	21	24	27	31	35	52	73	88	96	99	.	.
12	6	8	10	12	14	16	19	22	26	30	34	38	56	77	91	98	.	.	.
13	6	9	10	12	15	18	21	24	28	32	37	41	60	81	94	98	.	.	.
14	6	9	11	13	16	19	22	26	30	35	39	44	64	85	95	99	.	.	.
15	7	9	12	14	17	20	24	28	32	37	42	47	68	87	97	99	.	.	.
20	7	11	14	18	22	26	32	37	43	49	55	61	82	96	99	.	.	.	.
25	8	14	17	22	27	33	39	46	53	60	66	72	90	99	.	.	.	.	.
30	9	16	20	26	32	39	47	54	62	69	75	81	95	.	.	.	.	.	.
35	10	18	23	30	37	45	54	62	69	76	82	87	98	.	.	.	.	.	.
40	11	20	27	34	42	51	60	68	76	82	88	92	99	.	.	.	.	.	.
45	12	22	30	38	47	57	66	74	81	87	91	95	.	.	.	.	.	.	.
50	13	25	33	42	52	62	71	79	85	90	94	97	.	.	.	.	.	.	.
55	14	27	36	46	56	66	75	83	89	93	96	98	.	.	.	.	.	.	.
60	14	29	39	50	61	71	79	86	92	95	97	99	.	.	.	.	.	.	.
65	15	32	42	53	64	74	83	89	94	97	98	99	.	.	.	.	.	.	.
70	16	34	45	57	68	78	86	91	95	98	99	.	.	.	.	.	.	.	.
75	17	36	48	60	71	81	88	93	97	98	99	.	.	.	.	.	.	.	.
80	18	39	51	63	74	84	90	95	97	99	.	.	.	.	.	.	.	.	.
90	20	43	56	69	80	88	94	97	99	99	.	.	.	.	.	.	.	.	.
100	22	47	61	74	84	91	96	98	99	.	.	.	.	.	.	.	.	.	.
110	25	51	66	78	88	94	97	99	.	.	.	.	.	.	.	.	.	.	.
120	27	55	70	82	91	96	98	99	.	.	.	.	.	.	.	.	.	.	.
130	29	59	74	85	93	97	99	.	.	.	.	.	.	.	.	.	.	.	.
140	31	63	77	88	95	98	99	.	.	.	.	.	.	.	.	.	.	.	.
150	33	66	80	90	96	99	.	.	.	.	.	.	.	.	.	.	.	.	.
175	38	73	87	94	98	99	.	.	.	.	.	.	.	.	.	.	.	.	.
200	43	80	91	97	99	.	.	.	.	.	.	.	.	.	.	.	.	.	.
225	48	84	94	98	.	.	.	.	.	.	.	.	.	.	.	.	.	.	.
250	52	88	96	99	.	.	.	.	.	.	.	.	.	.	.	.	.	.	.
300	61	94	98	.	.	.	.	.	.	.	.	.	.	.	.	.	.	.	.
350	68	97	99	.	.	.	.	.	.	.	.	.	.	.	.	.	.	.	.
400	74	98	.	.	.	.	.	.	.	.	.	.	.	.	.	.	.	.	.
450	80	99	.	.	.	.	.	.	.	.	.	.	.	.	.	.	.	.	.
500	84	.	.	.	.	.	.	.	.	.	.	.	.	.	.	.	.	.	.
600	90	.	.	.	.	.	.	.	.	.	.	.	.	.	.	.	.	.	.
700	94	.	.	.	.	.	.	.	.	.	.	.	.	.	.	.	.	.	.
800	97	.	.	.	.	.	.	.	.	.	.	.	.	.	.	.	.	.	.
900	98	.	.	.	.	.	.	.	.	.	.	.	.	.	.	.	.	.	.
1000	99	.	.	.	.	.	.	.	.	.	.	.	.	.	.	.	.	.	.

Table 7.8. Power table for ANCOVA F-ratio; $r = 0.60$, pattern L, 4 groups at alpha $= 0.05$

n	Hypothesized ES																		
	0.20	0.30	0.35	0.40	0.45	0.50	0.55	0.60	0.65	0.70	0.75	0.80	1.00	1.25	1.50	1.75	2.00	2.50	3.00
5	6	7	7	8	9	10	11	12	14	15	17	19	28	42	58	72	84	97	.
6	6	7	8	9	10	11	13	14	16	18	21	23	35	52	69	83	92	99	.
7	6	7	8	10	11	13	15	17	19	22	24	28	42	61	78	90	96	.	.
8	6	8	9	10	12	14	16	19	22	25	28	32	48	69	85	94	98	.	.
9	6	8	10	11	13	16	18	21	25	28	32	36	54	75	90	97	99	.	.
10	6	9	10	12	15	17	20	24	28	32	36	40	60	80	93	98	.	.	.
11	6	9	11	13	16	19	22	26	30	35	40	45	65	85	95	99	.	.	.
12	7	10	12	14	17	21	24	29	33	38	43	49	69	88	97	.	.	.	.
13	7	10	12	15	19	22	26	31	36	41	47	52	73	91	98	.	.	.	.
14	7	11	13	16	20	24	29	34	39	45	50	56	77	93	99	.	.	.	.
15	7	11	14	17	21	26	31	36	42	48	54	60	80	95	99	.	.	.	.
20	8	14	18	23	28	34	41	48	55	62	68	74	92	99	.	.	.	.	.
25	9	17	22	28	35	42	50	58	66	73	79	84	97	.	.	.	.	.	.
30	11	20	26	33	42	50	59	67	75	81	87	91	99	.	.	.	.	.	.
35	12	23	30	39	48	57	66	75	82	87	92	95	.	.	.	.	.	.	.
40	13	26	34	44	54	64	73	81	87	92	95	97	.	.	.	.	.	.	.
45	14	29	38	49	60	70	78	86	91	95	97	99	.	.	.	.	.	.	.
50	16	32	42	54	65	75	83	89	94	97	98	99	.	.	.	.	.	.	.
55	17	35	46	58	69	79	87	92	96	98	99	.	.	.	.	.	.	.	.
60	18	38	50	62	74	83	90	94	97	99	99	.	.	.	.	.	.	.	.
65	19	41	54	66	77	86	92	96	98	99	.	.	.	.	.	.	.	.	.
70	21	44	57	70	81	89	94	97	99	.	.	.	.	.	.	.	.	.	.
75	22	47	60	73	83	91	95	98	99	.	.	.	.	.	.	.	.	.	.
80	23	49	63	76	86	93	97	99	.	.	.	.	.	.	.	.	.	.	.
90	26	55	69	81	90	95	98	99	.	.	.	.	.	.	.	.	.	.	.
100	29	60	74	86	93	97	99	.	.	.	.	.	.	.	.	.	.	.	.
110	32	64	79	89	95	98	99	.	.	.	.	.	.	.	.	.	.	.	.
120	34	68	82	92	97	99	.	.	.	.	.	.	.	.	.	.	.	.	.
130	37	72	85	94	98	99	.	.	.	.	.	.	.	.	.	.	.	.	.
140	40	76	88	95	99	.	.	.	.	.	.	.	.	.	.	.	.	.	.
150	42	79	90	97	99	.	.	.	.	.	.	.	.	.	.	.	.	.	.
175	48	85	94	98	.	.	.	.	.	.	.	.	.	.	.	.	.	.	.
200	54	90	97	99	.	.	.	.	.	.	.	.	.	.	.	.	.	.	.
225	60	93	98	.	.	.	.	.	.	.	.	.	.	.	.	.	.	.	.
250	65	95	99	.	.	.	.	.	.	.	.	.	.	.	.	.	.	.	.
300	74	98	.	.	.	.	.	.	.	.	.	.	.	.	.	.	.	.	.
350	81	99	.	.	.	.	.	.	.	.	.	.	.	.	.	.	.	.	.
400	86	.	.	.	.	.	.	.	.	.	.	.	.	.	.	.	.	.	.
450	90	.	.	.	.	.	.	.	.	.	.	.	.	.	.	.	.	.	.
500	93	.	.	.	.	.	.	.	.	.	.	.	.	.	.	.	.	.	.
600	97	.	.	.	.	.	.	.	.	.	.	.	.	.	.	.	.	.	.
700	98	.	.	.	.	.	.	.	.	.	.	.	.	.	.	.	.	.	.
800	99	.	.	.	.	.	.	.	.	.	.	.	.	.	.	.	.	.	.
900	.	.	.	.	.	.	.	.	.	.	.	.	.	.	.	.	.	.	.
1000	.	.	.	.	.	.	.	.	.	.	.	.	.	.	.	.	.	.	.

Table 7.9. Power table for ANCOVA F-ratio; $r = 0.40$, pattern M, 4 groups at alpha = 0.05

n	Hypothesized ES																		
	0.20	0.30	0.35	0.40	0.45	0.50	0.55	0.60	0.65	0.70	0.75	0.80	1.00	1.25	1.50	1.75	2.00	2.50	3.00
5	6	6	7	7	8	9	10	11	12	14	15	17	24	36	50	65	77	93	99
6	6	7	7	8	9	10	11	13	14	16	18	20	30	45	61	76	87	98	.
7	6	7	8	9	10	11	13	15	17	19	21	24	36	53	70	84	93	99	.
8	6	7	8	10	11	13	14	17	19	21	24	27	41	61	78	90	96	.	.
9	6	8	9	10	12	14	16	19	21	24	28	31	47	67	84	94	98	.	.
10	6	8	9	11	13	15	18	21	24	27	31	35	52	73	88	96	99	.	.
11	6	8	10	12	14	16	19	23	26	30	34	38	57	78	91	98	.	.	.
12	6	9	11	13	15	18	21	25	28	33	37	42	61	82	94	99	.	.	.
13	6	9	11	13	16	19	23	27	31	35	40	45	66	85	96	99	.	.	.
14	7	10	12	14	17	21	24	29	33	38	43	49	69	88	97	.	.	.	.
15	7	10	12	15	18	22	26	31	36	41	46	52	73	91	98	.	.	.	.
20	8	12	16	19	24	29	35	41	47	54	60	66	86	97	.	.	.	.	.
25	9	15	19	24	30	36	43	50	58	65	71	77	93	99	.	.	.	.	.
30	10	17	22	29	36	43	51	59	67	74	80	85	97	.	.	.	.	.	.
35	11	20	26	33	41	50	58	67	74	81	86	91	99	.	.	.	.	.	.
40	12	22	29	38	47	56	65	73	80	86	91	94	99	.	.	.	.	.	.
45	13	25	33	42	52	62	71	79	85	90	94	97	.	.	.	.	.	.	.
50	14	27	36	46	57	67	76	83	89	93	96	98	.	.	.	.	.	.	.
55	15	30	40	51	61	71	80	87	92	95	98	99	.	.	.	.	.	.	.
60	16	32	43	54	66	76	84	90	94	97	99	99	.	.	.	.	.	.	.
65	17	35	46	58	69	79	87	92	96	98	99	.	.	.	.	.	.	.	.
70	18	38	50	62	73	82	89	94	97	99	99	.	.	.	.	.	.	.	.
75	19	40	53	65	76	85	92	96	98	99	.	.	.	.	.	.	.	.	.
80	20	43	56	68	79	88	93	97	99	99	.	.	.	.	.	.	.	.	.
90	22	47	61	74	84	91	96	98	99	.	.	.	.	.	.	.	.	.	.
100	25	52	66	79	88	94	97	99	.	.	.	.	.	.	.	.	.	.	.
110	27	56	71	83	91	96	98	99	.	.	.	.	.	.	.	.	.	.	.
120	29	60	75	86	93	97	99	.	.	.	.	.	.	.	.	.	.	.	.
130	32	64	79	89	95	98	99	.	.	.	.	.	.	.	.	.	.	.	.
140	34	68	82	91	97	99	.	.	.	.	.	.	.	.	.	.	.	.	.
150	36	71	85	93	98	99	.	.	.	.	.	.	.	.	.	.	.	.	.
175	42	78	90	96	99	.	.	.	.	.	.	.	.	.	.	.	.	.	.
200	47	84	94	98	.	.	.	.	.	.	.	.	.	.	.	.	.	.	.
225	52	88	96	99	.	.	.	.	.	.	.	.	.	.	.	.	.	.	.
250	57	92	98	.	.	.	.	.	.	.	.	.	.	.	.	.	.	.	.
300	66	96	99	.	.	.	.	.	.	.	.	.	.	.	.	.	.	.	.
350	73	98	.	.	.	.	.	.	.	.	.	.	.	.	.	.	.	.	.
400	79	99	.	.	.	.	.	.	.	.	.	.	.	.	.	.	.	.	.
450	84	.	.	.	.	.	.	.	.	.	.	.	.	.	.	.	.	.	.
500	88	.	.	.	.	.	.	.	.	.	.	.	.	.	.	.	.	.	.
600	93	.	.	.	.	.	.	.	.	.	.	.	.	.	.	.	.	.	.
700	96	.	.	.	.	.	.	.	.	.	.	.	.	.	.	.	.	.	.
800	98	.	.	.	.	.	.	.	.	.	.	.	.	.	.	.	.	.	.
900	99	.	.	.	.	.	.	.	.	.	.	.	.	.	.	.	.	.	.
1000	.	.	.	.	.	.	.	.	.	.	.	.	.	.	.	.	.	.	.

Table 7.10. Power table for ANCOVA F-ratio; $r = 0.60$, pattern M, 4 groups at alpha $= 0.05$

n	Hypothesized ES																		
	0.20	0.30	0.35	0.40	0.45	0.50	0.55	0.60	0.65	0.70	0.75	0.80	1.00	1.25	1.50	1.75	2.00	2.50	3.00
5	6	7	8	8	9	10	12	13	15	17	19	21	31	46	63	77	88	98	.
6	6	7	8	9	11	12	14	16	18	20	23	25	38	57	74	87	95	.	.
7	6	8	9	10	12	14	16	18	21	24	27	30	46	66	83	93	98	.	.
8	6	8	10	11	13	15	18	21	24	27	31	35	53	74	89	96	99	.	.
9	6	9	10	12	15	17	20	23	27	31	35	40	59	80	93	98	.	.	.
10	7	9	11	13	16	19	22	26	30	35	40	44	65	85	95	99	.	.	.
11	7	10	12	14	17	21	25	29	34	38	44	49	70	89	97	.	.	.	.
12	7	10	13	15	19	23	27	32	37	42	48	53	74	92	98	.	.	.	.
13	7	11	13	17	20	24	29	34	40	45	51	57	78	94	99	.	.	.	.
14	7	11	14	18	22	26	31	37	43	49	55	61	82	96	99	.	.	.	.
15	8	12	15	19	23	28	34	40	46	52	58	65	85	97	.	.	.	.	.
20	9	15	20	25	31	38	45	52	59	67	73	79	94	99	.	.	.	.	.
25	10	18	24	31	38	47	55	63	71	78	83	88	98	.	.	.	.	.	.
30	11	22	29	37	46	55	64	72	79	85	90	94	99	.	.	.	.	.	.
35	13	25	33	43	52	62	71	79	86	91	94	97	.	.	.	.	.	.	.
40	14	28	38	48	59	69	78	85	91	94	97	98	.	.	.	.	.	.	.
45	15	32	42	54	65	75	83	89	94	97	98	99	.	.	.	.	.	.	.
50	17	35	47	58	70	79	87	92	96	98	99	.	.	.	.	.	.	.	.
55	18	38	51	63	74	83	90	95	97	99	.	.	.	.	.	.	.	.	.
60	20	42	55	67	78	87	93	96	98	99	.	.	.	.	.	.	.	.	.
65	21	45	58	71	82	90	95	98	99	.	.	.	.	.	.	.	.	.	.
70	23	48	62	75	85	92	96	98	99	.	.	.	.	.	.	.	.	.	.
75	24	51	65	78	87	94	97	99	.	.	.	.	.	.	.	.	.	.	.
80	26	54	69	81	90	95	98	99	.	.	.	.	.	.	.	.	.	.	.
90	29	59	74	86	93	97	99	.	.	.	.	.	.	.	.	.	.	.	.
100	32	65	79	89	95	98	99	.	.	.	.	.	.	.	.	.	.	.	.
110	35	69	83	92	97	99	.	.	.	.	.	.	.	.	.	.	.	.	.
120	38	73	86	94	98	99	.	.	.	.	.	.	.	.	.	.	.	.	.
130	41	77	89	96	99	.	.	.	.	.	.	.	.	.	.	.	.	.	.
140	44	80	91	97	99	.	.	.	.	.	.	.	.	.	.	.	.	.	.
150	46	83	93	98	.	.	.	.	.	.	.	.	.	.	.	.	.	.	.
175	53	89	96	99	.	.	.	.	.	.	.	.	.	.	.	.	.	.	.
200	59	93	98	.	.	.	.	.	.	.	.	.	.	.	.	.	.	.	.
225	65	95	99	.	.	.	.	.	.	.	.	.	.	.	.	.	.	.	.
250	70	97	.	.	.	.	.	.	.	.	.	.	.	.	.	.	.	.	.
300	78	99	.	.	.	.	.	.	.	.	.	.	.	.	.	.	.	.	.
350	85	.	.	.	.	.	.	.	.	.	.	.	.	.	.	.	.	.	.
400	90	.	.	.	.	.	.	.	.	.	.	.	.	.	.	.	.	.	.
450	93	.	.	.	.	.	.	.	.	.	.	.	.	.	.	.	.	.	.
500	95	.	.	.	.	.	.	.	.	.	.	.	.	.	.	.	.	.	.
600	98	.	.	.	.	.	.	.	.	.	.	.	.	.	.	.	.	.	.
700	99	.	.	.	.	.	.	.	.	.	.	.	.	.	.	.	.	.	.
800	.	.	.	.	.	.	.	.	.	.	.	.	.	.	.	.	.	.	.
900	.	.	.	.	.	.	.	.	.	.	.	.	.	.	.	.	.	.	.
1000	.	.	.	.	.	.	.	.	.	.	.	.	.	.	.	.	.	.	.

Table 7.11. Power table for ANCOVA *F*-ratio; $r = 0.40$, pattern H, 4 groups at alpha = 0.05

n	Hypothesized ES																		
	0.20	0.30	0.35	0.40	0.45	0.50	0.55	0.60	0.65	0.70	0.75	0.80	1.00	1.25	1.50	1.75	2.00	2.50	3.00
5	6	8	9	10	11	13	15	17	19	22	24	27	41	60	78	90	96	.	.
6	6	8	9	11	13	15	18	20	23	27	30	34	51	72	87	96	99	.	.
7	6	9	10	12	15	17	21	24	28	32	36	41	60	81	93	98	.	.	.
8	7	10	12	14	17	20	24	28	32	37	42	47	67	87	96	99	.	.	.
9	7	10	13	15	19	22	27	31	36	42	47	53	74	91	98	.	.	.	.
10	7	11	14	17	21	25	30	35	41	46	52	58	79	94	99	.	.	.	.
11	8	12	15	19	23	28	33	39	45	51	57	63	84	96	.	.	.	.	.
12	8	13	16	20	25	30	36	42	49	55	62	68	87	98	.	.	.	.	.
13	8	14	17	22	27	33	39	46	53	60	66	72	90	99	.	.	.	.	.
14	9	14	19	23	29	35	42	49	56	63	70	76	93	99	.	.	.	.	.
15	9	15	20	25	31	38	45	52	60	67	74	79	94	99	.	.	.	.	.
20	11	20	26	33	41	50	59	67	75	81	87	91	99	.	.	.	.	.	.
25	12	24	32	41	51	61	70	78	85	90	94	96	.	.	.	.	.	.	.
30	14	29	39	49	60	70	79	86	91	95	97	99	.	.	.	.	.	.	.
35	16	34	45	56	67	77	85	91	95	97	99	99	.	.	.	.	.	.	.
40	18	38	50	63	74	83	90	95	97	99	.	.	.	.	.	.	.	.	.
45	20	43	56	68	79	88	93	97	99	99	.	.	.	.	.	.	.	.	.
50	22	47	61	73	84	91	96	98	99	.	.	.	.	.	.	.	.	.	.
55	24	51	65	78	87	94	97	99	.	.	.	.	.	.	.	.	.	.	.
60	26	55	70	82	90	96	98	99	.	.	.	.	.	.	.	.	.	.	.
65	28	59	74	85	93	97	99	.	.	.	.	.	.	.	.	.	.	.	.
70	30	62	77	88	94	98	99	.	.	.	.	.	.	.	.	.	.	.	.
75	33	66	80	90	96	99	.	.	.	.	.	.	.	.	.	.	.	.	.
80	35	69	83	92	97	99	.	.	.	.	.	.	.	.	.	.	.	.	.
90	39	75	87	95	98	.	.	.	.	.	.	.	.	.	.	.	.	.	.
100	43	79	91	97	99	.	.	.	.	.	.	.	.	.	.	.	.	.	.
110	46	83	93	98	.	.	.	.	.	.	.	.	.	.	.	.	.	.	.
120	50	87	95	99	.	.	.	.	.	.	.	.	.	.	.	.	.	.	.
130	54	89	97	99	.	.	.	.	.	.	.	.	.	.	.	.	.	.	.
140	57	92	98	.	.	.	.	.	.	.	.	.	.	.	.	.	.	.	.
150	60	93	98	.	.	.	.	.	.	.	.	.	.	.	.	.	.	.	.
175	68	97	99	.	.	.	.	.	.	.	.	.	.	.	.	.	.	.	.
200	74	98	.	.	.	.	.	.	.	.	.	.	.	.	.	.	.	.	.
225	80	99	.	.	.	.	.	.	.	.	.	.	.	.	.	.	.	.	.
250	84	.	.	.	.	.	.	.	.	.	.	.	.	.	.	.	.	.	.
300	90	.	.	.	.	.	.	.	.	.	.	.	.	.	.	.	.	.	.
350	94	.	.	.	.	.	.	.	.	.	.	.	.	.	.	.	.	.	.
400	97	.	.	.	.	.	.	.	.	.	.	.	.	.	.	.	.	.	.
450	98	.	.	.	.	.	.	.	.	.	.	.	.	.	.	.	.	.	.
500	99	.	.	.	.	.	.	.	.	.	.	.	.	.	.	.	.	.	.
600	.	.	.	.	.	.	.	.	.	.	.	.	.	.	.	.	.	.	.
700	.	.	.	.	.	.	.	.	.	.	.	.	.	.	.	.	.	.	.
800	.	.	.	.	.	.	.	.	.	.	.	.	.	.	.	.	.	.	.
900	.	.	.	.	.	.	.	.	.	.	.	.	.	.	.	.	.	.	.
1000	.	.	.	.	.	.	.	.	.	.	.	.	.	.	.	.	.	.	.

Table 7.12. Power table for ANCOVA F-ratio; $r = 0.60$, pattern H, 4 groups at alpha = 0.05

n	Hypothesized ES																		
	0.20	0.30	0.35	0.40	0.45	0.50	0.55	0.60	0.65	0.70	0.75	0.80	1.00	1.25	1.50	1.75	2.00	2.50	3.00
5	6	8	10	11	13	16	18	21	24	28	31	35	52	73	89	96	99	.	.
6	7	9	11	13	16	19	22	26	30	34	39	44	64	84	95	99	.	.	.
7	7	10	13	15	18	22	26	31	36	41	46	52	73	91	98	.	.	.	.
8	7	11	14	17	21	25	30	36	41	47	53	59	80	95	99	.	.	.	.
9	8	12	15	19	24	29	34	40	47	53	59	66	86	97	.	.	.	.	.
10	8	13	17	21	26	32	38	45	52	59	65	71	90	98	.	.	.	.	.
11	9	15	19	24	29	36	42	49	57	64	70	76	93	99	.	.	.	.	.
12	9	16	20	26	32	39	46	54	61	68	75	81	95	.	.	.	.	.	.
13	9	17	22	28	35	42	50	58	65	73	79	84	97	.	.	.	.	.	.
14	10	18	23	30	37	45	54	62	69	76	82	87	98	.	.	.	.	.	.
15	10	19	25	32	40	48	57	65	73	80	85	90	98	.	.	.	.	.	.
20	13	25	33	43	53	62	72	80	86	91	94	97	.	.	.	.	.	.	.
25	15	31	42	53	64	74	82	89	93	96	98	99	.	.	.	.	.	.	.
30	18	37	49	61	73	82	89	94	97	99	99	.	.	.	.	.	.	.	.
35	20	43	56	69	80	88	94	97	99	99	.	.	.	.	.	.	.	.	.
40	23	49	63	76	85	92	96	98	99	.	.	.	.	.	.	.	.	.	.
45	26	54	69	81	90	95	98	99	.	.	.	.	.	.	.	.	.	.	.
50	29	59	74	85	93	97	99	.	.	.	.	.	.	.	.	.	.	.	.
55	31	64	78	89	95	98	99	.	.	.	.	.	.	.	.	.	.	.	.
60	34	68	82	91	97	99	.	.	.	.	.	.	.	.	.	.	.	.	.
65	37	72	85	94	98	99	.	.	.	.	.	.	.	.	.	.	.	.	.
70	39	75	88	95	98	.	.	.	.	.	.	.	.	.	.	.	.	.	.
75	42	78	90	96	99	.	.	.	.	.	.	.	.	.	.	.	.	.	.
80	44	81	92	97	99	.	.	.	.	.	.	.	.	.	.	.	.	.	.
90	49	86	95	99	.	.	.	.	.	.	.	.	.	.	.	.	.	.	.
100	54	90	97	99	.	.	.	.	.	.	.	.	.	.	.	.	.	.	.
110	59	92	98	.	.	.	.	.	.	.	.	.	.	.	.	.	.	.	.
120	63	95	99	.	.	.	.	.	.	.	.	.	.	.	.	.	.	.	.
130	67	96	99	.	.	.	.	.	.	.	.	.	.	.	.	.	.	.	.
140	70	97	.	.	.	.	.	.	.	.	.	.	.	.	.	.	.	.	.
150	73	98	.	.	.	.	.	.	.	.	.	.	.	.	.	.	.	.	.
175	80	99	.	.	.	.	.	.	.	.	.	.	.	.	.	.	.	.	.
200	86	.	.	.	.	.	.	.	.	.	.	.	.	.	.	.	.	.	.
225	90	.	.	.	.	.	.	.	.	.	.	.	.	.	.	.	.	.	.
250	93	.	.	.	.	.	.	.	.	.	.	.	.	.	.	.	.	.	.
300	97	.	.	.	.	.	.	.	.	.	.	.	.	.	.	.	.	.	.
350	98	.	.	.	.	.	.	.	.	.	.	.	.	.	.	.	.	.	.
400	99	.	.	.	.	.	.	.	.	.	.	.	.	.	.	.	.	.	.
450	.	.	.	.	.	.	.	.	.	.	.	.	.	.	.	.	.	.	.
500	.	.	.	.	.	.	.	.	.	.	.	.	.	.	.	.	.	.	.
600	.	.	.	.	.	.	.	.	.	.	.	.	.	.	.	.	.	.	.
700	.	.	.	.	.	.	.	.	.	.	.	.	.	.	.	.	.	.	.
800	.	.	.	.	.	.	.	.	.	.	.	.	.	.	.	.	.	.	.
900	.	.	.	.	.	.	.	.	.	.	.	.	.	.	.	.	.	.	.
1000	.	.	.	.	.	.	.	.	.	.	.	.	.	.	.	.	.	.	.

Table 7.13. Power table for ANCOVA F-ratio; $r = 0.40$, pattern L, 5 groups at alpha = 0.05

n	Hypothesized ES																		
	0.20	0.30	0.35	0.40	0.45	0.50	0.55	0.60	0.65	0.70	0.75	0.80	1.00	1.25	1.50	1.75	2.00	2.50	3.00
5	5	6	6	7	7	8	9	10	11	12	13	14	20	30	42	55	68	88	97
6	5	6	7	7	8	9	10	11	12	13	15	17	24	37	52	67	80	95	99
7	5	6	7	8	9	10	11	12	14	15	17	19	29	44	61	76	87	98	.
8	6	7	7	8	9	11	12	14	16	18	20	22	34	51	69	83	92	99	.
9	6	7	8	9	10	12	13	15	17	20	22	25	38	57	75	88	95	.	.
10	6	7	8	9	11	13	15	17	19	22	25	28	43	63	81	92	97	.	.
11	6	7	9	10	12	14	16	18	21	24	27	31	47	68	85	95	99	.	.
12	6	8	9	11	13	15	17	20	23	26	30	34	52	73	89	96	99	.	.
13	6	8	10	11	13	16	18	22	25	29	33	37	56	77	91	98	.	.	.
14	6	8	10	12	14	17	20	23	27	31	35	40	60	81	94	99	.	.	.
15	6	9	10	13	15	18	21	25	29	33	38	43	63	84	95	99	.	.	.
20	7	10	13	16	19	24	28	33	39	44	50	56	78	94	99	.	.	.	.
25	8	12	15	19	24	29	35	42	48	55	61	68	88	98	.	.	.	.	.
30	8	14	18	23	29	35	42	49	57	64	71	77	93	99	.	.	.	.	.
35	9	16	21	27	33	41	49	57	65	72	78	84	97	.	.	.	.	.	.
40	10	18	24	30	38	46	55	63	71	78	84	89	98	.	.	.	.	.	.
45	11	20	27	34	43	52	61	69	77	84	89	93	99	.	.	.	.	.	.
50	11	22	29	38	47	57	66	75	82	88	92	95	.	.	.	.	.	.	.
55	12	24	32	42	52	61	71	79	86	91	94	97	.	.	.	.	.	.	.
60	13	26	35	45	56	66	75	83	89	93	96	98	.	.	.	.	.	.	.
65	14	28	38	49	60	70	79	86	91	95	97	99	.	.	.	.	.	.	.
70	15	30	41	52	63	73	82	89	94	97	98	99	.	.	.	.	.	.	.
75	16	32	43	55	67	77	85	91	95	98	99	.	.	.	.	.	.	.	.
80	16	35	46	58	70	80	87	93	96	98	99	.	.	.	.	.	.	.	.
90	18	39	51	64	76	85	91	96	98	99	.	.	.	.	.	.	.	.	.
100	20	43	56	69	80	89	94	97	99	.	.	.	.	.	.	.	.	.	.
110	22	47	61	74	84	92	96	98	99	.	.	.	.	.	.	.	.	.	.
120	24	51	65	78	88	94	97	99	.	.	.	.	.	.	.	.	.	.	.
130	25	54	69	82	90	96	98	99	.	.	.	.	.	.	.	.	.	.	.
140	27	58	73	85	93	97	99	.	.	.	.	.	.	.	.	.	.	.	.
150	29	61	76	87	94	98	99	.	.	.	.	.	.	.	.	.	.	.	.
175	34	69	83	92	97	99	.	.	.	.	.	.	.	.	.	.	.	.	.
200	38	75	88	95	99	.	.	.	.	.	.	.	.	.	.	.	.	.	.
225	43	81	92	97	99	.	.	.	.	.	.	.	.	.	.	.	.	.	.
250	47	85	95	99	.	.	.	.	.	.	.	.	.	.	.	.	.	.	.
300	56	91	98	.	.	.	.	.	.	.	.	.	.	.	.	.	.	.	.
350	63	95	99	.	.	.	.	.	.	.	.	.	.	.	.	.	.	.	.
400	70	97	.	.	.	.	.	.	.	.	.	.	.	.	.	.	.	.	.
450	75	99	.	.	.	.	.	.	.	.	.	.	.	.	.	.	.	.	.
500	80	99	.	.	.	.	.	.	.	.	.	.	.	.	.	.	.	.	.
600	88	.	.	.	.	.	.	.	.	.	.	.	.	.	.	.	.	.	.
700	92	.	.	.	.	.	.	.	.	.	.	.	.	.	.	.	.	.	.
800	96	.	.	.	.	.	.	.	.	.	.	.	.	.	.	.	.	.	.
900	97	.	.	.	.	.	.	.	.	.	.	.	.	.	.	.	.	.	.
1000	99	.	.	.	.	.	.	.	.	.	.	.	.	.	.	.	.	.	.

Table 7.14. Power table for ANCOVA F-ratio; $r = 0.60$, pattern L, 5 groups at alpha = 0.05

n	Hypothesized ES																		
	0.20	0.30	0.35	0.40	0.45	0.50	0.55	0.60	0.65	0.70	0.75	0.80	1.00	1.25	1.50	1.75	2.00	2.50	3.00
5	5	6	7	8	8	9	10	11	13	14	15	17	25	38	54	69	81	96	99
6	6	7	7	8	9	10	12	13	15	17	19	21	31	48	65	80	90	99	.
7	6	7	8	9	10	12	13	15	17	19	22	25	38	56	74	87	95	.	.
8	6	7	8	10	11	13	15	17	20	22	25	29	44	64	81	92	98	.	.
9	6	8	9	10	12	14	16	19	22	25	29	32	49	71	87	96	99	.	.
10	6	8	9	11	13	16	18	21	25	28	32	36	55	76	91	97	.	.	.
11	6	8	10	12	14	17	20	23	27	31	36	40	60	81	94	99	.	.	.
12	6	9	11	13	15	18	22	26	30	34	39	44	65	85	96	99	.	.	.
13	7	9	11	14	17	20	24	28	32	37	42	48	69	88	97	.	.	.	.
14	7	10	12	15	18	21	25	30	35	40	46	51	73	91	98	.	.	.	.
15	7	10	13	15	19	23	27	32	37	43	49	55	76	93	99	.	.	.	.
20	8	13	16	20	25	30	36	43	50	57	63	70	89	98	.	.	.	.	.
25	9	15	19	25	31	38	45	53	61	68	75	81	95	.	.	.	.	.	.
30	10	18	23	30	37	45	54	62	70	77	83	88	98	.	.	.	.	.	.
35	11	20	27	35	43	52	61	70	78	84	89	93	99	.	.	.	.	.	.
40	12	23	31	39	49	59	68	77	84	89	93	96	.	.	.	.	.	.	.
45	13	26	34	44	55	65	74	82	88	93	96	98	.	.	.	.	.	.	.
50	14	28	38	49	60	70	79	86	92	95	98	99	.	.	.	.	.	.	.
55	15	31	42	53	64	75	83	90	94	97	99	99	.	.	.	.	.	.	.
60	16	34	45	57	69	79	87	92	96	98	99	.	.	.	.	.	.	.	.
65	17	37	49	61	73	82	90	94	97	99	.	.	.	.	.	.	.	.	.
70	18	39	52	65	76	85	92	96	98	99	.	.	.	.	.	.	.	.	.
75	20	42	55	68	80	88	94	97	99	.	.	.	.	.	.	.	.	.	.
80	21	45	59	72	82	90	95	98	99	.	.	.	.	.	.	.	.	.	.
90	23	50	64	77	87	94	97	99	.	.	.	.	.	.	.	.	.	.	.
100	26	55	70	82	91	96	98	99	.	.	.	.	.	.	.	.	.	.	.
110	28	59	74	86	93	97	99	.	.	.	.	.	.	.	.	.	.	.	.
120	31	63	78	89	95	98	.	.	.	.	.	.	.	.	.	.	.	.	.
130	33	67	82	92	97	99	.	.	.	.	.	.	.	.	.	.	.	.	.
140	35	71	85	94	98	99	.	.	.	.	.	.	.	.	.	.	.	.	.
150	38	74	88	95	98	.	.	.	.	.	.	.	.	.	.	.	.	.	.
175	44	81	92	98	99	.	.	.	.	.	.	.	.	.	.	.	.	.	.
200	49	87	96	99	.	.	.	.	.	.	.	.	.	.	.	.	.	.	.
225	55	91	97	99	.	.	.	.	.	.	.	.	.	.	.	.	.	.	.
250	60	94	99	.	.	.	.	.	.	.	.	.	.	.	.	.	.	.	.
300	69	97	.	.	.	.	.	.	.	.	.	.	.	.	.	.	.	.	.
350	76	99	.	.	.	.	.	.	.	.	.	.	.	.	.	.	.	.	.
400	82	.	.	.	.	.	.	.	.	.	.	.	.	.	.	.	.	.	.
450	87	.	.	.	.	.	.	.	.	.	.	.	.	.	.	.	.	.	.
500	91	.	.	.	.	.	.	.	.	.	.	.	.	.	.	.	.	.	.
600	95	.	.	.	.	.	.	.	.	.	.	.	.	.	.	.	.	.	.
700	98	.	.	.	.	.	.	.	.	.	.	.	.	.	.	.	.	.	.
800	99	.	.	.	.	.	.	.	.	.	.	.	.	.	.	.	.	.	.
900	.	.	.	.	.	.	.	.	.	.	.	.	.	.	.	.	.	.	.
1000	.	.	.	.	.	.	.	.	.	.	.	.	.	.	.	.	.	.	.

Table 7.15. Power table for ANCOVA *F*-ratio; $r = 0.40$, pattern M, 5 groups at alpha = 0.05

n	Hypothesized ES																		
	0.20	0.30	0.35	0.40	0.45	0.50	0.55	0.60	0.65	0.70	0.75	0.80	1.00	1.25	1.50	1.75	2.00	2.50	3.00
5	5	6	7	7	8	9	10	11	12	13	15	16	24	37	51	66	79	94	99
6	6	7	7	8	9	10	11	13	14	16	18	20	30	46	63	78	88	98	.
7	6	7	8	9	10	11	13	14	16	19	21	24	36	54	72	86	94	99	.
8	6	7	8	9	11	12	14	16	19	21	24	27	42	62	79	91	97	.	.
9	6	8	9	10	12	14	16	18	21	24	27	31	47	68	85	95	99	.	.
10	6	8	9	11	13	15	17	20	23	27	31	35	53	74	89	97	99	.	.
11	6	8	10	12	14	16	19	22	26	30	34	38	58	79	93	98	.	.	.
12	6	9	10	12	15	18	21	24	28	33	37	42	62	83	95	99	.	.	.
13	6	9	11	13	16	19	23	26	31	36	40	46	67	87	96	99	.	.	.
14	7	9	12	14	17	20	24	29	33	38	44	49	70	90	98	.	.	.	.
15	7	10	12	15	18	22	26	31	36	41	47	52	74	92	98	.	.	.	.
20	8	12	15	19	24	29	35	41	48	54	61	67	87	98	.	.	.	.	.
25	8	14	19	24	30	36	43	51	58	66	72	78	94	99	.	.	.	.	.
30	9	17	22	28	36	43	52	60	68	75	81	86	98	.	.	.	.	.	.
35	10	19	26	33	41	50	59	68	75	82	87	92	99	.	.	.	.	.	.
40	11	22	29	38	47	57	66	74	82	87	92	95	.	.	.	.	.	.	.
45	12	24	33	42	52	62	72	80	86	91	95	97	.	.	.	.	.	.	.
50	13	27	36	47	57	68	77	84	90	94	97	98	.	.	.	.	.	.	.
55	14	30	40	51	62	72	81	88	93	96	98	99	.	.	.	.	.	.	.
60	15	32	43	55	66	77	85	91	95	97	99	.	.	.	.	.	.	.	.
65	17	35	47	59	70	80	88	93	97	98	99	.	.	.	.	.	.	.	.
70	18	38	50	63	74	84	90	95	98	99	.	.	.	.	.	.	.	.	.
75	19	40	53	66	77	86	92	96	98	99	.	.	.	.	.	.	.	.	.
80	20	43	56	69	80	89	94	97	99	.	.	.	.	.	.	.	.	.	.
90	22	48	62	75	85	92	96	99	99	.	.	.	.	.	.	.	.	.	.
100	24	52	67	80	89	95	98	99	.	.	.	.	.	.	.	.	.	.	.
110	27	57	72	84	92	97	99	.	.	.	.	.	.	.	.	.	.	.	.
120	29	61	76	87	94	98	99	.	.	.	.	.	.	.	.	.	.	.	.
130	31	65	80	90	96	99	.	.	.	.	.	.	.	.	.	.	.	.	.
140	34	69	83	92	97	99	.	.	.	.	.	.	.	.	.	.	.	.	.
150	36	72	86	94	98	99	.	.	.	.	.	.	.	.	.	.	.	.	.
175	42	79	91	97	99	.	.	.	.	.	.	.	.	.	.	.	.	.	.
200	47	85	95	99	.	.	.	.	.	.	.	.	.	.	.	.	.	.	.
225	53	89	97	99	.	.	.	.	.	.	.	.	.	.	.	.	.	.	.
250	58	92	98	.	.	.	.	.	.	.	.	.	.	.	.	.	.	.	.
300	66	96	99	.	.	.	.	.	.	.	.	.	.	.	.	.	.	.	.
350	74	98	.	.	.	.	.	.	.	.	.	.	.	.	.	.	.	.	.
400	80	99	.	.	.	.	.	.	.	.	.	.	.	.	.	.	.	.	.
450	85	.	.	.	.	.	.	.	.	.	.	.	.	.	.	.	.	.	.
500	89	.	.	.	.	.	.	.	.	.	.	.	.	.	.	.	.	.	.
600	94	.	.	.	.	.	.	.	.	.	.	.	.	.	.	.	.	.	.
700	97	.	.	.	.	.	.	.	.	.	.	.	.	.	.	.	.	.	.
800	99	.	.	.	.	.	.	.	.	.	.	.	.	.	.	.	.	.	.
900	99	.	.	.	.	.	.	.	.	.	.	.	.	.	.	.	.	.	.
1000	.	.	.	.	.	.	.	.	.	.	.	.	.	.	.	.	.	.	.

Table 7.16. Power table for ANCOVA *F*-ratio; $r = 0.60$, pattern M, 5 groups at alpha = 0.05

n	Hypothesized ES																		
	0.20	0.30	0.35	0.40	0.45	0.50	0.55	0.60	0.65	0.70	0.75	0.80	1.00	1.25	1.50	1.75	2.00	2.50	3.00
5	6	7	7	8	9	10	12	13	15	17	19	21	31	47	64	79	90	99	.
6	6	7	8	9	10	12	14	15	18	20	23	25	39	58	76	89	96	.	.
7	6	8	9	10	12	13	16	18	21	24	27	30	46	67	84	94	98	.	.
8	6	8	9	11	13	15	18	21	24	27	31	35	53	75	90	97	99	.	.
9	6	9	10	12	14	17	20	23	27	31	36	40	60	81	94	99	.	.	.
10	6	9	11	13	16	19	22	26	30	35	40	45	66	86	96	99	.	.	.
11	7	10	12	14	17	21	24	29	34	39	44	49	71	90	98	.	.	.	.
12	7	10	12	15	19	22	27	32	37	42	48	54	75	93	99	.	.	.	.
13	7	11	13	16	20	24	29	34	40	46	52	58	80	95	99	.	.	.	.
14	7	11	14	18	22	26	31	37	43	49	56	62	83	96	.	.	.	.	.
15	8	12	15	19	23	28	34	40	46	53	59	66	86	97	.	.	.	.	.
20	9	15	19	25	31	38	45	53	60	68	74	80	95	.	.	.	.	.	.
25	10	18	24	31	38	47	55	64	72	79	85	89	98	.	.	.	.	.	.
30	11	21	29	37	46	55	65	73	81	87	91	95	.	.	.	.	.	.	.
35	12	25	33	43	53	63	72	80	87	92	95	97	.	.	.	.	.	.	.
40	14	28	38	49	59	70	79	86	91	95	97	99	.	.	.	.	.	.	.
45	15	32	42	54	65	76	84	90	95	97	99	99	.	.	.	.	.	.	.
50	17	35	47	59	71	81	88	93	97	98	99	.	.	.	.	.	.	.	.
55	18	39	51	64	75	85	91	95	98	99	.	.	.	.	.	.	.	.	.
60	20	42	55	68	79	88	94	97	99	.	.	.	.	.	.	.	.	.	.
65	21	45	59	72	83	91	95	98	99	.	.	.	.	.	.	.	.	.	.
70	23	48	63	76	86	93	97	99	.	.	.	.	.	.	.	.	.	.	.
75	24	51	66	79	88	94	98	99	.	.	.	.	.	.	.	.	.	.	.
80	26	54	69	82	91	96	98	99	.	.	.	.	.	.	.	.	.	.	.
90	29	60	75	87	94	98	99	.	.	.	.	.	.	.	.	.	.	.	.
100	32	65	80	90	96	99	.	.	.	.	.	.	.	.	.	.	.	.	.
110	35	70	84	93	98	99	.	.	.	.	.	.	.	.	.	.	.	.	.
120	38	74	87	95	98	.	.	.	.	.	.	.	.	.	.	.	.	.	.
130	41	78	90	97	99	.	.	.	.	.	.	.	.	.	.	.	.	.	.
140	44	81	92	98	99	.	.	.	.	.	.	.	.	.	.	.	.	.	.
150	47	84	94	98	.	.	.	.	.	.	.	.	.	.	.	.	.	.	.
175	53	90	97	99	.	.	.	.	.	.	.	.	.	.	.	.	.	.	.
200	60	94	99	.	.	.	.	.	.	.	.	.	.	.	.	.	.	.	.
225	66	96	99	.	.	.	.	.	.	.	.	.	.	.	.	.	.	.	.
250	71	98	.	.	.	.	.	.	.	.	.	.	.	.	.	.	.	.	.
300	79	99	.	.	.	.	.	.	.	.	.	.	.	.	.	.	.	.	.
350	86	.	.	.	.	.	.	.	.	.	.	.	.	.	.	.	.	.	.
400	91	.	.	.	.	.	.	.	.	.	.	.	.	.	.	.	.	.	.
450	94	.	.	.	.	.	.	.	.	.	.	.	.	.	.	.	.	.	.
500	96	.	.	.	.	.	.	.	.	.	.	.	.	.	.	.	.	.	.
600	98	.	.	.	.	.	.	.	.	.	.	.	.	.	.	.	.	.	.
700	99	.	.	.	.	.	.	.	.	.	.	.	.	.	.	.	.	.	.
800	.	.	.	.	.	.	.	.	.	.	.	.	.	.	.	.	.	.	.
900	.	.	.	.	.	.	.	.	.	.	.	.	.	.	.	.	.	.	.
1000	.	.	.	.	.	.	.	.	.	.	.	.	.	.	.	.	.	.	.

Table 7.17. Power table for ANCOVA F-ratio; $r = 0.40$, pattern H, 5 groups at alpha = 0.05

n	Hypothesized ES																		
	0.20	0.30	0.35	0.40	0.45	0.50	0.55	0.60	0.65	0.70	0.75	0.80	1.00	1.25	1.50	1.75	2.00	2.50	3.00
5	6	8	9	10	11	13	15	18	20	23	26	29	45	65	82	93	98	.	.
6	6	8	10	11	13	16	18	21	25	28	32	37	55	77	91	98	.	.	.
7	6	9	11	13	15	18	22	25	29	34	39	44	64	85	96	99	.	.	.
8	7	10	12	14	17	21	25	29	34	39	45	50	72	90	98	.	.	.	.
9	7	11	13	16	20	24	28	33	39	45	51	57	78	94	99	.	.	.	.
10	7	11	14	18	22	26	32	37	43	50	56	62	83	96	.	.	.	.	.
11	8	12	15	19	24	29	35	41	48	55	61	68	88	98	.	.	.	.	.
12	8	13	17	21	26	32	38	45	52	59	66	72	91	99	.	.	.	.	.
13	8	14	18	23	28	35	42	49	56	64	70	77	93	99	.	.	.	.	.
14	9	15	19	25	31	38	45	53	60	67	74	80	95	.	.	.	.	.	.
15	9	16	21	26	33	40	48	56	64	71	78	83	96	.	.	.	.	.	.
20	11	21	27	35	44	53	62	71	79	85	90	94	99	.	.	.	.	.	.
25	13	26	34	44	54	65	74	82	88	93	96	98	.	.	.	.	.	.	.
30	15	31	41	52	64	74	83	89	94	97	98	99	.	.	.	.	.	.	.
35	17	36	48	60	71	81	89	94	97	99	99	.	.	.	.	.	.	.	.
40	19	41	54	67	78	87	93	96	98	99	.	.	.	.	.	.	.	.	.
45	21	45	59	72	83	91	96	98	99	.	.	.	.	.	.	.	.	.	.
50	23	50	65	78	87	94	97	99	.	.	.	.	.	.	.	.	.	.	.
55	26	54	69	82	91	96	98	99	.	.	.	.	.	.	.	.	.	.	.
60	28	59	74	85	93	97	99	.	.	.	.	.	.	.	.	.	.	.	.
65	30	63	78	88	95	98	99	.	.	.	.	.	.	.	.	.	.	.	.
70	32	66	81	91	96	99	.	.	.	.	.	.	.	.	.	.	.	.	.
75	35	70	84	93	97	99	.	.	.	.	.	.	.	.	.	.	.	.	.
80	37	73	86	94	98	.	.	.	.	.	.	.	.	.	.	.	.	.	.
90	41	79	91	97	99	.	.	.	.	.	.	.	.	.	.	.	.	.	.
100	45	83	93	98	.	.	.	.	.	.	.	.	.	.	.	.	.	.	.
110	50	87	96	99	.	.	.	.	.	.	.	.	.	.	.	.	.	.	.
120	54	90	97	99	.	.	.	.	.	.	.	.	.	.	.	.	.	.	.
130	57	92	98	.	.	.	.	.	.	.	.	.	.	.	.	.	.	.	.
140	61	94	99	.	.	.	.	.	.	.	.	.	.	.	.	.	.	.	.
150	64	96	99	.	.	.	.	.	.	.	.	.	.	.	.	.	.	.	.
175	72	98	.	.	.	.	.	.	.	.	.	.	.	.	.	.	.	.	.
200	78	99	.	.	.	.	.	.	.	.	.	.	.	.	.	.	.	.	.
225	83	.	.	.	.	.	.	.	.	.	.	.	.	.	.	.	.	.	.
250	87	.	.	.	.	.	.	.	.	.	.	.	.	.	.	.	.	.	.
300	93	.	.	.	.	.	.	.	.	.	.	.	.	.	.	.	.	.	.
350	96	.	.	.	.	.	.	.	.	.	.	.	.	.	.	.	.	.	.
400	98	.	.	.	.	.	.	.	.	.	.	.	.	.	.	.	.	.	.
450	99	.	.	.	.	.	.	.	.	.	.	.	.	.	.	.	.	.	.
500	.	.	.	.	.	.	.	.	.	.	.	.	.	.	.	.	.	.	.
600	.	.	.	.	.	.	.	.	.	.	.	.	.	.	.	.	.	.	.
700	.	.	.	.	.	.	.	.	.	.	.	.	.	.	.	.	.	.	.
800	.	.	.	.	.	.	.	.	.	.	.	.	.	.	.	.	.	.	.
900	.	.	.	.	.	.	.	.	.	.	.	.	.	.	.	.	.	.	.
1000	.	.	.	.	.	.	.	.	.	.	.	.	.	.	.	.	.	.	.

Table 7.18. Power table for ANCOVA F-ratio; $r = 0.60$, pattern H, 5 groups at alpha = 0.05

n	Hypothesized ES																		
	0.20	0.30	0.35	0.40	0.45	0.50	0.55	0.60	0.65	0.70	0.75	0.80	1.00	1.25	1.50	1.75	2.00	2.50	3.00
5	6	9	10	12	14	16	19	22	26	29	34	38	57	78	92	98	.	.	.
6	7	9	11	14	16	20	23	27	32	37	42	47	68	88	97	.	.	.	.
7	7	10	13	16	19	23	28	33	38	44	50	56	77	94	99	.	.	.	.
8	7	12	14	18	22	27	32	38	44	51	57	63	84	97	.	.	.	.	.
9	8	13	16	20	25	31	37	43	50	57	64	70	89	98	.	.	.	.	.
10	8	14	18	22	28	34	41	48	55	63	69	76	93	99	.	.	.	.	.
11	9	15	19	25	31	38	45	53	61	68	75	81	95	.	.	.	.	.	.
12	9	16	21	27	34	41	49	57	65	73	79	85	97	.	.	.	.	.	.
13	10	17	23	29	37	45	53	62	70	77	83	88	98	.	.	.	.	.	.
14	10	19	25	32	40	48	57	66	74	80	86	91	99	.	.	.	.	.	.
15	11	20	27	34	43	52	61	69	77	84	89	93	99	.	.	.	.	.	.
20	13	26	36	46	56	67	76	83	89	94	96	98	.	.	.	.	.	.	.
25	16	33	44	56	68	78	86	92	96	98	99	.	.	.	.	.	.	.	.
30	19	40	53	65	77	86	92	96	98	99	.	.	.	.	.	.	.	.	.
35	21	46	60	73	84	91	96	98	99	.	.	.	.	.	.	.	.	.	.
40	24	52	67	80	89	95	98	99	.	.	.	.	.	.	.	.	.	.	.
45	27	58	73	85	93	97	99	.	.	.	.	.	.	.	.	.	.	.	.
50	30	63	78	89	95	98	99	.	.	.	.	.	.	.	.	.	.	.	.
55	33	68	82	92	97	99	.	.	.	.	.	.	.	.	.	.	.	.	.
60	36	72	86	94	98	99	.	.	.	.	.	.	.	.	.	.	.	.	.
65	39	76	89	96	99	.	.	.	.	.	.	.	.	.	.	.	.	.	.
70	42	79	91	97	99	.	.	.	.	.	.	.	.	.	.	.	.	.	.
75	45	82	93	98	.	.	.	.	.	.	.	.	.	.	.	.	.	.	.
80	47	85	95	99	.	.	.	.	.	.	.	.	.	.	.	.	.	.	.
90	53	89	97	99	.	.	.	.	.	.	.	.	.	.	.	.	.	.	.
100	58	93	98	.	.	.	.	.	.	.	.	.	.	.	.	.	.	.	.
110	62	95	99	.	.	.	.	.	.	.	.	.	.	.	.	.	.	.	.
120	67	96	99	.	.	.	.	.	.	.	.	.	.	.	.	.	.	.	.
130	71	98	.	.	.	.	.	.	.	.	.	.	.	.	.	.	.	.	.
140	74	98	.	.	.	.	.	.	.	.	.	.	.	.	.	.	.	.	.
150	77	99	.	.	.	.	.	.	.	.	.	.	.	.	.	.	.	.	.
175	84	.	.	.	.	.	.	.	.	.	.	.	.	.	.	.	.	.	.
200	89	.	.	.	.	.	.	.	.	.	.	.	.	.	.	.	.	.	.
225	93	.	.	.	.	.	.	.	.	.	.	.	.	.	.	.	.	.	.
250	95	.	.	.	.	.	.	.	.	.	.	.	.	.	.	.	.	.	.
300	98	.	.	.	.	.	.	.	.	.	.	.	.	.	.	.	.	.	.
350	99	.	.	.	.	.	.	.	.	.	.	.	.	.	.	.	.	.	.
400	.	.	.	.	.	.	.	.	.	.	.	.	.	.	.	.	.	.	.
450	.	.	.	.	.	.	.	.	.	.	.	.	.	.	.	.	.	.	.
500	.	.	.	.	.	.	.	.	.	.	.	.	.	.	.	.	.	.	.
600	.	.	.	.	.	.	.	.	.	.	.	.	.	.	.	.	.	.	.
700	.	.	.	.	.	.	.	.	.	.	.	.	.	.	.	.	.	.	.
800	.	.	.	.	.	.	.	.	.	.	.	.	.	.	.	.	.	.	.
900	.	.	.	.	.	.	.	.	.	.	.	.	.	.	.	.	.	.	.
1000	.	.	.	.	.	.	.	.	.	.	.	.	.	.	.	.	.	.	.

Table 7.19. Sample size table for one-way ANCOVA; alpha = 0.05, $r = 0.40$

Chart A. 2 groups																			
Power	Hypothesized ES																		
	0.20	0.30	0.35	0.40	0.45	0.50	0.55	0.60	0.65	0.70	0.75	0.80	1.00	1.25	1.50	1.75	2.00	2.50	3.00
0.80	323	144	106	82	65	53	44	37	32	28	24	22	14	10	7	6	5	4	4
0.90	433	193	142	109	87	71	59	49	42	37	32	29	19	13	9	7	6	5	4

Chart B. 3 groups, pattern L/M																			
Power	Hypothesized ES																		
	0.20	0.30	0.35	0.40	0.45	0.50	0.55	0.60	0.65	0.70	0.75	0.80	1.00	1.25	1.50	1.75	2.00	2.50	3.00
0.80	400	179	132	101	80	65	54	46	39	34	30	26	18	12	9	7	6	4	4
0.90	527	235	173	133	105	86	71	60	51	44	39	34	23	15	11	9	7	5	4

Chart C. 3 groups, pattern H																			
Power	Hypothesized ES																		
	0.20	0.30	0.35	0.40	0.45	0.50	0.55	0.60	0.65	0.70	0.75	0.80	1.00	1.25	1.50	1.75	2.00	2.50	3.00
0.80	300	134	99	76	61	49	41	35	30	26	23	20	14	9	7	6	5	4	3
0.90	395	177	130	100	79	65	54	45	39	34	30	26	17	12	9	7	6	4	4

Chart D. 4 groups, pattern L																			
Power	Hypothesized ES																		
	0.20	0.30	0.35	0.40	0.45	0.50	0.55	0.60	0.65	0.70	0.75	0.80	1.00	1.25	1.50	1.75	2.00	2.50	3.00
0.80	454	203	149	115	91	74	61	52	44	38	34	30	20	13	10	8	6	5	4
0.90	592	264	194	149	118	96	80	67	57	50	44	38	25	17	12	9	8	5	4

Chart E. 4 groups, pattern M																			
Power	Hypothesized ES																		
	0.20	0.30	0.35	0.40	0.45	0.50	0.55	0.60	0.65	0.70	0.75	0.80	1.00	1.25	1.50	1.75	2.00	2.50	3.00
0.80	409	183	135	103	82	67	55	47	40	35	31	27	18	12	9	7	6	4	4
0.90	533	238	175	134	107	87	72	61	52	45	39	35	23	15	11	9	7	5	4

Chart F. 4 groups, pattern H																			
Power	Hypothesized ES																		
	0.20	0.30	0.35	0.40	0.45	0.50	0.55	0.60	0.65	0.70	0.75	0.80	1.00	1.25	1.50	1.75	2.00	2.50	3.00
0.80	228	102	75	58	46	38	31	27	23	20	18	16	11	7	6	5	4	3	3
0.90	297	133	98	75	60	49	41	34	29	26	23	20	13	9	7	6	5	4	3

Table 7.19. (*cont.*)

Chart G. 5 groups, pattern L																			
Power	Hypothesized ES																		
	0.20	0.30	0.35	0.40	0.45	0.50	0.55	0.60	0.65	0.70	0.75	0.80	1.00	1.25	1.50	1.75	2.00	2.50	3.00
0.80	498	222	164	126	100	81	67	57	49	42	37	33	21	14	10	8	7	5	4
0.90	645	288	212	162	129	104	87	73	62	54	47	42	27	18	13	10	8	6	5

Chart H. 5 groups, pattern M																			
Power	Hypothesized ES																		
	0.20	0.30	0.35	0.40	0.45	0.50	0.55	0.60	0.65	0.70	0.75	0.80	1.00	1.25	1.50	1.75	2.00	2.50	3.00
0.80	399	178	131	101	80	65	54	46	39	34	30	26	17	12	9	7	6	4	3
0.90	517	230	170	130	103	84	70	59	50	44	38	34	22	15	11	8	7	5	4

Chart I. 5 groups, pattern H																			
Power	Hypothesized ES																		
	0.20	0.30	0.35	0.40	0.45	0.50	0.55	0.60	0.65	0.70	0.75	0.80	1.00	1.25	1.50	1.75	2.00	2.50	3.00
0.80	208	93	69	53	42	35	29	24	21	18	16	14	10	7	5	4	4	3	3
0.90	270	121	89	69	54	44	37	31	27	23	21	18	12	8	6	5	4	3	3

Table 7.20. Sample size table for one-way ANCOVA; alpha = 0.05, $r = 0.60$

Chart A. 2 groups

Power	Hypothesized ES																		
	0.20	0.30	0.35	0.40	0.45	0.50	0.55	0.60	0.65	0.70	0.75	0.80	1.00	1.25	1.50	1.75	2.00	2.50	3.00
0.80	246	110	81	63	50	41	34	29	25	22	19	17	11	8	6	5	4	4	3
0.90	330	148	109	84	66	54	45	38	33	28	25	22	15	10	8	6	5	4	4

Chart B. 3 groups, pattern L/M

Power	Hypothesized ES																		
	0.20	0.30	0.35	0.40	0.45	0.50	0.55	0.60	0.65	0.70	0.75	0.80	1.00	1.25	1.50	1.75	2.00	2.50	3.00
0.80	305	136	101	77	62	50	42	35	30	26	23	21	14	9	7	6	5	4	3
0.90	402	179	132	102	81	66	54	46	39	34	30	27	18	12	9	7	6	4	4

Chart C. 3 groups, pattern H

Power	Hypothesized ES																		
	0.20	0.30	0.35	0.40	0.45	0.50	0.55	0.60	0.65	0.70	0.75	0.80	1.00	1.25	1.50	1.75	2.00	2.50	3.00
0.80	229	103	76	58	47	38	32	27	23	20	18	16	11	7	6	5	4	3	3
0.90	302	135	100	77	61	50	41	35	30	26	23	20	14	9	7	6	5	4	3

Chart D. 4 groups, pattern L

Power	Hypothesized ES																		
	0.20	0.30	0.35	0.40	0.45	0.50	0.55	0.60	0.65	0.70	0.75	0.80	1.00	1.25	1.50	1.75	2.00	2.50	3.00
0.80	346	155	114	88	70	57	47	40	34	30	26	23	15	10	8	6	5	4	3
0.90	452	202	148	114	90	74	61	52	44	38	34	30	20	13	10	7	6	5	4

Chart E. 4 groups, pattern M

Power	Hypothesized ES																		
	0.20	0.30	0.35	0.40	0.45	0.50	0.55	0.60	0.65	0.70	0.75	0.80	1.00	1.25	1.50	1.75	2.00	2.50	3.00
0.80	312	139	103	79	63	51	43	37	31	27	24	21	14	10	7	6	5	4	3
0.90	407	182	134	103	82	66	55	48	40	35	30	27	18	12	9	7	6	4	4

Chart F. 4 groups, pattern H

Power	Hypothesized ES																		
	0.20	0.30	0.35	0.40	0.45	0.50	0.55	0.60	0.65	0.70	0.75	0.80	1.00	1.25	1.50	1.75	2.00	2.50	3.00
0.80	174	78	58	45	36	29	24	21	18	16	14	12	9	6	5	4	4	3	3
0.90	227	102	75	59	46	38	31	27	23	20	18	16	11	7	6	5	4	3	3

Table 7.20. (*cont.*)

Chart G. 5 groups, pattern L																			
Power	Hypothesized ES																		
	0.20	0.30	0.35	0.40	0.45	0.50	0.55	0.60	0.65	0.70	0.75	0.80	1.00	1.25	1.50	1.75	2.00	2.50	3.00
0.80	380	170	125	96	75	62	52	44	37	32	28	25	17	11	8	7	5	4	3
0.90	492	220	162	124	98	80	66	56	48	42	36	32	21	14	10	8	6	5	4

Chart H. 5 groups, pattern M																			
Power	Hypothesized ES																		
	0.20	0.30	0.35	0.40	0.45	0.50	0.55	0.60	0.65	0.70	0.75	0.80	1.00	1.25	1.50	1.75	2.00	2.50	3.00
0.80	304	136	100	77	61	50	42	35	30	26	23	20	14	9	7	6	5	4	3
0.90	394	176	130	100	79	64	53	45	39	34	29	26	17	12	9	7	6	4	3

Chart I. 5 groups, pattern H																			
Power	Hypothesized ES																		
	0.20	0.30	0.35	0.40	0.45	0.50	0.55	0.60	0.65	0.70	0.75	0.80	1.00	1.25	1.50	1.75	2.00	2.50	3.00
0.80	159	72	53	41	33	27	22	19	16	14	13	11	8	6	4	4	3	3	3
0.90	206	92	68	53	42	34	29	24	21	18	16	14	10	7	5	4	4	3	3

Table 7.21. Sample size table for ANCOVA; alpha = 0.01, r = 0.40

Chart A. 2 groups																			
Power	Hypothesized ES																		
	0.20	0.30	0.35	0.40	0.45	0.50	0.55	0.60	0.65	0.70	0.75	0.80	1.00	1.25	1.50	1.75	2.00	2.50	3.00
0.80	486	217	160	123	98	80	66	56	48	42	37	33	22	15	11	9	7	6	5
0.90	619	276	204	156	124	101	84	71	61	53	46	41	27	18	13	10	9	7	5
Chart B. 3 groups, pattern L/M																			
Power	Hypothesized ES																		
	0.20	0.30	0.35	0.40	0.45	0.50	0.55	0.60	0.65	0.70	0.75	0.80	1.00	1.25	1.50	1.75	2.00	2.50	3.00
0.80	580	259	191	146	116	95	78	66	57	49	43	38	25	17	12	10	8	6	5
0.90	728	325	239	184	146	118	98	83	71	61	54	47	31	21	15	12	9	7	6
Chart C. 3 groups, pattern H																			
Power	Hypothesized ES																		
	0.20	0.30	0.35	0.40	0.45	0.50	0.55	0.60	0.65	0.70	0.75	0.80	1.00	1.25	1.50	1.75	2.00	2.50	3.00
0.80	435	195	144	110	88	71	59	50	43	37	33	29	19	13	10	8	7	5	4
0.90	457	244	180	138	110	89	74	63	54	47	41	36	24	16	12	9	8	6	5
Chart D. 4 groups, pattern L																			
Power	Hypothesized ES																		
	0.20	0.30	0.35	0.40	0.45	0.50	0.55	0.60	0.65	0.70	0.75	0.80	1.00	1.25	1.50	1.75	2.00	2.50	3.00
0.80	646	288	212	163	129	105	87	74	63	55	48	42	28	29	13	10	9	6	5
0.90	806	359	264	203	161	131	108	91	78	68	59	52	34	23	16	13	10	7	6
Chart E. 4 groups, pattern M																			
Power	Hypothesized ES																		
	0.20	0.30	0.35	0.40	0.45	0.50	0.55	0.60	0.65	0.70	0.75	0.80	1.00	1.25	1.50	1.75	2.00	2.50	3.00
0.80	582	260	191	147	117	95	79	66	57	49	43	38	25	17	12	10	8	6	5
0.90	726	324	238	183	145	118	98	82	70	61	53	47	31	21	15	11	9	7	5
Chart F. 4 groups, pattern H																			
Power	Hypothesized ES																		
	0.20	0.30	0.35	0.40	0.45	0.50	0.55	0.60	0.65	0.70	0.75	0.80	1.00	1.25	1.50	1.75	2.00	2.50	3.00
0.80	324	145	107	83	55	54	45	38	32	28	25	22	15	10	8	6	5	4	4
0.90	404	181	133	102	81	66	55	47	40	35	31	27	18	12	9	7	6	5	4

Table 7.21. (*cont.*)

Chart G. 5 groups, pattern L																			
Power	Hypothesized ES																		
	0.20	0.30	0.35	0.40	0.45	0.50	0.55	0.60	0.65	0.70	0.75	0.80	1.00	1.25	1.50	1.75	2.00	2.50	3.00
0.80	701	313	230	177	140	114	94	80	68	59	52	46	30	20	14	11	9	6	5
0.90	869	387	285	219	173	141	117	98	84	73	64	56	37	24	17	13	11	8	6

Chart H. 5 groups, pattern M																			
Power	Hypothesized ES																		
	0.20	0.30	0.35	0.40	0.45	0.50	0.55	0.60	0.65	0.70	0.75	0.80	1.00	1.25	1.50	1.75	2.00	2.50	3.00
0.80	561	250	185	142	112	91	76	64	55	48	42	37	24	16	12	9	8	6	5
0.90	696	310	228	175	139	113	94	79	68	59	51	45	30	20	14	11	9	6	5

Chart I. 5 groups, pattern H																			
Power	Hypothesized ES																		
	0.20	0.30	0.35	0.40	0.45	0.50	0.55	0.60	0.65	0.70	0.75	0.80	1.00	1.25	1.50	1.75	2.00	2.50	3.00
0.80	293	131	97	75	59	48	40	34	29	26	23	20	14	9	7	6	5	4	3
0.90	363	163	120	92	73	60	50	42	36	31	28	24	16	11	8	7	6	4	4

Table 7.22. Sample size table for ANCOVA; alpha = 0.01, $r = 0.60$

Chart A. 2 groups

Power	Hypothesized ES																		
	0.20	0.30	0.35	0.40	0.45	0.50	0.55	0.60	0.65	0.70	0.75	0.80	1.00	1.25	1.50	1.75	2.00	2.50	3.00
0.80	371	166	123	94	75	61	51	43	37	32	29	25	17	12	9	7	6	5	4
0.90	472	211	156	120	95	77	64	54	47	41	36	32	21	14	11	9	7	6	5

Chart B. 3 groups, pattern L/M

Power	Hypothesized ES																		
	0.20	0.30	0.35	0.40	0.45	0.50	0.55	0.60	0.65	0.70	0.75	0.80	1.00	1.25	1.50	1.75	2.00	2.50	3.00
0.80	442	198	146	112	89	73	60	51	44	38	33	30	20	13	10	8	7	5	4
0.90	555	248	183	140	111	91	75	64	54	47	41	37	24	16	12	9	8	6	5

Chart C. 3 groups, pattern H

Power	Hypothesized ES																		
	0.20	0.30	0.35	0.40	0.45	0.50	0.55	0.60	0.65	0.70	0.75	0.80	1.00	1.25	1.50	1.75	2.00	2.50	3.00
0.80	332	149	110	85	67	55	46	39	33	29	26	23	15	11	8	7	6	4	4
0.90	417	186	138	106	84	68	57	48	41	36	32	28	19	13	10	8	6	5	4

Chart D. 4 groups, pattern L

Power	Hypothesized ES																		
	0.20	0.30	0.35	0.40	0.45	0.50	0.55	0.60	0.65	0.70	0.75	0.80	1.00	1.25	1.50	1.75	2.00	2.50	3.00
0.80	493	220	162	125	99	81	67	57	48	42	37	33	22	15	11	8	7	5	4
0.90	615	274	202	155	123	100	83	70	60	52	46	40	26	18	13	10	8	6	5

Chart E. 4 groups, pattern M

Power	Hypothesized ES																		
	0.20	0.30	0.35	0.40	0.45	0.50	0.55	0.60	0.65	0.70	0.75	0.80	1.00	1.25	1.50	1.75	2.00	2.50	3.00
0.80	444	198	146	112	89	73	60	51	44	38	33	30	20	13	10	8	7	5	4
0.90	553	247	182	140	111	90	75	63	54	47	41	36	24	16	12	9	8	6	5

Chart F. 4 groups, pattern H

Power	Hypothesized ES																		
	0.20	0.30	0.35	0.40	0.45	0.50	0.55	0.60	0.65	0.70	0.75	0.80	1.00	1.25	1.50	1.75	2.00	2.50	3.00
0.80	247	111	82	63	50	41	34	29	25	22	19	17	12	8	6	5	5	4	3
0.90	308	138	102	79	62	51	42	36	31	27	24	21	14	10	8	6	5	4	4

Table 7.22. (*cont.*)

Chart G. 5 groups, pattern L																			
Power	Hypothesized ES																		
	0.20	0.30	0.35	0.40	0.45	0.50	0.55	0.60	0.65	0.70	0.75	0.80	1.00	1.25	1.50	1.75	2.00	2.50	3.00
0.80	535	239	176	135	107	87	72	61	52	45	40	35	23	16	11	9	7	5	4
0.90	663	296	218	167	132	108	89	75	64	56	49	43	28	19	14	11	9	6	5

Chart H. 5 groups, pattern M																			
Power	Hypothesized ES																		
	0.20	0.30	0.35	0.40	0.45	0.50	0.55	0.60	0.65	0.70	0.75	0.80	1.00	1.25	1.50	1.75	2.00	2.50	3.00
0.80	428	191	141	108	86	70	58	49	42	37	32	29	19	13	10	8	6	5	4
0.90	531	237	175	134	106	86	72	61	52	45	39	35	23	15	11	9	7	5	4

Chart I. 5 groups, pattern H																			
Power	Hypothesized ES																		
	0.20	0.30	0.35	0.40	0.45	0.50	0.55	0.60	0.65	0.70	0.75	0.80	1.00	1.25	1.50	1.75	2.00	2.50	3.00
0.80	224	101	74	57	46	37	31	27	23	20	18	16	11	8	6	5	4	4	3
0.90	277	124	92	71	56	46	38	32	28	24	21	19	13	9	7	6	5	4	3

Table 7.23. Sample size table for ANCOVA; alpha = 0.10, $r = 0.40$

Chart A. 2 groups																			
Power	Hypothesized ES																		
	0.20	0.30	0.35	0.40	0.45	0.50	0.55	0.60	0.65	0.70	0.75	0.80	1.00	1.25	1.50	1.75	2.00	2.50	3.00
0.80	251	112	83	64	51	41	34	29	25	22	19	17	11	8	6	5	4	3	3
0.90	350	156	115	88	70	57	47	40	34	30	26	23	15	10	8	6	5	4	3

Chart B. 3 groups, pattern L/M																			
Power	Hypothesized ES																		
	0.20	0.30	0.35	0.40	0.45	0.50	0.55	0.60	0.65	0.70	0.75	0.80	1.00	1.25	1.50	1.75	2.00	2.50	3.00
0.80	319	142	105	81	64	52	43	37	31	27	24	21	14	10	7	6	5	4	3
0.90	434	194	143	109	87	71	59	49	42	37	32	28	19	12	9	7	6	4	4

Chart C. 3 groups, pattern H																			
Power	Hypothesized ES																		
	0.20	0.30	0.35	0.40	0.45	0.50	0.55	0.60	0.65	0.70	0.75	0.80	1.00	1.25	1.50	1.75	2.00	2.50	3.00
0.80	239	107	79	61	48	39	33	28	24	21	18	16	11	8	6	5	4	3	3
0.90	326	146	107	82	65	53	44	37	32	28	24	22	14	10	7	6	5	4	3

Chart D. 4 groups, pattern L																			
Power	Hypothesized ES																		
	0.20	0.30	0.35	0.40	0.45	0.50	0.55	0.60	0.65	0.70	0.75	0.80	1.00	1.25	1.50	1.75	2.00	2.50	3.00
0.80	365	163	120	92	73	60	49	42	36	31	27	24	16	11	8	6	5	4	3
0.90	493	220	162	124	98	80	66	56	48	41	36	32	21	14	10	8	6	5	4

Chart E. 4 groups, pattern M																			
Power	Hypothesized ES																		
	0.20	0.30	0.35	0.40	0.45	0.50	0.55	0.60	0.65	0.70	0.75	0.80	1.00	1.25	1.50	1.75	2.00	2.50	3.00
0.80	329	147	108	83	66	54	45	38	32	28	25	22	14	10	7	6	5	4	3
0.90	443	198	146	112	89	72	60	50	43	37	33	29	19	13	9	7	6	4	4

Chart F. 4 groups, pattern H																			
Power	Hypothesized ES																		
	0.20	0.30	0.35	0.40	0.45	0.50	0.55	0.60	0.65	0.70	0.75	0.80	1.00	1.25	1.50	1.75	2.00	2.50	3.00
0.80	183	82	61	47	37	30	25	22	19	16	14	13	9	6	5	4	3	3	3
0.90	247	110	82	63	50	41	34	29	25	21	19	17	11	8	6	5	4	3	3

Table 7.23. (*cont.*)

Chart G. 5 groups, pattern L																			
Power	Hypothesized ES																		
	0.20	0.30	0.35	0.40	0.45	0.50	0.55	0.60	0.65	0.70	0.75	0.80	1.00	1.25	1.50	1.75	2.00	2.50	3.00
0.80	403	180	133	102	81	66	54	46	39	34	30	26	17	12	9	7	5	4	3
0.90	540	241	177	136	108	87	72	61	52	45	40	35	23	15	11	8	7	5	4
Chart H. 5 groups, pattern M																			
Power	Hypothesized ES																		
	0.20	0.30	0.35	0.40	0.45	0.50	0.55	0.60	0.65	0.70	0.75	0.80	1.00	1.25	1.50	1.75	2.00	2.50	3.00
0.80	323	144	106	82	65	53	44	37	32	28	24	21	14	10	7	6	5	4	3
0.90	432	193	142	109	86	70	58	49	42	36	32	28	19	12	9	7	6	4	3
Chart I. 5 groups, pattern H																			
Power	Hypothesized ES																		
	0.20	0.30	0.35	0.40	0.45	0.50	0.55	0.60	0.65	0.70	0.75	0.80	1.00	1.25	1.50	1.75	2.00	2.50	3.00
0.80	169	76	56	43	34	28	23	20	17	15	13	12	8	6	4	4	3	3	2
0.90	226	101	75	57	46	37	31	26	23	20	17	15	10	7	5	4	4	3	3

Table 7.24. Sample size table for ANCOVA; alpha = 0.10, $r = 0.60$

Chart A. 2 groups																			
Power	Hypothesized ES																		
	0.20	0.30	0.35	0.40	0.45	0.50	0.55	0.60	0.65	0.70	0.75	0.80	1.00	1.25	1.50	1.75	2.00	2.50	3.00
0.80	192	86	64	49	39	32	26	22	19	17	15	13	9	6	5	4	4	3	3
0.90	267	119	88	68	54	44	36	31	26	23	20	18	12	8	6	5	4	4	3

Chart B. 3 groups, pattern L/M																			
Power	Hypothesized ES																		
	0.20	0.30	0.35	0.40	0.45	0.50	0.55	0.60	0.65	0.70	0.75	0.80	1.00	1.25	1.50	1.75	2.00	2.50	3.00
0.80	243	109	80	62	49	40	33	28	24	21	19	16	11	8	6	5	4	3	3
0.90	331	148	109	84	66	54	45	39	33	28	25	22	15	10	7	6	5	4	3

Chart C. 3 groups, pattern H																			
Power	Hypothesized ES																		
	0.20	0.30	0.35	0.40	0.45	0.50	0.55	0.60	0.65	0.70	0.75	0.80	1.00	1.25	1.50	1.75	2.00	2.50	3.00
0.80	183	82	60	47	37	30	25	21	18	16	14	13	9	6	5	4	3	3	2
0.90	249	111	82	63	50	41	34	29	25	22	19	17	11	8	6	5	4	3	3

Chart D. 4 groups, pattern L																			
Power	Hypothesized ES																		
	0.20	0.30	0.35	0.40	0.45	0.50	0.55	0.60	0.65	0.70	0.75	0.80	1.00	1.25	1.50	1.75	2.00	2.50	3.00
0.80	279	125	92	71	56	46	38	32	28	24	21	19	12	9	6	5	4	3	3
0.90	376	168	124	95	75	61	51	43	37	32	28	25	16	11	8	6	5	4	3

Chart E. 4 groups, pattern M																			
Power	Hypothesized ES																		
	0.20	0.30	0.35	0.40	0.45	0.50	0.55	0.60	0.65	0.70	0.75	0.80	1.00	1.25	1.50	1.75	2.00	2.50	3.00
0.80	251	112	83	64	51	41	34	29	25	22	19	17	11	8	6	5	4	3	3
0.90	338	151	111	86	68	55	46	39	33	29	25	22	15	10	7	6	5	4	3

Chart F. 4 groups, pattern H																			
Power	Hypothesized ES																		
	0.20	0.30	0.35	0.40	0.45	0.50	0.55	0.60	0.65	0.70	0.75	0.80	1.00	1.25	1.50	1.75	2.00	2.50	3.00
0.80	140	63	47	36	29	24	20	17	14	13	11	10	7	5	4	3	3	3	2
0.90	188	84	62	48	38	31	26	22	19	17	15	13	9	6	5	4	3	3	3

Table 7.24. (*cont.*)

Chart G. 5 groups, pattern L

Power	Hypothesized ES																	
	0.20	0.30	0.35	0.40	0.45	0.50	0.55	0.60	0.65	0.70	0.75	0.80	1.00	1.25	1.50	1.75	2.00	2.50	3.00
0.80	308	137	101	78	62	50	42	35	30	26	23	20	14	9	7	5	5	4	3
0.90	411	184	135	104	82	67	56	47	40	35	30	27	18	12	9	7	6	4	3

Chart H. 5 groups, pattern M

Power	Hypothesized ES																	
	0.20	0.30	0.35	0.40	0.45	0.50	0.55	0.60	0.65	0.70	0.75	0.80	1.00	1.25	1.50	1.75	2.00	2.50	3.00
0.80	246	110	81	63	50	41	34	29	25	21	19	17	11	8	6	5	4	3	3
0.90	329	147	108	83	66	54	45	38	32	28	25	22	14	10	7	6	5	4	3

Chart I. 5 groups, pattern H

Power	Hypothesized ES																	
	0.20	0.30	0.35	0.40	0.45	0.50	0.55	0.60	0.65	0.70	0.75	0.80	1.00	1.25	1.50	1.75	2.00	2.50	3.00
0.80	129	58	43	33	27	22	18	16	13	12	10	9	7	5	4	3	3	3	2
0.90	172	77	57	44	35	29	24	20	17	15	13	12	8	6	5	4	3	3	2

Table 7.25. ANCOVA power table for Tukey HSD; $r = 0.40$, 3 groups at alpha = 0.05

n	Hypothesized ES																		
	0.20	0.30	0.35	0.40	0.45	0.50	0.55	0.60	0.65	0.70	0.75	0.80	1.00	1.25	1.50	1.75	2.00	2.50	3.00
5	2	3	3	4	5	6	6	7	9	10	11	13	20	32	46	61	74	92	98
6	2	3	4	5	6	7	8	9	11	12	14	16	26	41	58	73	85	97	.
7	2	4	5	5	7	8	9	11	13	15	17	20	32	49	68	82	92	99	.
8	3	4	5	6	8	9	11	13	15	18	20	23	37	57	75	88	96	.	.
9	3	4	6	7	8	10	12	15	17	20	24	27	43	64	82	93	98	.	.
10	3	5	6	8	9	12	14	17	20	23	27	31	48	70	87	95	99	.	.
11	3	5	7	8	10	13	16	19	22	26	30	34	53	76	90	97	99	.	.
12	3	6	7	9	12	14	17	21	25	29	33	38	58	80	93	98	.	.	.
13	3	6	8	10	13	16	19	23	27	32	36	42	63	84	95	99	.	.	.
14	4	6	8	11	14	17	21	25	29	34	40	45	67	87	97	99	.	.	.
15	4	7	9	12	15	18	22	27	32	37	43	48	70	90	98	.	.	.	.
20	5	9	12	16	20	25	31	37	44	50	57	63	85	97	.	.	.	.	.
25	6	11	15	20	26	32	39	47	54	62	69	75	92	99	.	.	.	.	.
30	6	14	19	25	32	39	48	56	64	71	78	84	96	.	.	.	.	.	.
35	7	16	22	29	37	46	55	64	72	79	85	90	98	.	.	.	.	.	.
40	8	18	26	34	43	53	62	71	78	85	90	93	99	.	.	.	.	.	.
45	9	21	29	38	48	59	68	77	84	89	93	96	.	.	.	.	.	.	.
50	10	23	32	43	53	64	73	82	88	92	96	98	.	.	.	.	.	.	.
55	11	26	36	47	58	69	78	86	91	95	97	99	.	.	.	.	.	.	.
60	12	29	39	51	63	73	82	89	93	96	98	99	.	.	.	.	.	.	.
65	13	31	43	55	67	77	85	91	95	98	99	.	.	.	.	.	.	.	.
70	14	34	46	59	71	81	88	93	97	98	99	.	.	.	.	.	.	.	.
75	15	36	49	62	74	84	90	95	98	99	.	.	.	.	.	.	.	.	.
80	17	39	52	66	77	86	92	96	98	99	.	.	.	.	.	.	.	.	.
90	19	44	58	71	82	90	95	98	99	.	.	.	.	.	.	.	.	.	.
100	21	48	63	77	87	93	97	99	.	.	.	.	.	.	.	.	.	.	.
110	23	53	68	81	90	95	98	99	.	.	.	.	.	.	.	.	.	.	.
120	25	57	73	85	93	97	99	.	.	.	.	.	.	.	.	.	.	.	.
130	28	61	77	88	95	98	99	.	.	.	.	.	.	.	.	.	.	.	.
140	30	65	80	90	96	99	.	.	.	.	.	.	.	.	.	.	.	.	.
150	32	68	83	92	97	99	.	.	.	.	.	.	.	.	.	.	.	.	.
175	38	76	89	96	99	.	.	.	.	.	.	.	.	.	.	.	.	.	.
200	43	82	93	98	99	.	.	.	.	.	.	.	.	.	.	.	.	.	.
225	49	87	96	99	.	.	.	.	.	.	.	.	.	.	.	.	.	.	.
250	54	90	97	99	.	.	.	.	.	.	.	.	.	.	.	.	.	.	.
300	63	95	99	.	.	.	.	.	.	.	.	.	.	.	.	.	.	.	.
350	71	98	.	.	.	.	.	.	.	.	.	.	.	.	.	.	.	.	.
400	77	99	.	.	.	.	.	.	.	.	.	.	.	.	.	.	.	.	.
450	82	99	.	.	.	.	.	.	.	.	.	.	.	.	.	.	.	.	.
500	87	.	.	.	.	.	.	.	.	.	.	.	.	.	.	.	.	.	.
600	92	.	.	.	.	.	.	.	.	.	.	.	.	.	.	.	.	.	.
700	96	.	.	.	.	.	.	.	.	.	.	.	.	.	.	.	.	.	.
800	98	.	.	.	.	.	.	.	.	.	.	.	.	.	.	.	.	.	.
900	99	.	.	.	.	.	.	.	.	.	.	.	.	.	.	.	.	.	.
1000	99	.	.	.	.	.	.	.	.	.	.	.	.	.	.	.	.	.	.

Table 7.26. Power table for Tukey HSD; $r = 0.60$, 3 groups at alpha = 0.05

n	Hypothesized ES																		
	0.20	0.30	0.35	0.40	0.45	0.50	0.55	0.60	0.65	0.70	0.75	0.80	1.00	1.25	1.50	1.75	2.00	2.50	3.00
5	2	3	4	5	6	7	8	9	11	13	15	17	27	42	59	74	86	97	.
6	3	4	5	6	7	8	10	12	14	16	19	21	34	53	71	85	94	99	.
7	3	4	5	7	8	10	12	14	17	20	23	26	42	63	81	92	97	.	.
8	3	5	6	8	10	12	14	17	20	23	27	31	49	71	87	96	99	.	.
9	3	5	7	9	11	13	16	20	23	27	31	36	55	78	92	98	.	.	.
10	3	6	8	10	12	15	19	22	26	31	36	41	62	83	95	99	.	.	.
11	4	7	8	11	14	17	21	25	30	35	40	45	67	87	97	99	.	.	.
12	4	7	9	12	15	19	23	28	33	38	44	50	72	90	98	.	.	.	.
13	4	8	10	13	17	21	25	30	36	42	48	54	76	93	99	.	.	.	.
14	4	8	11	14	18	22	28	33	39	45	52	58	80	95	99	.	.	.	.
15	5	9	12	15	19	24	30	36	42	49	55	62	83	96	.	.	.	.	.
20	6	12	16	21	27	34	41	49	56	64	71	77	93	99	.	.	.	.	.
25	7	15	20	27	35	43	51	60	68	75	82	87	98	.	.	.	.	.	.
30	8	18	25	33	42	51	61	70	77	84	89	93	99	.	.	.	.	.	.
35	9	21	29	39	49	59	69	77	84	90	94	96	.	.	.	.	.	.	.
40	11	25	34	45	56	66	76	83	89	94	96	98	.	.	.	.	.	.	.
45	12	28	39	50	62	72	81	88	93	96	98	99	.	.	.	.	.	.	.
50	13	31	43	55	67	77	86	91	95	98	99	.	.	.	.	.	.	.	.
55	15	35	47	60	72	82	89	94	97	99	99	.	.	.	.	.	.	.	.
60	16	38	51	65	76	85	92	96	98	99	.	.	.	.	.	.	.	.	.
65	18	41	55	69	80	88	94	97	99	.	.	.	.	.	.	.	.	.	.
70	19	44	59	72	83	91	95	98	99	.	.	.	.	.	.	.	.	.	.
75	21	47	62	76	86	93	97	99	.	.	.	.	.	.	.	.	.	.	.
80	22	51	66	79	88	94	98	99	.	.	.	.	.	.	.	.	.	.	.
90	25	56	72	84	92	97	99	.	.	.	.	.	.	.	.	.	.	.	.
100	28	62	77	88	95	98	99	.	.	.	.	.	.	.	.	.	.	.	.
110	31	66	81	91	96	99	.	.	.	.	.	.	.	.	.	.	.	.	.
120	34	71	85	93	98	99	.	.	.	.	.	.	.	.	.	.	.	.	.
130	37	75	88	95	99	.	.	.	.	.	.	.	.	.	.	.	.	.	.
140	40	78	90	97	99	.	.	.	.	.	.	.	.	.	.	.	.	.	.
150	43	81	92	98	99	.	.	.	.	.	.	.	.	.	.	.	.	.	.
175	50	88	96	99	.	.	.	.	.	.	.	.	.	.	.	.	.	.	.
200	56	92	98	.	.	.	.	.	.	.	.	.	.	.	.	.	.	.	.
225	62	95	99	.	.	.	.	.	.	.	.	.	.	.	.	.	.	.	.
250	67	97	99	.	.	.	.	.	.	.	.	.	.	.	.	.	.	.	.
300	76	99	.	.	.	.	.	.	.	.	.	.	.	.	.	.	.	.	.
350	83	.	.	.	.	.	.	.	.	.	.	.	.	.	.	.	.	.	.
400	88	.	.	.	.	.	.	.	.	.	.	.	.	.	.	.	.	.	.
450	92	.	.	.	.	.	.	.	.	.	.	.	.	.	.	.	.	.	.
500	95	.	.	.	.	.	.	.	.	.	.	.	.	.	.	.	.	.	.
600	98	.	.	.	.	.	.	.	.	.	.	.	.	.	.	.	.	.	.
700	99	.	.	.	.	.	.	.	.	.	.	.	.	.	.	.	.	.	.
800	.	.	.	.	.	.	.	.	.	.	.	.	.	.	.	.	.	.	.
900	.	.	.	.	.	.	.	.	.	.	.	.	.	.	.	.	.	.	.
1000	.	.	.	.	.	.	.	.	.	.	.	.	.	.	.	.	.	.	.

Table 7.27. Power table for Tukey HSD; $r = 0.40$, 4 groups at $\alpha = 0.05$

n	Hypothesized ES																		
	0.20	0.30	0.35	0.40	0.45	0.50	0.55	0.60	0.65	0.70	0.75	0.80	1.00	1.25	1.50	1.75	2.00	2.50	3.00
5	1	2	2	3	3	4	4	5	6	7	8	9	15	26	40	55	69	90	98
6	1	2	3	3	4	4	5	6	7	9	10	12	20	34	51	67	81	96	.
7	1	2	3	4	4	5	6	8	9	11	13	15	25	42	61	77	89	98	.
8	2	3	3	4	5	6	7	9	11	13	15	18	30	50	69	84	94	99	.
9	2	3	4	5	6	7	9	11	13	15	18	21	35	57	76	90	96	.	.
10	2	3	4	5	6	8	10	12	15	17	20	24	40	63	82	93	98	.	.
11	2	3	4	6	7	9	11	14	16	20	23	27	45	69	87	96	99	.	.
12	2	4	5	6	8	10	12	15	18	22	26	30	50	74	90	97	99	.	.
13	2	4	5	7	9	11	14	17	20	24	29	34	55	78	93	98	.	.	.
14	2	4	6	7	9	12	15	19	23	27	32	37	59	82	95	99	.	.	.
15	2	4	6	8	10	13	16	20	25	29	35	40	63	85	96	99	.	.	.
20	3	6	8	11	15	19	24	29	35	42	48	55	79	95	99	.	.	.	.
25	4	8	11	15	19	25	31	38	46	53	61	68	89	98	.	.	.	.	.
30	4	9	13	18	24	31	39	47	55	64	71	78	94	.	.	.	.	.	.
35	5	11	16	22	29	38	46	55	64	72	79	85	97	.	.	.	.	.	.
40	5	13	19	26	35	44	53	63	72	79	85	90	99	.	.	.	.	.	.
45	6	15	22	30	40	50	60	69	78	85	90	94	99	.	.	.	.	.	.
50	7	17	25	34	45	55	66	75	83	89	93	96	.	.	.	.	.	.	.
55	8	19	28	38	49	61	71	80	87	92	95	98	.	.	.	.	.	.	.
60	8	22	31	42	54	65	76	84	90	94	97	99	.	.	.	.	.	.	.
65	9	24	34	46	58	70	80	87	93	96	98	99	.	.	.	.	.	.	.
70	10	26	37	50	62	74	83	90	95	97	99	99	.	.	.	.	.	.	.
75	11	28	40	54	66	77	86	92	96	98	99	.	.	.	.	.	.	.	.
80	12	31	43	57	70	81	89	94	97	99	.	.	.	.	.	.	.	.	.
90	13	35	49	63	76	86	92	96	98	99	.	.	.	.	.	.	.	.	.
100	15	40	55	69	81	90	95	98	99	.	.	.	.	.	.	.	.	.	.
110	17	44	60	74	86	93	97	99	.	.	.	.	.	.	.	.	.	.	.
120	19	48	65	79	89	95	98	99	.	.	.	.	.	.	.	.	.	.	.
130	21	52	69	83	92	97	99	.	.	.	.	.	.	.	.	.	.	.	.
140	23	56	73	86	94	98	99	.	.	.	.	.	.	.	.	.	.	.	.
150	25	60	77	88	95	98	.	.	.	.	.	.	.	.	.	.	.	.	.
175	30	69	84	93	98	99	.	.	.	.	.	.	.	.	.	.	.	.	.
200	35	76	89	96	99	.	.	.	.	.	.	.	.	.	.	.	.	.	.
225	40	81	93	98	.	.	.	.	.	.	.	.	.	.	.	.	.	.	.
250	45	86	95	99	.	.	.	.	.	.	.	.	.	.	.	.	.	.	.
300	54	92	98	.	.	.	.	.	.	.	.	.	.	.	.	.	.	.	.
350	62	96	99	.	.	.	.	.	.	.	.	.	.	.	.	.	.	.	.
400	70	98	.	.	.	.	.	.	.	.	.	.	.	.	.	.	.	.	.
450	76	99	.	.	.	.	.	.	.	.	.	.	.	.	.	.	.	.	.
500	81	.	.	.	.	.	.	.	.	.	.	.	.	.	.	.	.	.	.
600	89	.	.	.	.	.	.	.	.	.	.	.	.	.	.	.	.	.	.
700	93	.	.	.	.	.	.	.	.	.	.	.	.	.	.	.	.	.	.
800	96	.	.	.	.	.	.	.	.	.	.	.	.	.	.	.	.	.	.
900	98	.	.	.	.	.	.	.	.	.	.	.	.	.	.	.	.	.	.
1000	99	.	.	.	.	.	.	.	.	.	.	.	.	.	.	.	.	.	.

Table 7.28. Power table for Tukey HSD; $r = 0.60$, 4 groups at alpha = 0.05

n	Hypothesized ES																		
	0.20	0.30	0.35	0.40	0.45	0.50	0.55	0.60	0.65	0.70	0.75	0.80	1.00	1.25	1.50	1.75	2.00	2.50	3.00
5	1	2	3	3	4	5	6	7	8	9	11	12	21	36	53	70	83	97	.
6	2	2	3	4	5	6	7	8	10	12	14	16	28	46	65	81	92	99	.
7	2	3	4	4	6	7	9	10	12	15	17	20	35	56	75	89	96	.	.
8	2	3	4	5	7	8	10	12	15	18	21	25	41	64	83	94	98	.	.
9	2	3	5	6	8	9	12	14	17	21	25	29	48	71	88	97	99	.	.
10	2	4	5	7	9	11	13	17	20	24	28	33	54	78	92	98	.	.	.
11	2	4	6	7	10	12	15	19	23	27	32	37	59	83	95	99	.	.	.
12	2	5	6	8	11	14	17	21	26	31	36	42	65	87	97	.	.	.	.
13	3	5	7	9	12	15	19	23	28	34	40	46	69	90	98	.	.	.	.
14	3	5	7	10	13	17	21	26	31	37	43	50	74	92	99	.	.	.	.
15	3	6	8	11	14	18	23	28	34	40	47	53	77	94	99	.	.	.	.
20	4	8	11	15	20	26	33	40	48	55	63	70	90	99	.	.	.	.	.
25	4	10	15	20	27	35	43	51	60	68	75	82	96	.	.	.	.	.	.
30	5	13	18	25	34	43	52	62	70	78	84	89	99	.	.	.	.	.	.
35	6	15	22	31	40	51	61	70	78	85	90	94	.	.	.	.	.	.	.
40	7	18	26	36	47	58	68	77	85	90	94	97	.	.	.	.	.	.	.
45	8	21	30	41	53	64	75	83	89	94	97	98	.	.	.	.	.	.	.
50	9	24	34	46	59	70	80	87	93	96	98	99	.	.	.	.	.	.	.
55	10	27	38	51	64	75	84	91	95	98	99	.	.	.	.	.	.	.	.
60	11	30	42	56	69	80	88	93	97	99	99	.	.	.	.	.	.	.	.
65	12	33	46	60	73	83	91	95	98	99	.	.	.	.	.	.	.	.	.
70	14	36	50	64	77	87	93	97	99	.	.	.	.	.	.	.	.	.	.
75	15	39	54	68	80	89	95	98	99	.	.	.	.	.	.	.	.	.	.
80	16	42	57	72	83	91	96	98	99	.	.	.	.	.	.	.	.	.	.
90	18	47	64	78	88	95	98	99	.	.	.	.	.	.	.	.	.	.	.
100	21	53	70	83	92	97	99	.	.	.	.	.	.	.	.	.	.	.	.
110	24	58	75	87	94	98	99	.	.	.	.	.	.	.	.	.	.	.	.
120	26	63	79	90	96	99	.	.	.	.	.	.	.	.	.	.	.	.	.
130	29	67	83	93	97	99	.	.	.	.	.	.	.	.	.	.	.	.	.
140	31	71	86	95	98	.	.	.	.	.	.	.	.	.	.	.	.	.	.
150	34	75	89	96	99	.	.	.	.	.	.	.	.	.	.	.	.	.	.
175	41	82	93	98	.	.	.	.	.	.	.	.	.	.	.	.	.	.	.
200	47	88	96	99	.	.	.	.	.	.	.	.	.	.	.	.	.	.	.
225	53	92	98	.	.	.	.	.	.	.	.	.	.	.	.	.	.	.	.
250	59	95	99	.	.	.	.	.	.	.	.	.	.	.	.	.	.	.	.
300	69	98	.	.	.	.	.	.	.	.	.	.	.	.	.	.	.	.	.
350	77	99	.	.	.	.	.	.	.	.	.	.	.	.	.	.	.	.	.
400	83	.	.	.	.	.	.	.	.	.	.	.	.	.	.	.	.	.	.
450	88	.	.	.	.	.	.	.	.	.	.	.	.	.	.	.	.	.	.
500	92	.	.	.	.	.	.	.	.	.	.	.	.	.	.	.	.	.	.
600	96	.	.	.	.	.	.	.	.	.	.	.	.	.	.	.	.	.	.
700	98	.	.	.	.	.	.	.	.	.	.	.	.	.	.	.	.	.	.
800	99	.	.	.	.	.	.	.	.	.	.	.	.	.	.	.	.	.	.
900	.	.	.	.	.	.	.	.	.	.	.	.	.	.	.	.	.	.	.
1000	.	.	.	.	.	.	.	.	.	.	.	.	.	.	.	.	.	.	.

Table 7.29. Power table for Tukey HSD; $r = 0.40$, 5 groups at alpha = 0.05

n	Hypothesized ES																		
	0.20	0.30	0.35	0.40	0.45	0.50	0.55	0.60	0.65	0.70	0.75	0.80	1.00	1.25	1.50	1.75	2.00	2.50	3.00
5	1	1	2	2	2	3	3	4	5	5	6	7	12	22	35	50	65	88	97
6	1	1	2	2	3	3	4	5	6	7	8	9	17	30	46	63	78	95	99
7	1	2	2	3	3	4	5	6	7	8	10	12	21	37	56	73	86	98	.
8	1	2	2	3	4	5	6	7	8	10	12	14	26	44	64	81	92	99	.
9	1	2	3	3	4	5	7	8	10	12	14	17	30	51	72	87	95	.	.
10	1	2	3	4	5	6	8	9	11	14	17	20	35	58	78	91	97	.	.
11	1	2	3	4	5	7	9	11	13	16	19	23	40	64	83	94	99	.	.
12	1	3	3	5	6	8	10	12	15	18	22	25	44	69	87	96	99	.	.
13	1	3	4	5	7	8	11	13	17	20	24	28	49	74	91	98	.	.	.
14	2	3	4	5	7	9	12	15	18	22	27	31	53	78	93	99	.	.	.
15	2	3	4	6	8	10	13	16	20	25	29	34	57	82	95	99	.	.	.
20	2	4	6	8	11	15	19	24	30	36	42	49	74	93	99	.	.	.	.
25	2	6	8	11	15	20	26	33	40	47	55	62	86	98	.	.	.	.	.
30	3	7	10	14	20	26	33	41	49	58	66	73	93	99	.	.	.	.	.
35	3	8	13	18	24	32	40	49	58	67	74	81	96	.	.	.	.	.	.
40	4	10	15	21	29	38	47	57	66	74	82	87	98	.	.	.	.	.	.
45	4	12	18	25	34	44	54	64	73	81	87	92	99	.	.	.	.	.	.
50	5	13	20	29	39	49	60	70	79	86	91	95	.	.	.	.	.	.	.
55	6	15	23	32	43	55	65	75	83	89	94	97	.	.	.	.	.	.	.
60	6	17	26	36	48	60	70	80	87	92	96	98	.	.	.	.	.	.	.
65	7	19	29	40	52	64	75	84	90	95	97	99	.	.	.	.	.	.	.
70	7	21	32	44	56	69	79	87	93	96	98	99	.	.	.	.	.	.	.
75	8	23	34	47	60	72	82	90	94	97	99	.	.	.	.	.	.	.	.
80	9	25	37	51	64	76	85	92	96	98	99	.	.	.	.	.	.	.	.
90	10	29	43	57	71	82	90	95	98	99	.	.	.	.	.	.	.	.	.
100	12	34	49	64	77	87	93	97	99	.	.	.	.	.	.	.	.	.	.
110	13	38	54	69	82	90	96	98	99	.	.	.	.	.	.	.	.	.	.
120	15	42	59	74	86	93	97	99	.	.	.	.	.	.	.	.	.	.	.
130	17	46	63	78	89	95	98	99	.	.	.	.	.	.	.	.	.	.	.
140	18	50	68	82	91	97	99	.	.	.	.	.	.	.	.	.	.	.	.
150	20	54	72	85	93	98	99	.	.	.	.	.	.	.	.	.	.	.	.
175	24	63	80	91	97	99	.	.	.	.	.	.	.	.	.	.	.	.	.
200	29	71	86	95	99	.	.	.	.	.	.	.	.	.	.	.	.	.	.
225	34	77	91	97	99	.	.	.	.	.	.	.	.	.	.	.	.	.	.
250	39	82	94	98	.	.	.	.	.	.	.	.	.	.	.	.	.	.	.
300	48	90	97	.	.	.	.	.	.	.	.	.	.	.	.	.	.	.	.
350	56	94	99	.	.	.	.	.	.	.	.	.	.	.	.	.	.	.	.
400	64	97	.	.	.	.	.	.	.	.	.	.	.	.	.	.	.	.	.
450	71	99	.	.	.	.	.	.	.	.	.	.	.	.	.	.	.	.	.
500	76	99	.	.	.	.	.	.	.	.	.	.	.	.	.	.	.	.	.
600	85	.	.	.	.	.	.	.	.	.	.	.	.	.	.	.	.	.	.
700	91	.	.	.	.	.	.	.	.	.	.	.	.	.	.	.	.	.	.
800	95	.	.	.	.	.	.	.	.	.	.	.	.	.	.	.	.	.	.
900	97	.	.	.	.	.	.	.	.	.	.	.	.	.	.	.	.	.	.
1000	98	.	.	.	.	.	.	.	.	.	.	.	.	.	.	.	.	.	.

Table 7.30. Power table for Tukey HSD; $r = 0.60$, 5 groups at alpha $= 0.05$

n	Hypothesized ES																		
	0.20	0.30	0.35	0.40	0.45	0.50	0.55	0.60	0.65	0.70	0.75	0.80	1.00	1.25	1.50	1.75	2.00	2.50	3.00
5	1	2	2	2	3	3	4	5	6	7	9	10	18	31	48	66	80	96	.
6	1	2	2	3	3	4	5	6	8	9	11	13	24	41	61	78	90	99	.
7	1	2	3	3	4	5	7	8	10	12	14	17	30	51	71	87	95	.	.
8	1	2	3	4	5	6	8	10	12	14	17	20	36	59	79	92	98	.	.
9	1	2	3	4	6	7	9	11	14	17	20	24	42	67	86	95	99	.	.
10	1	3	4	5	6	8	11	13	16	20	24	28	48	73	90	97	.	.	.
11	2	3	4	6	7	9	12	15	19	23	27	32	54	79	93	99	.	.	.
12	2	3	5	6	8	11	14	17	21	26	31	36	59	83	96	99	.	.	.
13	2	4	5	7	9	12	15	19	24	29	34	40	64	87	97	.	.	.	.
14	2	4	5	7	10	13	17	21	26	32	38	44	69	90	98	.	.	.	.
15	2	4	6	8	11	14	19	23	29	35	41	48	73	92	99	.	.	.	.
20	3	6	9	12	16	22	28	34	42	49	57	65	87	98	.	.	.	.	.
25	3	8	11	16	22	29	37	45	54	62	70	77	95	.	.	.	.	.	.
30	4	10	15	21	28	37	46	56	65	73	80	86	98	.	.	.	.	.	.
35	5	12	18	26	34	44	55	65	74	81	88	92	99	.	.	.	.	.	.
40	5	14	21	30	41	52	62	72	81	87	92	96	.	.	.	.	.	.	.
45	6	17	25	35	47	58	69	79	86	92	95	98	.	.	.	.	.	.	.
50	7	19	29	40	53	65	75	84	90	95	97	99	.	.	.	.	.	.	.
55	8	22	33	45	58	70	80	88	93	97	98	99	.	.	.	.	.	.	.
60	9	25	36	50	63	75	84	91	95	98	99	.	.	.	.	.	.	.	.
65	10	27	40	54	68	79	88	94	97	99	.	.	.	.	.	.	.	.	.
70	10	30	44	58	72	83	91	95	98	99	.	.	.	.	.	.	.	.	.
75	11	33	48	63	76	86	93	97	99	.	.	.	.	.	.	.	.	.	.
80	12	36	51	66	79	89	95	98	99	.	.	.	.	.	.	.	.	.	.
90	15	41	58	73	85	93	97	99	.	.	.	.	.	.	.	.	.	.	.
100	17	47	64	79	89	95	98	99	.	.	.	.	.	.	.	.	.	.	.
110	19	52	69	83	92	97	99	.	.	.	.	.	.	.	.	.	.	.	.
120	21	57	74	87	95	98	.	.	.	.	.	.	.	.	.	.	.	.	.
130	24	61	79	90	96	99	.	.	.	.	.	.	.	.	.	.	.	.	.
140	26	66	82	93	98	99	.	.	.	.	.	.	.	.	.	.	.	.	.
150	29	70	85	94	98	.	.	.	.	.	.	.	.	.	.	.	.	.	.
175	35	78	91	97	99	.	.	.	.	.	.	.	.	.	.	.	.	.	.
200	41	85	95	99	.	.	.	.	.	.	.	.	.	.	.	.	.	.	.
225	47	89	97	99	.	.	.	.	.	.	.	.	.	.	.	.	.	.	.
250	53	93	98	.	.	.	.	.	.	.	.	.	.	.	.	.	.	.	.
300	63	97	.	.	.	.	.	.	.	.	.	.	.	.	.	.	.	.	.
350	72	99	.	.	.	.	.	.	.	.	.	.	.	.	.	.	.	.	.
400	79	99	.	.	.	.	.	.	.	.	.	.	.	.	.	.	.	.	.
450	85	.	.	.	.	.	.	.	.	.	.	.	.	.	.	.	.	.	.
500	89	.	.	.	.	.	.	.	.	.	.	.	.	.	.	.	.	.	.
600	95	.	.	.	.	.	.	.	.	.	.	.	.	.	.	.	.	.	.
700	97	.	.	.	.	.	.	.	.	.	.	.	.	.	.	.	.	.	.
800	99	.	.	.	.	.	.	.	.	.	.	.	.	.	.	.	.	.	.
900	99	.	.	.	.	.	.	.	.	.	.	.	.	.	.	.	.	.	.
1000	.	.	.	.	.	.	.	.	.	.	.	.	.	.	.	.	.	.	.

Table 7.31. Sample size table for ANCOVA multiple comparison; $p = 0.05$, $r = 0.40$

Chart A. 3 groups, $r = 0.40$ at alpha = 0.05; NK no intervening groups

Power	Hypothesized ES																		
	0.20	0.30	0.35	0.40	0.45	0.50	0.55	0.60	0.65	0.70	0.75	0.80	1.00	1.25	1.50	1.75	2.00	2.50	3.00
0.80	331	148	109	84	66	54	45	38	32	28	25	22	15	10	7	6	5	4	3
0.90	443	197	145	112	88	72	60	50	43	37	33	29	19	13	9	7	6	4	4

Chart B. 3 groups, $r = 0.40$ at alpha = 0.05; NK 1 intervening group, Tukey 3 groups

Power	Hypothesized ES																		
	0.20	0.30	0.35	0.40	0.45	0.50	0.55	0.60	0.65	0.70	0.75	0.80	1.00	1.25	1.50	1.75	2.00	2.50	3.00
0.80	428	191	141	108	86	70	58	49	42	36	32	28	19	12	9	7	6	5	4
0.90	553	247	182	139	111	90	74	63	54	47	41	36	24	16	11	9	7	5	4

Chart C. 4 groups, $r = 0.40$ at alpha = 0.05; NK no intervening groups

Power	Hypothesized ES																		
	0.20	0.30	0.35	0.40	0.45	0.50	0.55	0.60	0.65	0.70	0.75	0.80	1.00	1.25	1.50	1.75	2.00	2.50	3.00
0.80	331	148	109	83	66	54	45	38	32	28	25	22	14	10	7	6	5	4	3
0.90	442	197	145	111	88	72	59	50	43	37	32	29	19	12	9	7	6	4	3

Chart D. 4 groups, $r = 0.40$ at alpha = 0.05; NK 1 intervening group

Power	Hypothesized ES																		
	0.20	0.30	0.35	0.40	0.45	0.50	0.55	0.60	0.65	0.70	0.75	0.80	1.00	1.25	1.50	1.75	2.00	2.50	3.00
0.80	427	191	140	108	85	69	58	49	42	36	32	28	18	12	9	7	6	4	4
0.90	553	247	181	139	110	90	74	63	54	46	40	37	23	15	11	9	7	5	4

Chart E. 4 groups, $r = 0.40$ at alpha = 0.05; NK 2 intervening groups, Tukey 4 groups

Power	Hypothesized ES																		
	0.20	0.30	0.35	0.40	0.45	0.50	0.55	0.60	0.65	0.70	0.75	0.80	1.00	1.25	1.50	1.75	2.00	2.50	3.00
0.80	490	218	161	123	98	80	66	56	48	41	36	32	21	14	10	8	6	5	4
0.90	624	278	205	157	124	101	84	71	60	52	46	40	26	17	12	10	8	6	4

Chart F. 5 groups, $r = 0.40$ at alpha = 0.05; NK no intervening groups

Power	Hypothesized ES																		
	0.20	0.30	0.35	0.40	0.45	0.50	0.55	0.60	0.65	0.70	0.75	0.80	1.00	1.25	1.50	1.75	2.00	2.50	3.00
0.80	331	147	109	83	66	54	45	38	32	28	24	22	14	9	7	5	4	3	3
0.90	442	197	145	111	88	72	59	50	43	37	32	29	19	12	9	7	5	4	3

Table 7.31. (*cont.*)

Chart G. 5 groups, $r = 0.40$ at alpha = 0.05; NK 1 intervening group																			
Power	Hypothesized ES																		
	0.20	0.30	0.35	0.40	0.45	0.50	0.55	0.60	0.65	0.70	0.75	0.80	1.00	1.25	1.50	1.75	2.00	2.50	3.00
0.80	427	190	140	108	85	69	57	48	41	36	31	28	18	12	9	7	6	4	3
0.90	553	246	181	139	110	89	74	62	53	46	40	36	23	15	11	8	7	5	4

Chart H. 5 groups, $r = 0.40$ at alpha = 0.05; NK 2 intervening groups																			
Power	Hypothesized ES																		
	0.20	0.30	0.35	0.40	0.45	0.50	0.55	0.60	0.65	0.70	0.75	0.80	1.00	1.25	1.50	1.75	2.00	2.50	3.00
0.80	490	218	161	123	98	79	66	55	47	41	36	32	21	14	10	8	6	5	4
0.90	624	278	205	157	124	101	84	70	60	52	45	40	26	17	12	9	8	5	4

Chart I. 5 groups, $r = 0.40$ at alpha = 0.05; NK 3 intervening groups, Tukey 5 groups																			
Power	Hypothesized ES																		
	0.20	0.30	0.35	0.40	0.45	0.50	0.55	0.60	0.65	0.70	0.75	0.80	1.00	1.25	1.50	1.75	2.00	2.50	3.00
0.80	536	239	176	135	107	87	72	61	52	45	39	35	23	15	11	8	7	5	4
0.90	676	301	222	170	135	109	91	76	65	56	49	43	28	19	13	10	8	6	4

Table 7.32. Sample size table for ANCOVA multiple comparison; $p = 0.05$, $r = 0.60$

Chart A. 3 groups, $r = 0.60$ at alpha = 0.05; NK no intervening groups																			
Power	Hypothesized ES																		
	0.20	0.30	0.35	0.40	0.45	0.50	0.55	0.60	0.65	0.70	0.75	0.80	1.00	1.25	1.50	1.75	2.00	2.50	3.00
0.80	252	113	83	64	51	41	34	29	25	22	19	17	11	8	6	5	4	3	3
0.90	337	151	111	85	68	55	46	39	33	29	25	22	15	10	7	6	5	4	3
Chart B. 3 groups, $r = 0.60$ at alpha = 0.05; NK 1 intervening group, Tukey 3 groups																			
Power	Hypothesized ES																		
	0.20	0.30	0.35	0.40	0.45	0.50	0.55	0.60	0.65	0.70	0.75	0.80	1.00	1.25	1.50	1.75	2.00	2.50	3.00
0.80	326	146	108	83	66	53	44	38	32	28	25	22	15	10	7	6	5	4	3
0.90	422	188	139	107	85	69	57	48	41	36	31	28	18	12	9	7	6	5	4
Chart C. 4 groups, $r = 0.60$ at alpha = 0.05; NK no intervening groups																			
Power	Hypothesized ES																		
	0.20	0.30	0.35	0.40	0.45	0.50	0.55	0.60	0.65	0.70	0.75	0.80	1.00	1.25	1.50	1.75	2.00	2.50	3.00
0.80	252	113	83	64	51	41	34	29	25	22	19	17	11	8	6	5	4	3	3
0.90	337	150	111	85	67	55	46	38	33	29	25	22	15	10	7	6	5	4	3
Chart D. 4 groups, $r = 0.60$ at alpha = 0.05; NK 1 intervening group																			
Power	Hypothesized ES																		
	0.20	0.30	0.35	0.40	0.45	0.50	0.55	0.60	0.65	0.70	0.75	0.80	1.00	1.25	1.50	1.75	2.00	2.50	3.00
0.80	326	146	107	82	65	53	44	37	32	28	24	22	14	10	7	6	5	4	3
0.90	422	188	139	106	84	69	57	48	41	36	31	28	18	12	9	7	6	4	3
Chart E. 4 groups, $r = 0.60$ at alpha = 0.05; NK 2 intervening groups, Tukey 4 groups																			
Power	Hypothesized ES																		
	0.20	0.30	0.35	0.40	0.45	0.50	0.55	0.60	0.65	0.70	0.75	0.80	1.00	1.25	1.50	1.75	2.00	2.50	3.00
0.80	374	167	123	94	75	61	51	43	37	32	28	25	16	11	8	6	5	4	3
0.90	476	212	156	120	95	77	64	54	46	40	35	31	20	14	10	8	6	5	4
Chart F. 5 groups, $r = 0.60$ at alpha = 0.05; NK no intervening groups																			
Power	Hypothesized ES																		
	0.20	0.30	0.35	0.40	0.45	0.50	0.55	0.60	0.65	0.70	0.75	0.80	1.00	1.25	1.50	1.75	2.00	2.50	3.00
0.80	252	113	83	64	51	41	34	29	25	21	19	17	11	7	6	4	4	3	3
0.90	337	150	111	85	67	55	45	38	33	28	25	22	14	10	7	5	4	3	3

Table 7.32. (*cont.*)

Chart G. 5 groups, $r = 0.60$ at alpha = 0.05; NK 1 intervening group																			
Power	Hypothesized ES																		
	0.20	0.30	0.35	0.40	0.45	0.50	0.55	0.60	0.65	0.70	0.75	0.80	1.00	1.25	1.50	1.75	2.00	2.50	3.00
0.80	326	145	107	82	65	53	44	37	32	28	24	21	14	9	7	5	5	3	3
0.90	422	188	138	106	106	68	57	48	41	35	31	27	18	12	9	7	5	4	3

Chart H. 5 groups, $r = 0.60$ at alpha = 0.05; NK 2 intervening groups																			
Power	Hypothesized ES																		
	0.20	0.30	0.35	0.40	0.45	0.50	0.55	0.60	0.65	0.70	0.75	0.80	1.00	1.25	1.50	1.75	2.00	2.50	3.00
0.80	373	167	123	94	75	61	50	43	36	32	28	24	16	11	8	6	5	4	3
0.90	476	212	156	120	95	77	64	54	46	40	35	31	20	13	10	7	6	4	4

Chart I. 5 groups, $r = 0.60$ at alpha = 0.05; NK 3 intervening groups, Tukey 5 groups																			
Power	Hypothesized ES																		
	0.20	0.30	0.35	0.40	0.45	0.50	0.55	0.60	0.65	0.70	0.75	0.80	1.00	1.25	1.50	1.75	2.00	2.50	3.00
0.80	409	182	134	103	82	66	55	47	40	35	30	27	18	12	9	7	5	4	3
0.90	516	230	169	130	103	84	69	58	50	43	38	33	22	14	10	8	7	5	4

Table 7.33. Sample size table for ANCOVA multiple comparison; $p = 0.01$, $r = 0.40$

Chart A. 3 groups, $r = 0.40$ at alpha = 0.01; NK no intervening groups

Power	Hypothesized ES																		
	0.20	0.30	0.35	0.40	0.45	0.50	0.55	0.60	0.65	0.70	0.75	0.80	1.00	1.25	1.50	1.75	2.00	2.50	3.00
0.80	492	220	162	124	99	80	67	56	48	42	37	32	21	14	11	8	7	5	4
0.90	627	280	206	158	125	102	84	71	61	53	46	41	27	18	13	10	8	6	5

Chart B. 3 groups, $r = 0.40$ at alpha = 0.01; NK 1 intervening group, Tukey 3 groups

Power	Hypothesized ES																		
	0.20	0.30	0.35	0.40	0.45	0.50	0.55	0.60	0.65	0.70	0.75	0.80	1.00	1.25	1.50	1.75	2.00	2.50	3.00
0.80	594	265	195	150	119	97	80	68	58	50	44	39	26	17	13	10	8	6	5
0.90	741	331	243	187	148	120	100	84	72	62	55	48	32	21	15	12	10	7	6

Chart C. 4 groups, $r = 0.40$ at alpha = 0.01; NK no intervening groups

Power	Hypothesized ES																		
	0.20	0.30	0.35	0.40	0.45	0.50	0.55	0.60	0.65	0.70	0.75	0.80	1.00	1.25	1.50	1.75	2.00	2.50	3.00
0.80	492	219	162	124	98	80	66	56	48	41	36	32	21	14	10	8	7	5	4
0.90	626	276	206	158	125	101	84	71	61	52	46	40	26	17	13	10	8	6	5

Chart D. 4 groups, $r = 0.40$ at alpha = 0.01; NK 1 intervening group

Power	Hypothesized ES																		
	0.20	0.30	0.35	0.40	0.45	0.50	0.55	0.60	0.65	0.70	0.75	0.80	1.00	1.25	1.50	1.75	2.00	2.50	3.00
0.80	594	265	195	150	119	96	80	67	58	50	44	39	25	17	12	9	8	6	5
0.90	741	330	243	186	148	120	99	84	72	62	54	48	31	21	15	11	9	7	5

Chart E. 4 groups, $r = 0.40$ at alpha = 0.01; NK 2 intervening groups, Tukey 4 groups

Power	Hypothesized ES																		
	0.20	0.30	0.35	0.40	0.45	0.50	0.55	0.60	0.65	0.70	0.75	0.80	1.00	1.25	1.50	1.75	2.00	2.50	3.00
0.80	659	294	216	166	132	107	89	75	64	55	48	43	28	19	14	10	8	6	5
0.90	813	362	267	205	162	132	109	92	79	68	59	52	34	23	16	12	10	7	6

Chart F. 5 groups, $r = 0.40$ at alpha = 0.01; NK no intervening groups

Power	Hypothesized ES																		
	0.20	0.30	0.35	0.40	0.45	0.50	0.55	0.60	0.65	0.70	0.75	0.80	1.00	1.25	1.50	1.75	2.00	2.50	3.00
0.80	492	219	161	124	98	80	66	56	48	41	36	32	21	14	10	8	6	5	4
0.90	626	279	205	157	125	101	84	71	60	52	46	40	26	17	12	9	8	5	4

Table 7.33. (*cont.*)

Chart G. 5 groups, $r = 0.40$ at alpha = 0.01; NK 1 intervening group																			
Power	Hypothesized ES																		
	0.20	0.30	0.35	0.40	0.45	0.50	0.55	0.60	0.65	0.70	0.75	0.80	1.00	1.25	1.50	1.75	2.00	2.50	3.00
0.80	594	265	195	149	118	96	80	67	57	50	44	38	25	17	12	9	7	5	4
0.90	741	330	243	186	147	120	99	84	71	62	54	48	31	20	15	11	9	6	5

Chart H. 5 groups, $r = 0.40$ at alpha = 0.01; NK 2 intervening groups																			
Power	Hypothesized ES																		
	0.20	0.30	0.35	0.40	0.45	0.50	0.55	0.60	0.65	0.70	0.75	0.80	1.00	1.25	1.50	1.75	2.00	2.50	3.00
0.80	658	293	216	166	131	107	88	75	64	55	48	43	28	18	13	10	8	6	5
0.90	813	362	266	204	162	131	109	92	78	68	59	52	34	22	16	12	10	7	5

Chart I. 5 groups, $r = 0.40$ at alpha = 0.01; NK 3 intervening groups, Tukey 5 groups																			
Power	Hypothesized ES																		
	0.20	0.30	0.35	0.40	0.45	0.50	0.55	0.60	0.65	0.70	0.75	0.80	1.00	1.25	1.50	1.75	2.00	2.50	3.00
0.80	706	315	232	178	141	114	95	80	68	59	52	46	30	20	14	11	9	6	5
0.90	866	386	284	218	172	140	116	98	83	72	63	56	36	24	17	13	10	7	6

Table 7.34. Sample size table for ANCOVA multiple comparison; $p = 0.01$, $r = 0.60$

Chart A. 3 groups, $r = 0.60$ at alpha = 0.01; NK no intervening groups

Power	Hypothesized ES																		
	0.20	0.30	0.35	0.40	0.45	0.50	0.55	0.60	0.65	0.70	0.75	0.80	1.00	1.25	1.50	1.75	2.00	2.50	3.00
0.80	376	168	124	95	76	62	51	43	39	32	28	25	17	11	9	7	6	5	4
0.90	478	213	157	121	96	78	65	55	49	41	36	32	21	14	10	8	7	5	4

Chart B. 3 groups, $r = 0.60$ at alpha = 0.01; NK 1 intervening group, Tukey 3 groups

Power	Hypothesized ES																		
	0.20	0.30	0.35	0.40	0.45	0.50	0.55	0.60	0.65	0.70	0.75	0.80	1.00	1.25	1.50	1.75	2.00	2.50	3.00
0.80	453	203	149	115	91	74	62	52	45	39	34	30	20	14	10	8	7	5	4
0.90	565	252	186	143	113	92	76	65	55	48	42	37	25	17	12	10	8	6	5

Chart C. 4 groups, $r = 0.60$ at alpha = 0.01; NK no intervening groups

Power	Hypothesized ES																		
	0.20	0.30	0.35	0.40	0.45	0.50	0.55	0.60	0.65	0.70	0.75	0.80	1.00	1.25	1.50	1.75	2.00	2.50	3.00
0.80	375	168	123	95	75	61	51	43	37	32	28	25	16	11	8	6	5	4	4
0.90	478	213	157	120	95	78	64	54	47	40	35	31	21	14	10	8	6	5	4

Chart D. 4 groups, $r = 0.60$ at alpha = 0.01; NK 1 intervening group

Power	Hypothesized ES																		
	0.20	0.30	0.35	0.40	0.45	0.50	0.55	0.60	0.65	0.70	0.75	0.80	1.00	1.25	1.50	1.75	2.00	2.50	3.00
0.80	453	202	149	114	91	74	61	52	44	38	34	30	20	13	10	8	6	5	4
0.90	565	252	186	142	113	92	76	64	55	48	42	37	24	16	12	9	7	5	4

Chart E. 4 groups, $r = 0.60$ at alpha = 0.01; NK 2 intervening groups, Tukey 4 groups

Power	Hypothesized ES																		
	0.20	0.30	0.35	0.40	0.45	0.50	0.55	0.60	0.65	0.70	0.75	0.80	1.00	1.25	1.50	1.75	2.00	2.50	3.00
0.80	502	224	165	127	101	82	68	57	49	43	37	33	22	15	11	8	7	5	4
0.90	620	276	204	156	124	101	83	70	60	52	46	40	27	18	13	10	8	6	5

Chart F. 5 groups, $r = 0.60$ at alpha = 0.01; NK no intervening groups

Power	Hypothesized ES																		
	0.20	0.30	0.35	0.40	0.45	0.50	0.55	0.60	0.65	0.70	0.75	0.80	1.00	1.25	1.50	1.75	2.00	2.50	3.00
0.80	375	167	123	95	75	61	51	43	37	32	28	25	16	11	8	6	5	4	3
0.90	477	213	157	120	95	77	64	54	46	40	35	31	20	13	10	8	6	5	4

Table 7.34. (*cont.*)

Chart G. 5 groups, $r = 0.60$ at alpha = 0.01; NK 1 intervening group																			
Power	Hypothesized ES																		
	0.20	0.30	0.35	0.40	0.45	0.50	0.55	0.60	0.65	0.70	0.75	0.80	1.00	1.25	1.50	1.75	2.00	2.50	3.00
0.80	453	202	149	114	91	74	61	52	44	38	34	30	19	13	10	7	6	5	4
0.90	565	262	185	142	113	92	76	64	55	47	41	37	24	16	11	9	7	5	4

Chart H. 5 groups, $r = 0.60$ at alpha = 0.01; NK 2 intervening groups																			
Power	Hypothesized ES																		
	0.20	0.30	0.35	0.40	0.45	0.50	0.55	0.60	0.65	0.70	0.75	0.80	1.00	1.25	1.50	1.75	2.00	2.50	3.00
0.80	502	224	165	127	100	82	68	57	49	42	37	33	22	14	10	8	7	5	4
0.90	620	276	203	156	124	100	83	70	60	52	45	40	26	17	13	10	8	6	5

Chart I. 5 groups, $r = 0.60$ at alpha = 0.01; NK 3 intervening groups, Tukey 5 groups																			
Power	Hypothesized ES																		
	0.20	0.30	0.35	0.40	0.45	0.50	0.55	0.60	0.65	0.70	0.75	0.80	1.00	1.25	1.50	1.75	2.00	2.50	3.00
0.80	539	240	177	136	108	88	73	61	52	45	40	35	23	15	11	9	7	5	4
0.90	660	294	217	166	132	107	89	75	64	55	48	43	28	18	13	10	8	6	5

Table 7.35. Sample size table for ANCOVA multiple comparison; $p = 0.10$, $r = 0.40$

Chart A. 3 groups, $r = 0.40$ at alpha = 0.10; NK no intervening groups																			
Power	Hypothesized ES																		
	0.20	0.30	0.35	0.40	0.45	0.50	0.55	0.60	0.65	0.70	0.75	0.80	1.00	1.25	1.50	1.75	2.00	2.50	3.00
0.80	261	116	86	66	52	43	35	30	26	22	20	17	12	8	6	5	4	3	3
0.90	361	161	118	91	72	59	49	41	35	30	27	24	15	10	8	6	5	4	3

Chart B. 3 groups, $r = 0.40$ at alpha = 0.10; NK 1 intervening group, Tukey 3 groups																			
Power	Hypothesized ES																		
	0.20	0.30	0.35	0.40	0.45	0.50	0.55	0.60	0.65	0.70	0.75	0.80	1.00	1.25	1.50	1.75	2.00	2.50	3.00
0.80	353	158	116	89	71	58	48	40	35	30	26	23	15	10	8	6	5	4	3
0.90	468	209	154	118	93	76	63	53	45	39	34	30	20	13	10	7	6	5	4

Chart C. 4 groups, $r = 0.40$ at alpha = 0.10; NK no intervening groups																			
Power	Hypothesized ES																		
	0.20	0.30	0.35	0.40	0.45	0.50	0.55	0.60	0.65	0.70	0.75	0.80	1.00	1.25	1.50	1.75	2.00	2.50	3.00
0.80	261	116	86	66	52	42	35	30	26	22	19	17	11	8	6	4	4	3	3
0.90	361	161	118	91	72	58	48	41	35	30	27	23	15	10	7	6	5	3	3

Chart D. 4 groups, $r = 0.40$ at alpha = 0.10; NK 1 intervening group																			
Power	Hypothesized ES																		
	0.20	0.30	0.35	0.40	0.45	0.50	0.55	0.60	0.65	0.70	0.75	0.80	1.00	1.25	1.50	1.75	2.00	2.50	3.00
0.80	353	157	116	89	71	57	48	40	34	30	26	23	15	10	7	6	5	4	3
0.90	468	209	153	118	93	76	63	53	45	39	34	30	20	13	9	7	6	4	3

Chart E. 4 groups, $r = 0.40$ at alpha = 0.10; NK 2 intervening groups, Tukey 4 groups																			
Power	Hypothesized ES																		
	0.20	0.30	0.35	0.40	0.45	0.50	0.55	0.60	0.65	0.70	0.75	0.80	1.00	1.25	1.50	1.75	2.00	2.50	3.00
0.80	413	184	136	104	83	67	56	47	40	35	30	27	18	12	9	7	5	4	3
0.90	537	239	176	135	107	87	72	61	52	45	39	35	23	15	11	8	7	5	4

Chart F. 5 groups, $r = 0.40$ at alpha = 0.10; NK no intervening groups																			
Power	Hypothesized ES																		
	0.20	0.30	0.35	0.40	0.45	0.50	0.55	0.60	0.65	0.70	0.75	0.80	1.00	1.25	1.50	1.75	2.00	2.50	3.00
0.80	261	116	86	66	52	42	35	30	25	22	19	17	11	8	6	4	4	3	2
0.90	361	161	118	91	72	58	48	41	35	30	26	23	15	10	7	6	5	3	3

Table 7.35. (*cont.*)

Chart G. 5 groups, $r = 0.40$ at alpha = 0.10; NK 1 intervening group																			
Power	Hypothesized ES																		
	0.20	0.30	0.35	0.40	0.45	0.50	0.55	0.60	0.65	0.70	0.75	0.80	1.00	1.25	1.50	1.75	2.00	2.50	3.00
0.80	353	157	116	89	70	57	47	40	34	30	26	23	15	10	7	6	5	3	3
0.90	468	208	153	118	93	76	63	53	45	39	34	30	20	13	9	7	6	4	3

Chart H. 5 groups, $r = 0.40$ at alpha = 0.10; NK 2 intervening groups																			
Power	Hypothesized ES																		
	0.20	0.30	0.35	0.40	0.45	0.50	0.55	0.60	0.65	0.70	0.75	0.80	1.00	1.25	1.50	1.75	2.00	2.50	3.00
0.80	413	184	136	104	82	67	56	47	40	35	30	27	18	12	8	7	5	4	3
0.90	537	239	176	135	107	87	72	61	52	45	39	35	22	15	11	8	6	5	4

Chart I. 5 groups, $r = 0.40$ at alpha = 0.10; NK 3 intervening groups, Tukey 5 groups																			
Power	Hypothesized ES																		
	0.20	0.30	0.35	0.40	0.45	0.50	0.55	0.60	0.65	0.70	0.75	0.80	1.00	1.25	1.50	1.75	2.00	2.50	3.00
0.80	459	205	151	116	91	74	62	52	44	38	34	30	19	13	9	7	6	4	3
0.90	589	262	193	148	117	95	79	66	57	49	43	38	25	16	12	9	7	5	4

Table 7.36. Sample size table for ANCOVA multiple comparison; $p = 0.10$, $r = 0.60$

Chart A. 3 groups, $r = 0.60$ at alpha $= 0.10$; NK no intervening groups

Power	Hypothesized ES																	
	0.20	0.30	0.35	0.40	0.45	0.50	0.55	0.60	0.65	0.70	0.75	0.80	1.00	1.25	1.50	1.75	2.00	2.50	3.00
0.80	199	89	66	51	40	33	27	23	20	17	15	13	9	6	5	4	3	3	3
0.90	275	123	91	70	55	45	37	32	27	23	21	18	12	8	6	5	4	3	3

Chart B. 3 groups, $r = 0.60$ at alpha $= 0.10$; NK 1 intervening group, Tukey 3 groups

Power	Hypothesized ES																	
	0.20	0.30	0.35	0.40	0.45	0.50	0.55	0.60	0.65	0.70	0.75	0.80	1.00	1.25	1.50	1.75	2.00	2.50	3.00
0.80	269	120	89	68	54	44	37	31	27	23	20	18	12	8	6	5	4	3	3
0.90	357	159	117	90	72	58	48	41	35	30	27	24	16	10	8	6	5	4	3

Chart C. 4 groups, $r = 0.60$ at alpha $= 0.10$; NK no intervening groups

Power	Hypothesized ES																	
	0.20	0.30	0.35	0.40	0.45	0.50	0.55	0.60	0.65	0.70	0.75	0.80	1.00	1.25	1.50	1.75	2.00	2.50	3.00
0.80	199	89	66	50	40	33	27	23	20	17	15	13	9	6	5	4	3	3	2
0.90	275	123	90	69	55	45	37	31	27	23	20	18	12	8	6	5	4	3	3

Chart D. 4 groups, $r = 0.60$ at alpha $= 0.10$; NK 1 intervening group

Power	Hypothesized ES																	
	0.20	0.30	0.35	0.40	0.45	0.50	0.55	0.60	0.65	0.70	0.75	0.80	1.00	1.25	1.50	1.75	2.00	2.50	3.00
0.80	269	120	89	68	54	44	37	31	26	23	20	18	12	8	6	5	4	3	3
0.90	357	159	117	90	71	58	48	41	35	30	26	23	15	10	7	6	5	4	3

Chart E. 4 groups, $r = 0.60$ at alpha $= 0.10$; NK 2 intervening groups, Tukey 4 groups

Power	Hypothesized ES																	
	0.20	0.30	0.35	0.40	0.45	0.50	0.55	0.60	0.65	0.70	0.75	0.80	1.00	1.25	1.50	1.75	2.00	2.50	3.00
0.80	315	141	104	80	63	51	43	36	31	27	24	21	14	9	7	5	5	4	3
0.90	410	183	135	103	82	67	55	47	40	35	30	27	18	12	9	7	5	4	3

Chart F. 5 groups, $r = 0.60$ at alpha $= 0.10$; NK no intervening groups

Power	Hypothesized ES																	
	0.20	0.30	0.35	0.40	0.45	0.50	0.55	0.60	0.65	0.70	0.75	0.80	1.00	1.25	1.50	1.75	2.00	2.50	3.00
0.80	199	89	65	50	40	33	27	23	20	17	15	13	9	6	4	4	3	3	2
0.90	275	123	90	69	55	45	37	31	27	23	20	18	12	8	6	5	4	3	2

Table 7.36. (*cont.*)

Chart G. 5 groups, $r=0.60$ at alpha = 0.10; NK 1 intervening group																			
Power	Hypothesized ES																		
	0.20	0.30	0.35	0.40	0.45	0.50	0.55	0.60	0.65	0.70	0.75	0.80	1.00	1.25	1.50	1.75	2.00	2.50	3.00
0.80	269	120	88	68	54	44	36	31	26	23	20	18	12	8	6	5	4	3	3
0.90	357	159	117	90	71	58	48	40	35	30	26	23	15	10	7	6	5	3	3

Chart H. 5 groups, $r=0.60$ at alpha = 0.10; NK 2 intervening groups																			
Power	Hypothesized ES																		
	0.20	0.30	0.35	0.40	0.45	0.50	0.55	0.60	0.65	0.70	0.75	0.80	1.00	1.25	1.50	1.75	2.00	2.50	3.00
0.80	315	141	104	80	63	51	43	36	31	27	23	21	14	9	7	5	4	3	3
0.90	410	183	134	103	82	66	55	46	40	34	30	27	17	12	8	6	5	4	3

Chart I. 5 groups, $r=0.60$ at alpha = 0.10; NK 3 intervening groups, Tukey 5 groups																			
Power	Hypothesized ES																		
	0.20	0.30	0.35	0.40	0.45	0.50	0.55	0.60	0.65	0.70	0.75	0.80	1.00	1.25	1.50	1.75	2.00	2.50	3.00
0.80	350	156	115	88	70	57	47	40	34	30	26	23	15	10	7	6	5	4	3
0.90	449	200	147	113	90	73	60	51	43	38	33	29	19	13	9	7	6	4	3

8 One-way repeated measures analysis of variance

Purpose of the statistic

The one-way, repeated measures analysis of variance is exactly analogous to the one-way between subjects ANOVA except that the groups contain either the same subjects or individuals who have been explicitly matched[1] in some way. It is used to ascertain how likely these within subject mean differences would be to occur by chance alone. Studies that might employ such a design include the multiple (i.e., three or more times) measurement of a single group of individuals across time (e.g., adding a long term follow-up assessment to a single group, pretest–posttest design) or, less commonly, a situation in which the same group of individuals is exposed to three or more different conditions.

A within subject design is extremely efficient in comparison to one which employs different subjects in each group, requiring far fewer subjects when even a moderate correlation can be obtained among its repeated observations. A one-way, repeated measures (RM) ANOVA, then, is used when:

(1) there is a single, independent variable which is defined as group membership in three or more groups *or* as three or more separate points in time (recalling that if only two groups are involved a paired *t*-test can be employed, which is inferentially identical to a two-group RM ANOVA),
(2) the dependent variable is measured in such a way that it can be described by a mean (i.e., it is continuous in nature and not categorical),
(3) the same or matched subjects are contained in the groups (or measured at the indicated time intervals),
(4) the hypothesis being tested is expressed in terms of a mean difference, and
(5) there is no other independent variable (e.g., a baseline/pretest measure, or covariate) present (if there are such measures, these tables may still be used although the ES must first be adjusted as described in Chapter 10).

This latter condition deserves some explication because it is an important one. For most cross-over designs (i.e., studies in which the same subjects receive different treatments), there is a washout period following each treatment, after which a baseline measure is obtained prior to implementing the next treatment. In these circumstances there are three acceptable alternatives:

(1) The data are analyzed as a two-factor (i.e., group × time) RM ANOVA in which case the contrast of interest is the interaction (and both factors are repeated measures).
(2) The data are analyzed as a one-way RM ANCOVA, where each unique baseline assessment serves as the covariate.
(3) Difference scores are computed between each unique baseline and each EOT assessment and these difference scores are used as the dependent variable in a one-way RM ANOVA.

From a purely statistical perspective we would normally recommend the second option, although it is becoming increasingly difficult to find software capable of performing this analysis easily. From a power analytic perspective, however, we recommend that for either of the first two options the researcher employ either (a) the tables in this chapter to produce a relatively conservative power/sample size estimate (since the analytic use of the baseline will normally result in an increase in statistical power over and above a one-way RM ANOVA) or (b) the ANCOVA strategy detailed in Chapter 10 that allows for adjusting the tables in this chapter. We normally do not recommend the use of difference scores (the third analytic option) because they are usually less reliable than either the baseline or EOT scores upon which they are based, but for practical purposes their use will also provide an acceptable power analytic result as long as the hypothesized ES values are framed in terms of difference scores rather than raw means. When this option is used, the tables in this chapter are directly applicable for studies employing three to five groups, while the tables in Chapter 5 are applicable for two-group studies.

Assuming that some variant of a one-way RM ANOVA or RM ANCOVA is employed, the primary assumption governing the use of this statistic is that the correlations among the groups with respect to the dependent variable are similar *and* that the variances within the groups are approximately equal. The combination of these two assumptions is defined as sphericity, which is routinely tested by most statistical packages. The primary implication of not meeting this assumption from the present perspective is that power will be overestimated, although violation of the assumption should have little effect upon the power of pairwise comparisons employing

MCPs as long as the estimated power analytic parameters (e.g., the pairwise ES values and correlations across the repeated observations) are appropriate. There is some evidence, however, that the *average* correlation among the repeated measures will produce reasonably accurate results from a power analytic perspective.

Like its between subjects analog, a one-way repeated measures ANOVA is a two-step process, involving (a) computing an overall *F*-ratio and, if statistically significant, (b) computing a multiple comparison procedure of some sort to indicate exactly which means differ significantly from which other means. The remainder of this chapter will therefore be divided into two parts as was done in the previous two chapters. The first will present power and sample size tables for the overall *F*-ratio. The second will present tables that allow the investigator to estimate the power (or sample size requirements) for individual contrasts using two common multiple comparison procedures.

Part I. The *F*-ratio tables

Tables 8.1 through 8.16 present estimated power for a wide range of values of ES, *N*/group, and patterns of group means for *F*-ratios involving three through five groups. (Tables 8.17 through 8.22 present the comparable sample size tables for powers of 0.80 and 0.90 relevant to three alpha levels, 0.05, 0.01, and 0.10.) These tables are used in the same way as those presented in Chapter 7 except that for an RM ANOVA the Pearson *r* that must be estimated involves the most likely correlation between the subjects' multiple measurements. The repeated measures model, however, results in considerably more power than does the ANCOVA (which in turn produced more power than an ANOVA without comparable control variables).

As with any analysis of variance model employing more than three groups per factor, a power analysis involving the overall *F*-ratio is of limited use due to the facts that (a) further pairwise comparisons among the group means are almost always required and (b) the hypothesized pattern/dispersion of these means exerts such a dramatic effect upon power.

As with ANCOVA, it is sometimes relatively difficult to estimate the exact correlation among subjects (or matched subjects) with respect to their repeated measures scores. As always the results of pilot work or previously published studies provide the best estimates for these values. In lieu of such estimates, the guidelines presented in Chapter 7 are also applicable here:

(1) If the dependent variable is a relatively stable attribute a correlation of 0.60 may be possible (although the length of the interval among the measurements is negatively correlated with r).

(2) Since the estimation of this parameter (like most of the others involved in a power analysis) is subject to error, it is always wise to model power/sample size estimates based upon different values of r. To facilitate this process, we have provided tables for r values of both 0.40 and 0.60 as well as Table 10.3 in Chapter 10 which permits the present tables to be converted to more exact values of r ranging from 0.20 to 0.80.

Example. To illustrate how these tables are employed, let us assume that a researcher wishes to test three different types of acupuncture (i.e., with no electrical stimulation, with electrical stimulation involving a very weak current, and with relatively strong electrical stimulation) with respect to one another and a sham acupuncture procedure involving the insertion of needles in non-active points (to control for the stress of handling and needle insertion over and above any analgesic effects of acupuncture itself). Employing the latency with which rats withdraw their paw from a thermal stimulus as the dependent variable, the investigator decided to subject each animal to each type of acupuncture (counterbalancing the order in which the treatments were introduced) following a suitable washout period in order to decrease the number of animals required for the experiment.

Let us further assume that no baseline latency measures were collected following each washout period because of the desire to avoid injecting this additional source of stress into the experiment. If a baseline measure had been administered prior to each condition, then the investigator could (a) employ difference scores, (b) first adjust the hypothesized ES based upon the estimated correlation between the baseline and end-of-period latency measure as described in Chapter 10, or (c) opt for a slightly more conservative power/sample size estimate by ignoring the effect of the baseline (hence employing each latency measure following the administration of the treatment as the dependent variable). Option (b) would normally provide the highest (and probably most accurate) estimate of power, but options (a) and (c) would probably provide sufficiently accurate results as well.

Assuming, however, that our investigator needed to determine how much power would be available for achieving statistical significance in an experiment employing 20 rats in which no unique baseline measure was available for each treatment, it would be necessary to specify the following parameters:

(1) The ES for the largest mean difference likely to occur among the four groups (which in this case would probably involve the control condition vs. acupuncture accompanied by a relatively strong current).
(2) The projected *N*/group (20, since in a repeated measures design the total *N* is the same as the *N*/group).
(3) A judgment regarding whether the spread of means is likely to reflect a low, medium, or high dispersion pattern (low/medium in this case since the researcher would probably hypothesize a fairly even spread among the groups).
(4) The most likely baseline to end-of-treatment dependent variable relationship.

At the risk of redundancy, this latter correlation can also be estimated from (a) pilot work, (b) previous published research (in our experience these relationships are seldom mentioned in published reports, although the requisite information can occasionally be obtained by contacting the investigators of these reports), or (c) modeling different values of *r*.

Let us assume in this example that the investigator did not have prior information on the most likely value of the relationship and therefore chose to model both correlations of 0.40 and 0.60. To facilitate the actual power calculations, two templates are provided that are relevant for from three to five groups (Templates 8.1 and 8.2) and are presented at the end of this chapter. (For two-group studies the tables presented in Chapter 5 may be used.)

Basically these templates are employed in exactly the same manner as described in Chapter 7. In way of illustration, let us therefore posit the necessary parameters for this four-group model as requested by the preliminary Template 8.1.

The first and most crucial step involves estimating the ES values among the four experimental conditions. For investigators uncomfortable with working directly with standardized means, this (as always) can be accomplished by (a) hypothesizing the means accruing from each experimental condition in the original metric, (b) dividing them by the pooled standard deviation, and (c) setting the lowest standardized mean to zero by simply subtracting it from itself and each of the remaining group means. (Please refer to Chart 7.1 in the previous chapter for a numerical review of this process.)

Specifically, our investigator would need to begin by rank ordering the four means with respect to their most likely outcome and then placing them on the provided ES line as follows:

Step 1. Write in the names/codes of the groups in the chart to the right in ascending order based upon their expected means (i.e., the name of the group expected to have the lowest mean or the weakest effect will be written next to ①, followed by the next strongest treatment and so forth).	Group definitions ① Sham control ② Low Hz acupuncture ③ Med. Hz acupuncture ④ High Hz acupuncture ⑤

(Note that the fifth number is not used since there are only four groups.)

Step 2. Plot these groups, using the circled numbers rather than their codes, on the ES line below. In the case of a tie between two groups with respect to their actual ES value, place one to the right of the other based upon the best theoretical evidence for which might be slightly superior.

ES line

①		②		③				④							
0	0.1	0.2	0.3	0.4	0.5	0.6	0.7	0.8	0.9	1.0	1.1	1.2	1.3	1.4	1.5

From this hypothesized spread of means, we would estimate the ES to be entered into the power table (step 3) to be 1.0 (i.e., the largest standardized mean difference or the control vs. acupuncture accompanied by a strong electrical current). Assuming that 20 rats were projected to be employed (each of which, it will be remembered, is subjected to all four conditions) this value is entered as the step 3 parameter. (If the sample size necessary for a fixed level of power were desired, the latter would be entered here.) The mean differences most closely approximate a medium dispersion pattern when compared to the options in step 4 and, not knowing exactly which correlation among the repeated observations to posit, our investigator decides to model both 0.40 and 0.60 (step 5) in the power analysis, at which point he/she is directed to Template 8.2.

Step 6 in the power template (8.2) requests the specification of the largest group ES, which is, by definition, the ES corresponding to group 4 from the ES line in step 2, or 1.0. Armed with these estimates, step 7 directs our investigator to find the intersection of the 1.0 ES column and the row N/group = 20 in Table 8.7 for r = 0.40 (medium mean pattern) and Table 8.8 for r = 0.60 (medium mean pattern), producing power estimates of 0.95 and 0.99 respectively. Heartened by this finding, our investigator would rightly conclude that he/she had an excellent chance of obtaining a statistically significant F-ratio for either of the projected correlation coefficients. The results of this power analysis, which would follow a justification for the hypothesized ES between the highest and lowest groups with respect to the dependent variable (perhaps based upon a smaller pilot study) and the most likely pattern of means to accrue (which could be justified rationally), might be presented as follows:

A power analysis indicated that this four-group experiment would have a 95% chance of yielding a statistically significant F-ratio given a medium dispersion of group means, an average correlation of 0.40 among the repeated measures for a difference between the most extreme groups (the control vs. acupuncture with strong electrical stimulation) of one standard deviation. Should a correlation of 0.60 accrue with these data, the available power would be 0.99.

Part II. The multiple comparison tables

The researcher would be *incorrect*, however, if he/she concluded that the probability of obtaining statistical significance between two specific groups would range between 0.95 and 0.99. The statistically correct way to ascertain the probability of obtaining statistical significance for individual group contrasts is to compute the power for a multiple comparison procedure, which can be accomplished by completing the remainder of the within group power template.

The first step in ascertaining the power of the six pairwise comparisons resulting from a four-group study is to compute their individual ES values. These have already been hypothesized via step 2 in the preliminary template; hence all that needs to be done is to perform the actual subtractions requested by step 9 of Template 8.2.

In most cases an investigator would probably not be interested in all of these contrasts from a scientific perspective, or even expect that they all would prove to be statistically significant. There is no harm in reviewing all of the individual pairwise ES values via the completion of step 9, however, since if nothing else this will force the researcher to consider explicitly his/her implicit expectations for the study. It may also suggest, upon reflection, that the initial ES line projections (step 2, Template 8.1) should be revised (in which case the power of the overall F might need to be recalculated).

In any case, let us assume that our investigator did not need to revise his/her original hypothesized pairwise ES values, which in turn would produce the following six ES values (note that any contrasts involving a fifth group are irrelevant for this study):

Step 9. Using the hypothesized values from step 2 in the preliminary template (8.1), fill out the hypothesized ES values by performing the indicated subtractions listed under the "Contrasts" column under the appropriate model.

ES ② − ① = 0.3	ES ④ − ① = 1.0	ES ⑤ − ① =
ES ③ − ① = 0.6	ES ④ − ② = 0.7	ES ⑤ − ② =
ES ③ − ② = 0.3	ES ④ − ③ = 0.4	ES ⑤ − ③ =
		ES ⑤ − ④ =

The final step in the process, once the MCP is chosen, is to access the power tables indicated in step 10. Assuming that the Tukey HSD procedure were chosen, the provided chart indicates that Tables 8.25 (for $r=0.40$) and 8.26 (for $r=0.60$) would be employed for all six contrasts. Locating the intersection of the *N*/group row of 20 and the five different ES values would produce the following range of results for $r=0.40$ and $r=0.60$:

Step 10. For power of the above pairwise contrasts (step 9) using the **Tukey HSD** procedure, use the tables indicated in the following chart. (Note that the table is chosen based *only* upon the contrasts involved and *not* the number of groups.) Locate the power in the indicated table at the intersection of the ES column (step 9) and the *N*/group row (preliminary step 3, Template 8.1). Interpolate as desired.

For contrasts ② − ① , ③ − ② , ④ − ③ , and/or ⑤ − ④
If $r=0.40$, use Table 8.23 — If $r=0.60$, use Table 8.24

For contrasts ③ − ① ,④ − ②, and/or ⑤ − ③
If $r=0.40$, use Table 8.25 — If $r=0.60$, use Table 8.26

For contrasts ④ − ①, and/or ⑤ − ②
If $r=0.40$, use Table 8.27 — If $r=0.60$, use Table 8.28

For contrast ⑤ − ①
If $r=0.40$, use Table 8.29 — If $r=0.60$, use Table 8.30

Power ② − ① = 0.08–0.13	Power ④ − ① = 0.92–0.99	Power ⑤ − ① =
Power ③ − ① = 0.42–0.63	Power ④ − ② = 0.58–0.80	Power ⑤ − ② =
Power ③ − ② = 0.08–0.13	Power ④ − ③ = 0.16–0.27	Power ⑤ − ③ =
		Power ⑤ − ④ =

Thus while the power of the overall *F* ranged from an impressive 0.95 to very close to 1.0, only two of the contrasts (group 4, acupuncture coupled with high frequency electrical stimulation, vs. group 1, the sham control, and group 4 vs. group 2 assuming a correlation of 0.60 between observations) possess an adequate level of power.

Computing the required *N*/group for a within group multiple comparison procedure. Let us assume that all six of the contrasts produced by this design were of sufficient scientific interest to question the wisdom of conducting the study as currently designed. The first step in deciding how to proceed would therefore probably entail determining how many subjects per group would be needed to yield an acceptable power, which in this case we will assume to be 0.80. Let us further assume that our investigator decided to employ the Newman–Keuls procedure and was willing to take a chance that a correlation among the repeated observations of 0.60 would be obtained.

To conduct this sample size analysis our investigator could simply proceed directly to step 10 of Template 8.3 since all of the preliminary steps have already been completed. There, he/she would be instructed to access Table 8.30 where Chart C would be used for three of the contrasts, Chart D for two, and Chart E for the high frequency acupuncture vs. control contrast. Following these instructions would produce the following six sample size requirements:

N/group ② – ① = 71	*N*/group ④ – ① = 11	*N*/group ⑤ – ① =
N/group ③ – ① = 24	*N*/group ④ – ② = 18	*N*/group ⑤ – ② =
N/group ③ – ② = 71	*N*/group ④ – ③ = 41	*N*/group ⑤ – ③ =
		N/group ⑤ – ④ =

Armed with this information, the investigator would need to make a number of decisions. If he/she had 71 rats available, then the experiment might be quite feasible to run, since sufficient power would be assured for all of the available contrasts. Otherwise, one or more of the options presented in Chapter 2 might be considered, such as reducing the number of groups and/or increasing the ES. One possibility, for example, might be to combine the low and medium frequency groups by selecting a frequency halfway in between, thereby producing the following revised ES line:

Group definitions
① Sham control
② Moderate Hz acupuncture
③ High Hz acupuncture

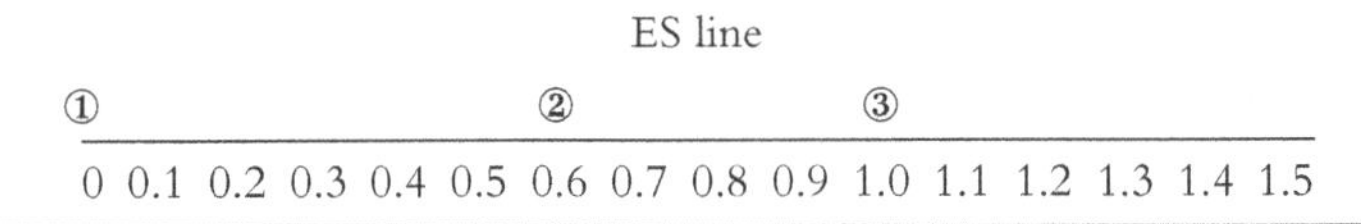

This strategy would increase the smallest pairwise ES to 0.40 which, using Table 8.30 Chart A would reduce the number of animals by more than half (71 to 33) for these contrasts, which in turn would provide even more power for the sham vs. high Hz acupuncture contrast. Other alternatives exist, but they would not be available if the problem were not identified prior to conducting the study, which is the primary value of conducting an a priori power analysis.

Computing the maximum detectable ES. Unfortunately it is not uncommon for both the *N*/group and the acceptable power level to be fixed for an investigator by practical constraints. Returning to our four-group example, let us assume that the minimum acceptable power level for the investigator's discipline is 0.80 and he/she had exactly 20 rats with which to work.

A natural question might then arise as to what type of ES would have to be obtained in this experiment in order actually to provide an 80% chance of obtaining statistical significance. (We will further assume that a hypothesized *r* of 0.60 between the repeated measures is still reasonable and that the Newman–Keuls procedure is still the MCP of choice.)

The simplest way to perform this analysis would be to access Table 8.30 Chart C and locate the nearest *N*/group value to 20 in the power = 0.80 row in the appropriate chart. For the adjacent groups (i.e., 2 vs. 1, 3 vs. 2, and 4 vs. 3) on the ES line (i.e., that possessed no intervening groups separating them), this value would be 20, which corresponds to the ES of 0.60. For groups 3 vs. 1 and 4 vs. 2, the closest detectable ES value (Chart D) would be 0.65 or 0.70 (corresponding to the *N*/group of 21 or 18, respectively). Our investigator would either interpolate or choose the more conservative figure, which would produce a detectable ES of 0.68 or 0.70 respectively. For the largest ES (4 vs. 1), the corresponding smallest detectable ES would be approximately 0.70 (Chart E). At this point, the obvious decision would become whether or not it would be practical and scientifically defensible to design an experiment capable of achieving an ES of this magnitude.

Template 8.1. Preliminary one-way within subjects ANOVA power and sample size template

---

This preliminary template is applicable to all one-way within subjects (repeated measures) ANOVA designs employing between three and five groups.

Step 1. Write in the names/codes of the groups in the chart to the right in ascending order based upon their expected means (i.e., the name of the group expected to have the lowest mean or the weakest effect will be written next to ①, followed by the next strongest treatment and so forth).

Group definitions
①
②
③
④
⑤

Step 2. Plot these groups, using the circled numbers rather than their codes, on the ES line below. In the case of a tie between two groups with respect to their actual ES value, place one to the right of the other based upon the best theoretical evidence for which might be slightly superior.

ES line

①

0 0.1 0.2 0.3 0.4 0.5 0.6 0.7 0.8 0.9 1.0 1.1 1.2 1.3 1.4 1.5

Step 3. For the estimation of power, specify the *N*/group available. For required sample size, enter the desired power level.

N/group =   or, desired power =

Step 4. Compare the low, medium, and high ES patterns below with the graphed ES line from step 2 above. Choose the pattern which most closely matches the hypothesized pattern of means (step 2).

(a) Three-group designs

ES pattern for low/medium *F* power	①　　　　②　　　　③
ES pattern for high *F* power	①②　　　　　　　　③
or	①　　　　　　　　②③

(b) Four-group designs

ES pattern for low *F* power	①　　②③　　　　④
ES pattern for medium *F* power	①　　②　　③　　④
ES pattern for high *F* power	①②　　　　　　③④

(c) Five-group designs

ES pattern for low *F* power	①　　②③④　　⑤
ES pattern for medium *F* power	①　②　③　④　⑤
ES pattern for high *F* power	①②③　　　　④⑤
or	①②　　　　③④⑤

Step 5. Select the most likely correlation between the repeated (or matched) dependent variable observations. (For values other than 0.40 or 0.60, see Chapter 10.)

$r = 0.40$ ___ $r = 0.60$ ___

To compute power, turn to Template 8.2. To determine *N*/group, turn to Template 8.3.

Template 8.2. One-way within subjects ANOVA power template

Power of the overall *F*-ratio

Steps 1–5, see preliminary Template 8.1.

Step 6. Use the largest hypothesized ES produced in preliminary step 2 (Template 8.1). For three-group studies this will be the ES for group ③, and so forth.

Largest ES from preliminary step 2:

Step 7. Use the chart below to find the appropriate power table.

(a) For three groups, choose the appropriate model below:

$r = 0.40$	Table	$r = 0.60$	Table
L/M pattern	8.1	L/M pattern	8.2
H pattern	8.3	H pattern	8.4

(b) For four groups, choose the appropriate model below:

$r = 0.40$	Table	$r = 0.60$	Table
L pattern	8.5	L pattern	8.6
M pattern	8.7	M pattern	8.8
H pattern	8.9	H pattern	8.10

(c) For five groups, choose the appropriate model below:

$r=0.40$	Table	$r=0.60$	Table
L pattern	8.11	L pattern	8.12
M pattern	8.13	M pattern	8.14
H pattern	8.15	H pattern	8.16

Step 8. Turn to the table identified in step 7 and find the power at the intersection between the ES column from step 6 and the *N*/group specified in step 3 of the preliminary template. Interpolate as desired.

Power =

Power of the pairwise contrasts

Step 9. Using the hypothesized values from step 2 in the preliminary template (8.1), fill out the hypothesized ES values by performing the indicated subtractions listed under the "Contrasts" column in the chart below.

ES ② − ① =	ES ④ − ① =	ES ⑤ − ① =
ES ③ − ① =	ES ④ − ② =	ES ⑤ − ② =
ES ③ − ② =	ES ④ − ③ =	ES ⑤ − ③ =
		ES ⑤ − ④ =

Step 10. For power of the above pairwise contrasts (step 9) using the **Tukey HSD** procedure, use the tables indicated in the "Power tables for the Tukey HSD procedure" chart (only sample size tables are available for the **Newman–Keuls** procedure, see Template 8.3). Locate the power in the indicated table at the intersection of the ES column (step 9) and the *N*/group row (preliminary step 3, Template 8.1). Interpolate as desired.

Power tables for the Tukey HSD procedure

Three-group studies	If $r=0.40$, use Table 8.23	If $r=0.60$, use Table 8.24
Four-group studies	If $r=0.40$, use Table 8.25	If $r=0.60$, use Table 8.26
Five-group studies	If $r=0.40$, use Table 8.27	If $r=0.60$, use Table 8.28

Power ② − ① =	Power ④ − ① =	Power ⑤ − ① =
Power ③ − ① =	Power ④ − ② =	Power ⑤ − ② =
Power ③ − ② =	Power ④ − ③ =	Power ⑤ − ③ =
		Power ⑤ − ④ =

Template 8.3. One-way within subjects ANOVA sample size template

Required sample size for a statistically significant overall *F*-ratio

Steps 1–5, see preliminary Template 8.1.

Step 6. Use the largest hypothesized ES produced in preliminary step 2 (Template 8.1). For three-group studies this will be the ES for group ③, and so forth.

Largest ES from preliminary step 2:

Step 7. For desired power values of 0.80 and 0.90, use the chart below to find the appropriate sample size table. Note that sample size tables are also provided for $p=0.01$ (Table 8.19 for $r=0.40$; Table 8.20 for $r=0.60$) and $p=0.10$ (Tables 8.21 and 8.22).

(a) For three groups, choose the appropriate model below:

$r=0.40$	Table	$r=0.60$	Table
L/M pattern	8.17 Chart A	L/M pattern	8.18 Chart A
H pattern	8.17 Chart B	H pattern	8.18 Chart B

(b) For four groups, choose the appropriate model below:

$r=0.40$	Table	$r=0.60$	Table
L pattern	8.17 Chart C	L pattern	8.18 Chart C
M pattern	8.17 Chart D	M pattern	8.18 Chart D
H pattern	8.17 Chart E	H pattern	8.18 Chart E

(c) For five groups, choose the appropriate model below:

$r=0.40$	Table	$r=0.60$	Table
L pattern	8.17 Chart F	L pattern	8.18 Chart F
M pattern	8.17 Chart G	M pattern	8.18 Chart G
H pattern	8.17 Chart H	H pattern	8.18 Chart H

Step 8. For desired powers of 0.80 and 0.90, turn to the table identified in step 7 and find the *N*/group at the intersection between the ES column from step 6 and the *N*/group specified in step 3 of the preliminary template. For powers other than 0.80 and 0.90, find the appropriate table from step 7 in Template 8.2. Locate the nearest power value in the ES row (step 6) and read the required *N*/group associated with that row. Interpolate as desired.

N/group =

Sample size for the pairwise contrasts

Step 9. Using the hypothesized values from step 2 in the preliminary template (8.1), fill out the hypothesized ES values by performing the indicated subtractions listed under the "Contrasts" column in the chart below.

ES ② − ① =	ES ④ − ① =	ES ⑤ − ① =
ES ③ − ① =	ES ④ − ② =	ES ⑤ − ② =
ES ③ − ② =	ES ④ − ③ =	ES ⑤ − ③ =
		ES ⑤ − ④ =

Step 10. For the required sample size for powers of 0.80 or 0.90 of the pairwise contrasts listed in step 9 for the **Tukey HSD** procedure, use the tables indicated in the "Tukey HSD" chart below. For powers of 0.80 or 0.90 for the **Newman–Keuls** procedure, use the tables indicated in the "Newman–Keuls" chart. (Note that the tables specified for the Tukey procedure are based *only* upon the number of groups and not the particular contrast involved while the tables used for Newman–Keuls are based *only* upon the contrasts involved and *not* the number of groups.) Locate the *N*/group in the indicated table at the intersection of the ES column (step 9) and the desired power row (preliminary step 3, Template 8.1). For powers other than 0.80 and 0.90 (for the Tukey HSD procedure), use the appropriate tables indicated in step 10 of Template 8.2 by finding the nearest power value in the ES column (step 9) and reading the *N*/group associated with that row. Interpolate as desired. (Note that sample size requirements for $p=0.10$ and $p=0.01$ are

provided in Tables 8.31 to 8.34 and are mirror images of the $p=0.05$ tables/charts presented below.)

Sample size tables for the Tukey HSD procedure

Three-group studies

If $r=0.40$, use Table 8.29 Chart B — If $r=0.60$, use Table 8.30 Chart B

Four-group studies

If $r=0.40$, use Table 8.29 Chart E — If $r=0.60$, use Table 8.30 Chart E

Five-group studies

If $r=0.40$, use Table 8.29 Chart I — If $r=0.60$, use Table 8.30 Chart I

Sample size tables for the Newman–Keuls procedure

Determining the sample size requirements for the Newman–Keuls procedure is slightly more complicated since the sample size is determined by both the number of groups and the number of intervening variables in Table 8.29 ($r=0.40$) and 8.30 ($r=0.60$) for $p=0.05$ and Tables 8.31 to 8.34 for $p=0.01$ and $p=0.10$:

Three groups

Contrast	Chart
② − ①	A
③ − ①	B
③ − ②	A

Four groups

Contrast	Chart	Contrast	Chart
② − ①	C	④ − ①	E
③ − ①	D	④ − ②	D
③ − ②	C	④ − ③	C

Five groups

Contrast	Chart	Contrast	Chart	Contrast	Chart
② − ①	F	④ − ①	H	⑤ − ①	I
③ − ①	G	③ − ①	G	⑤ − ②	H
③ − ②	F	④ − ③	F	⑤ − ③	G
				⑤ − ④	F

N/group ② − ① =	N/group ④ − ① =	N/group ⑤ − ① =
N/group ③ − ① =	N/group ④ − ② =	N/group ⑤ − ② =
N/group ③ − ② =	N/group ④ − ③ =	N/group ⑤ − ③ =
		N/group ⑤ − ④ =

Table 8.1. Power table for repeated measures ANOVA; $r = 0.40$, pattern L/M, 3 groups at alpha = 0.05

n	Hypothesized ES																		
	0.20	0.30	0.35	0.40	0.45	0.50	0.55	0.60	0.65	0.70	0.75	0.80	1.00	1.25	1.50	1.75	2.00	2.50	3.00
5	6	7	8	9	10	11	12	14	16	17	19	21	31	46	61	75	86	97	.
6	6	8	9	10	11	13	15	17	19	21	24	26	39	57	73	86	94	99	.
7	6	8	9	11	13	15	17	19	22	25	28	32	47	66	82	92	97	.	.
8	6	9	10	12	14	17	19	22	25	29	33	37	54	74	89	96	99	.	.
9	7	9	11	13	16	18	22	25	29	33	37	42	60	80	93	98	.	.	.
10	7	10	12	14	17	20	24	28	32	37	42	47	66	85	96	99	.	.	.
11	7	11	13	16	19	23	27	31	36	41	46	51	72	89	97	.	.	.	.
12	7	11	14	17	21	25	29	34	39	45	50	56	76	92	98	.	.	.	.
13	8	12	15	18	22	27	32	37	43	48	54	60	80	94	99	.	.	.	.
14	8	13	16	19	24	29	34	40	46	52	58	64	83	96	99	.	.	.	.
15	8	13	17	21	26	31	37	43	49	55	61	67	86	97	.	.	.	.	.
20	10	17	22	27	34	41	48	56	63	70	76	81	95	.	.	.	.	.	.
25	11	20	27	34	42	50	59	67	74	80	86	90	98	.	.	.	.	.	.
30	13	24	32	40	50	59	68	75	82	88	92	95	.	.	.	.	.	.	.
35	14	28	37	47	57	66	75	82	88	92	95	97	.	.	.	.	.	.	.
40	16	32	42	52	63	73	81	87	92	95	98	99	.	.	.	.	.	.	.
45	17	35	46	58	69	78	86	91	95	97	99	99	.	.	.	.	.	.	.
50	19	39	51	63	74	83	89	94	97	98	99	.	.	.	.	.	.	.	.
55	21	42	55	67	78	86	92	96	98	99	.	.	.	.	.	.	.	.	.
60	22	46	59	71	82	89	94	97	99	.	.	.	.	.	.	.	.	.	.
65	24	49	63	75	85	92	96	98	99	.	.	.	.	.	.	.	.	.	.
70	26	52	66	78	87	94	97	99	.	.	.	.	.	.	.	.	.	.	.
75	27	56	70	81	90	95	98	99	.	.	.	.	.	.	.	.	.	.	.
80	29	58	73	84	92	96	98	99	.	.	.	.	.	.	.	.	.	.	.
90	32	64	78	88	95	98	99	.	.	.	.	.	.	.	.	.	.	.	.
100	36	69	82	91	96	99	.	.	.	.	.	.	.	.	.	.	.	.	.
110	39	73	86	94	98	99	.	.	.	.	.	.	.	.	.	.	.	.	.
120	42	77	89	96	99	.	.	.	.	.	.	.	.	.	.	.	.	.	.
130	45	81	91	97	99	.	.	.	.	.	.	.	.	.	.	.	.	.	.
140	48	84	93	98	99	.	.	.	.	.	.	.	.	.	.	.	.	.	.
150	51	86	95	98	.	.	.	.	.	.	.	.	.	.	.	.	.	.	.
175	58	91	97	99	.	.	.	.	.	.	.	.	.	.	.	.	.	.	.
200	64	94	99	.	.	.	.	.	.	.	.	.	.	.	.	.	.	.	.
225	69	97	99	.	.	.	.	.	.	.	.	.	.	.	.	.	.	.	.
250	74	98	.	.	.	.	.	.	.	.	.	.	.	.	.	.	.	.	.
300	82	99	.	.	.	.	.	.	.	.	.	.	.	.	.	.	.	.	.
350	88	.	.	.	.	.	.	.	.	.	.	.	.	.	.	.	.	.	.
400	92	.	.	.	.	.	.	.	.	.	.	.	.	.	.	.	.	.	.
450	95	.	.	.	.	.	.	.	.	.	.	.	.	.	.	.	.	.	.
500	96	.	.	.	.	.	.	.	.	.	.	.	.	.	.	.	.	.	.
600	99	.	.	.	.	.	.	.	.	.	.	.	.	.	.	.	.	.	.
700	99	.	.	.	.	.	.	.	.	.	.	.	.	.	.	.	.	.	.
800	.	.	.	.	.	.	.	.	.	.	.	.	.	.	.	.	.	.	.
900	.	.	.	.	.	.	.	.	.	.	.	.	.	.	.	.	.	.	.
1000	.	.	.	.	.	.	.	.	.	.	.	.	.	.	.	.	.	.	.

Table 8.2. Power table for repeated measures ANOVA; $r = 0.60$, pattern L/M, 3 groups at alpha = 0.05

n	Hypothesized ES																		
	0.20	0.30	0.35	0.40	0.45	0.50	0.55	0.60	0.65	0.70	0.75	0.80	1.00	1.25	1.50	1.75	2.00	2.50	3.00
5	7	8	10	11	12	14	16	19	21	24	27	30	44	63	79	90	96	.	.
6	7	9	11	12	15	17	20	23	26	30	34	38	55	75	89	96	99	.	.
7	7	10	12	14	17	20	23	27	31	36	40	45	64	84	95	99	.	.	.
8	7	11	13	16	19	23	27	32	36	42	47	52	72	90	97	.	.	.	.
9	8	12	15	18	22	26	31	36	41	47	53	59	79	94	99	.	.	.	.
10	8	13	16	20	24	29	35	40	46	52	59	64	84	96	99	.	.	.	.
11	8	14	17	22	27	32	38	44	51	57	64	70	88	98	.	.	.	.	.
12	9	15	19	24	29	35	42	49	55	62	68	74	91	99	.	.	.	.	.
13	9	16	20	26	32	38	45	52	59	66	73	78	93	99	.	.	.	.	.
14	10	17	22	28	34	41	49	56	63	70	76	82	95	.	.	.	.	.	.
15	10	18	23	30	37	44	52	60	67	74	80	85	97	.	.	.	.	.	.
20	12	24	31	39	48	57	66	74	81	87	91	94	99	.	.	.	.	.	.
25	15	29	38	49	59	68	77	84	90	94	96	98	.	.	.	.	.	.	.
30	17	35	46	57	68	77	85	91	95	97	99	99	.	.	.	.	.	.	.
35	20	40	52	64	75	84	90	95	97	99	99	.	.	.	.	.	.	.	.
40	22	45	58	71	81	89	94	97	99	99	.	.	.	.	.	.	.	.	.
45	25	50	64	76	86	92	96	98	99	.	.	.	.	.	.	.	.	.	.
50	27	55	69	81	89	95	98	99	.	.	.	.	.	.	.	.	.	.	.
55	30	59	74	85	92	97	99	.	.	.	.	.	.	.	.	.	.	.	.
60	32	64	78	88	94	98	99	.	.	.	.	.	.	.	.	.	.	.	.
65	34	67	81	90	96	99	.	.	.	.	.	.	.	.	.	.	.	.	.
70	37	71	84	93	97	99	.	.	.	.	.	.	.	.	.	.	.	.	.
75	39	74	87	94	98	99	.	.	.	.	.	.	.	.	.	.	.	.	.
80	42	77	89	95	99	.	.	.	.	.	.	.	.	.	.	.	.	.	.
90	46	82	92	97	99	.	.	.	.	.	.	.	.	.	.	.	.	.	.
100	51	86	95	98	.	.	.	.	.	.	.	.	.	.	.	.	.	.	.
110	55	89	96	99	.	.	.	.	.	.	.	.	.	.	.	.	.	.	.
120	59	92	98	.	.	.	.	.	.	.	.	.	.	.	.	.	.	.	.
130	62	94	98	.	.	.	.	.	.	.	.	.	.	.	.	.	.	.	.
140	66	95	99	.	.	.	.	.	.	.	.	.	.	.	.	.	.	.	.
150	69	97	99	.	.	.	.	.	.	.	.	.	.	.	.	.	.	.	.
175	76	98	.	.	.	.	.	.	.	.	.	.	.	.	.	.	.	.	.
200	82	99	.	.	.	.	.	.	.	.	.	.	.	.	.	.	.	.	.
225	86	.	.	.	.	.	.	.	.	.	.	.	.	.	.	.	.	.	.
250	90	.	.	.	.	.	.	.	.	.	.	.	.	.	.	.	.	.	.
300	94	.	.	.	.	.	.	.	.	.	.	.	.	.	.	.	.	.	.
350	97	.	.	.	.	.	.	.	.	.	.	.	.	.	.	.	.	.	.
400	99	.	.	.	.	.	.	.	.	.	.	.	.	.	.	.	.	.	.
450	99	.	.	.	.	.	.	.	.	.	.	.	.	.	.	.	.	.	.
500	.	.	.	.	.	.	.	.	.	.	.	.	.	.	.	.	.	.	.
600	.	.	.	.	.	.	.	.	.	.	.	.	.	.	.	.	.	.	.
700	.	.	.	.	.	.	.	.	.	.	.	.	.	.	.	.	.	.	.
800	.	.	.	.	.	.	.	.	.	.	.	.	.	.	.	.	.	.	.
900	.	.	.	.	.	.	.	.	.	.	.	.	.	.	.	.	.	.	.
1000	.	.	.	.	.	.	.	.	.	.	.	.	.	.	.	.	.	.	.

Table 8.3. Power table for repeated measures ANOVA; $r = 0.40$, pattern H, 3 groups at alpha = 0.05

n	Hypothesized ES																		
	0.20	0.30	0.35	0.40	0.45	0.50	0.55	0.60	0.65	0.70	0.75	0.80	1.00	1.25	1.50	1.75	2.00	2.50	3.00
5	6	8	9	10	12	13	15	17	19	22	24	27	40	58	74	87	94	99	.
6	7	9	10	12	13	16	18	21	24	27	30	34	50	70	85	94	98	.	.
7	7	9	11	13	15	18	21	25	28	32	36	41	59	79	92	98	.	.	.
8	7	10	12	15	18	21	24	28	33	37	42	47	67	86	96	99	.	.	.
9	7	11	13	16	20	24	28	32	37	43	48	53	74	91	98	.	.	.	.
10	8	12	15	18	22	26	31	36	42	48	53	59	79	94	99	.	.	.	.
11	8	13	16	20	24	29	34	40	46	52	58	64	84	96	99	.	.	.	.
12	8	14	17	21	26	32	38	44	50	57	63	69	87	98	.	.	.	.	.
13	9	15	18	23	29	34	41	47	54	61	67	73	90	99	.	.	.	.	.
14	9	15	20	25	31	37	44	51	58	65	71	77	93	99	.	.	.	.	.
15	9	16	21	27	33	40	47	54	62	68	75	80	94	99	.	.	.	.	.
20	11	21	28	35	44	52	61	69	76	82	87	91	99	.	.	.	.	.	.
25	13	26	35	44	54	63	72	79	86	91	94	96	.	.	.	.	.	.	.
30	16	31	41	52	62	72	80	87	92	95	97	99	.	.	.	.	.	.	.
35	18	36	47	59	70	79	87	92	95	98	99	.	.	.	.	.	.	.	.
40	20	41	53	65	76	85	91	95	98	99	.	.	.	.	.	.	.	.	.
45	22	46	59	71	81	89	94	97	99	99	.	.	.	.	.	.	.	.	.
50	24	50	64	76	85	92	96	98	99	.	.	.	.	.	.	.	.	.	.
55	27	54	68	80	89	94	98	99	.	.	.	.	.	.	.	.	.	.	.
60	29	58	72	84	91	96	98	99	.	.	.	.	.	.	.	.	.	.	.
65	31	62	76	87	94	97	99	.	.	.	.	.	.	.	.	.	.	.	.
70	33	65	79	89	95	98	99	.	.	.	.	.	.	.	.	.	.	.	.
75	35	69	82	91	96	99	.	.	.	.	.	.	.	.	.	.	.	.	.
80	38	72	85	93	97	99	.	.	.	.	.	.	.	.	.	.	.	.	.
90	42	77	89	96	99	.	.	.	.	.	.	.	.	.	.	.	.	.	.
100	46	82	92	97	99	.	.	.	.	.	.	.	.	.	.	.	.	.	.
110	50	85	94	98	.	.	.	.	.	.	.	.	.	.	.	.	.	.	.
120	54	88	96	99	.	.	.	.	.	.	.	.	.	.	.	.	.	.	.
130	57	91	97	99	.	.	.	.	.	.	.	.	.	.	.	.	.	.	.
140	61	93	98	.	.	.	.	.	.	.	.	.	.	.	.	.	.	.	.
150	64	94	99	.	.	.	.	.	.	.	.	.	.	.	.	.	.	.	.
175	71	97	99	.	.	.	.	.	.	.	.	.	.	.	.	.	.	.	.
200	77	99	.	.	.	.	.	.	.	.	.	.	.	.	.	.	.	.	.
225	82	99	.	.	.	.	.	.	.	.	.	.	.	.	.	.	.	.	.
250	86	.	.	.	.	.	.	.	.	.	.	.	.	.	.	.	.	.	.
300	92	.	.	.	.	.	.	.	.	.	.	.	.	.	.	.	.	.	.
350	95	.	.	.	.	.	.	.	.	.	.	.	.	.	.	.	.	.	.
400	97	.	.	.	.	.	.	.	.	.	.	.	.	.	.	.	.	.	.
450	99	.	.	.	.	.	.	.	.	.	.	.	.	.	.	.	.	.	.
500	99	.	.	.	.	.	.	.	.	.	.	.	.	.	.	.	.	.	.
600	.	.	.	.	.	.	.	.	.	.	.	.	.	.	.	.	.	.	.
700	.	.	.	.	.	.	.	.	.	.	.	.	.	.	.	.	.	.	.
800	.	.	.	.	.	.	.	.	.	.	.	.	.	.	.	.	.	.	.
900	.	.	.	.	.	.	.	.	.	.	.	.	.	.	.	.	.	.	.
1000	.	.	.	.	.	.	.	.	.	.	.	.	.	.	.	.	.	.	.

Table 8.4. Power table for repeated measures ANOVA; $r = 0.60$, pattern H, 3 groups at alpha = 0.05

n	Hypothesized ES																		
	0.20	0.30	0.35	0.40	0.45	0.50	0.55	0.60	0.65	0.70	0.75	0.80	1.00	1.25	1.50	1.75	2.00	2.50	3.00
5	7	9	11	13	15	18	20	23	27	31	34	39	56	76	90	97	99	.	.
6	7	11	13	15	18	21	25	29	34	38	43	48	68	87	96	99	.	.	.
7	8	12	14	18	21	25	30	35	40	46	52	57	77	93	99	.	.	.	.
8	8	13	16	20	25	30	35	41	47	53	59	65	84	96	.	.	.	.	.
9	9	14	18	23	28	34	40	46	53	60	66	72	89	98	.	.	.	.	.
10	9	16	20	25	31	38	44	52	59	65	72	77	93	99	.	.	.	.	.
11	10	17	22	28	34	42	49	56	64	71	77	82	95	.	.	.	.	.	.
12	10	19	24	31	38	45	53	61	68	75	81	86	97	.	.	.	.	.	.
13	11	20	26	33	41	49	57	65	73	79	85	89	98	.	.	.	.	.	.
14	12	22	28	36	44	53	61	69	76	83	88	92	99	.	.	.	.	.	.
15	12	23	30	38	47	56	65	73	80	86	90	94	99	.	.	.	.	.	.
20	15	30	40	51	61	71	79	86	91	95	97	98	.	.	.	.	.	.	.
25	19	38	49	61	72	81	88	93	96	98	99	.	.	.	.	.	.	.	.
30	22	45	58	70	80	88	94	97	99	99	.	.	.	.	.	.	.	.	.
35	25	51	65	77	87	93	97	99	99	.	.	.	.	.	.	.	.	.	.
40	28	58	72	83	91	96	98	99	.	.	.	.	.	.	.	.	.	.	.
45	32	63	77	88	94	98	99	.	.	.	.	.	.	.	.	.	.	.	.
50	35	68	82	91	96	99	.	.	.	.	.	.	.	.	.	.	.	.	.
55	38	73	85	93	98	99	.	.	.	.	.	.	.	.	.	.	.	.	.
60	41	77	89	95	98	.	.	.	.	.	.	.	.	.	.	.	.	.	.
65	45	80	91	97	99	.	.	.	.	.	.	.	.	.	.	.	.	.	.
70	48	83	93	98	99	.	.	.	.	.	.	.	.	.	.	.	.	.	.
75	50	86	95	98	.	.	.	.	.	.	.	.	.	.	.	.	.	.	.
80	53	88	96	99	.	.	.	.	.	.	.	.	.	.	.	.	.	.	.
90	59	92	98	99	.	.	.	.	.	.	.	.	.	.	.	.	.	.	.
100	63	94	99	.	.	.	.	.	.	.	.	.	.	.	.	.	.	.	.
110	68	96	99	.	.	.	.	.	.	.	.	.	.	.	.	.	.	.	.
120	72	97	.	.	.	.	.	.	.	.	.	.	.	.	.	.	.	.	.
130	76	98	.	.	.	.	.	.	.	.	.	.	.	.	.	.	.	.	.
140	79	99	.	.	.	.	.	.	.	.	.	.	.	.	.	.	.	.	.
150	82	99	.	.	.	.	.	.	.	.	.	.	.	.	.	.	.	.	.
175	88	.	.	.	.	.	.	.	.	.	.	.	.	.	.	.	.	.	.
200	92	.	.	.	.	.	.	.	.	.	.	.	.	.	.	.	.	.	.
225	94	.	.	.	.	.	.	.	.	.	.	.	.	.	.	.	.	.	.
250	96	.	.	.	.	.	.	.	.	.	.	.	.	.	.	.	.	.	.
300	99	.	.	.	.	.	.	.	.	.	.	.	.	.	.	.	.	.	.
350	99	.	.	.	.	.	.	.	.	.	.	.	.	.	.	.	.	.	.
400	.	.	.	.	.	.	.	.	.	.	.	.	.	.	.	.	.	.	.
450	.	.	.	.	.	.	.	.	.	.	.	.	.	.	.	.	.	.	.
500	.	.	.	.	.	.	.	.	.	.	.	.	.	.	.	.	.	.	.
600	.	.	.	.	.	.	.	.	.	.	.	.	.	.	.	.	.	.	.
700	.	.	.	.	.	.	.	.	.	.	.	.	.	.	.	.	.	.	.
800	.	.	.	.	.	.	.	.	.	.	.	.	.	.	.	.	.	.	.
900	.	.	.	.	.	.	.	.	.	.	.	.	.	.	.	.	.	.	.
1000	.	.	.	.	.	.	.	.	.	.	.	.	.	.	.	.	.	.	.

Table 8.5. Power table for repeated measures ANOVA; $r = 0.40$, pattern L, 4 groups at alpha = 0.05

n	Hypothesized ES																		
	0.20	0.30	0.35	0.40	0.45	0.50	0.55	0.60	0.65	0.70	0.75	0.80	1.00	1.25	1.50	1.75	2.00	2.50	3.00
5	6	7	7	8	9	10	11	12	14	15	17	19	28	42	58	72	84	97	.
6	6	7	8	9	10	11	13	15	17	19	21	23	35	52	70	83	92	99	.
7	6	7	9	10	11	13	15	17	19	22	25	28	42	62	79	90	97	.	.
8	6	8	9	11	12	15	17	19	22	25	29	33	49	70	86	95	99	.	.
9	6	8	10	12	14	16	19	22	25	29	33	37	55	76	90	97	99	.	.
10	6	9	11	13	15	18	21	24	28	32	37	42	61	82	94	99	.	.	.
11	7	9	11	14	16	20	23	27	31	36	41	46	66	86	96	99	.	.	.
12	7	10	12	15	18	21	25	30	34	39	45	50	71	89	98	.	.	.	.
13	7	10	13	16	19	23	27	32	37	43	48	54	75	92	98	.	.	.	.
14	7	11	14	17	21	25	30	35	40	46	52	58	79	94	99	.	.	.	.
15	7	12	14	18	22	27	32	37	43	49	56	62	82	96	99	.	.	.	.
20	9	14	19	24	29	36	42	50	57	64	70	76	93	99	.	.	.	.	.
25	10	18	23	29	37	44	52	60	68	75	81	86	97	.	.	.	.	.	.
30	11	21	27	35	44	52	61	70	77	83	89	92	99	.	.	.	.	.	.
35	12	24	32	41	50	60	69	77	84	89	93	96	.	.	.	.	.	.	.
40	14	27	36	46	56	66	75	83	89	93	96	98	.	.	.	.	.	.	.
45	15	30	40	51	62	72	81	88	93	96	98	99	.	.	.	.	.	.	.
50	16	34	45	56	67	77	85	91	95	97	99	99	.	.	.	.	.	.	.
55	18	37	49	61	72	82	89	94	97	98	99	.	.	.	.	.	.	.	.
60	19	40	53	65	76	85	91	96	98	99	.	.	.	.	.	.	.	.	.
65	20	43	56	69	80	88	94	97	99	99	.	.	.	.	.	.	.	.	.
70	22	46	60	73	83	91	95	98	99	.	.	.	.	.	.	.	.	.	.
75	23	49	63	76	86	93	97	99	99	.	.	.	.	.	.	.	.	.	.
80	25	52	66	79	88	94	97	99	.	.	.	.	.	.	.	.	.	.	.
90	28	57	72	84	92	96	99	.	.	.	.	.	.	.	.	.	.	.	.
100	31	62	77	88	94	98	99	.	.	.	.	.	.	.	.	.	.	.	.
110	33	67	81	91	96	99	.	.	.	.	.	.	.	.	.	.	.	.	.
120	36	71	85	93	98	99	.	.	.	.	.	.	.	.	.	.	.	.	.
130	39	75	88	95	98	.	.	.	.	.	.	.	.	.	.	.	.	.	.
140	42	78	90	96	99	.	.	.	.	.	.	.	.	.	.	.	.	.	.
150	45	81	92	97	99	.	.	.	.	.	.	.	.	.	.	.	.	.	.
175	51	87	96	99	.	.	.	.	.	.	.	.	.	.	.	.	.	.	.
200	57	92	98	.	.	.	.	.	.	.	.	.	.	.	.	.	.	.	.
225	63	95	99	.	.	.	.	.	.	.	.	.	.	.	.	.	.	.	.
250	68	97	99	.	.	.	.	.	.	.	.	.	.	.	.	.	.	.	.
300	76	99	.	.	.	.	.	.	.	.	.	.	.	.	.	.	.	.	.
350	83	.	.	.	.	.	.	.	.	.	.	.	.	.	.	.	.	.	.
400	88	.	.	.	.	.	.	.	.	.	.	.	.	.	.	.	.	.	.
450	92	.	.	.	.	.	.	.	.	.	.	.	.	.	.	.	.	.	.
500	94	.	.	.	.	.	.	.	.	.	.	.	.	.	.	.	.	.	.
600	98	.	.	.	.	.	.	.	.	.	.	.	.	.	.	.	.	.	.
700	99	.	.	.	.	.	.	.	.	.	.	.	.	.	.	.	.	.	.
800	.	.	.	.	.	.	.	.	.	.	.	.	.	.	.	.	.	.	.
900	.	.	.	.	.	.	.	.	.	.	.	.	.	.	.	.	.	.	.
1000	.	.	.	.	.	.	.	.	.	.	.	.	.	.	.	.	.	.	.

Table 8.6. Power table for repeated measures ANOVA; $r = 0.60$, pattern L, 4 groups at alpha = 0.05

n	Hypothesized ES																		
	0.20	0.30	0.35	0.40	0.45	0.50	0.55	0.60	0.65	0.70	0.75	0.80	1.00	1.25	1.50	1.75	2.00	2.50	3.00
5	6	8	9	10	11	13	15	17	19	21	24	27	41	60	77	89	96	.	.
6	6	8	10	11	13	15	18	20	23	27	30	34	51	72	87	96	99	.	.
7	7	9	11	13	15	18	21	24	28	32	36	41	60	81	93	98	.	.	.
8	7	10	12	14	17	20	24	28	32	37	42	47	68	87	97	99	.	.	.
9	7	10	13	16	19	23	27	32	37	42	48	53	74	92	98	.	.	.	.
10	7	11	14	17	21	25	30	36	41	47	53	59	80	95	99	.	.	.	.
11	8	12	15	19	23	28	33	39	46	52	58	64	85	97	.	.	.	.	.
12	8	13	16	21	25	31	37	43	50	56	63	69	88	98	.	.	.	.	.
13	8	14	18	22	28	33	40	47	54	61	67	73	91	99	.	.	.	.	.
14	9	15	19	24	30	36	43	50	57	65	71	77	93	99	.	.	.	.	.
15	9	16	20	26	32	39	46	54	61	68	75	81	95	.	.	.	.	.	.
20	11	20	27	34	43	51	60	68	76	82	88	92	99	.	.	.	.	.	.
25	13	25	33	43	53	62	71	79	86	91	94	97	.	.	.	.	.	.	.
30	15	30	40	51	61	72	80	87	92	96	98	99	.	.	.	.	.	.	.
35	17	35	46	58	69	79	87	92	96	98	99	.	.	.	.	.	.	.	.
40	19	40	52	65	76	85	91	95	98	99	.	.	.	.	.	.	.	.	.
45	21	44	58	70	81	89	94	97	99	.	.	.	.	.	.	.	.	.	.
50	23	49	63	75	85	92	96	98	99	.	.	.	.	.	.	.	.	.	.
55	25	53	67	80	89	95	98	99	.	.	.	.	.	.	.	.	.	.	.
60	27	57	72	84	92	96	99	.	.	.	.	.	.	.	.	.	.	.	.
65	30	61	76	87	94	97	99	.	.	.	.	.	.	.	.	.	.	.	.
70	32	65	79	89	95	98	99	.	.	.	.	.	.	.	.	.	.	.	.
75	34	68	82	91	97	99	.	.	.	.	.	.	.	.	.	.	.	.	.
80	36	71	85	93	97	99	.	.	.	.	.	.	.	.	.	.	.	.	.
90	40	77	89	96	99	.	.	.	.	.	.	.	.	.	.	.	.	.	.
100	44	81	92	97	99	.	.	.	.	.	.	.	.	.	.	.	.	.	.
110	48	85	94	98	.	.	.	.	.	.	.	.	.	.	.	.	.	.	.
120	52	88	96	99	.	.	.	.	.	.	.	.	.	.	.	.	.	.	.
130	56	91	97	99	.	.	.	.	.	.	.	.	.	.	.	.	.	.	.
140	59	93	98	.	.	.	.	.	.	.	.	.	.	.	.	.	.	.	.
150	63	95	99	.	.	.	.	.	.	.	.	.	.	.	.	.	.	.	.
175	70	97	.	.	.	.	.	.	.	.	.	.	.	.	.	.	.	.	.
200	76	99	.	.	.	.	.	.	.	.	.	.	.	.	.	.	.	.	.
225	82	99	.	.	.	.	.	.	.	.	.	.	.	.	.	.	.	.	.
250	86	.	.	.	.	.	.	.	.	.	.	.	.	.	.	.	.	.	.
300	92	.	.	.	.	.	.	.	.	.	.	.	.	.	.	.	.	.	.
350	95	.	.	.	.	.	.	.	.	.	.	.	.	.	.	.	.	.	.
400	98	.	.	.	.	.	.	.	.	.	.	.	.	.	.	.	.	.	.
450	99	.	.	.	.	.	.	.	.	.	.	.	.	.	.	.	.	.	.
500	99	.	.	.	.	.	.	.	.	.	.	.	.	.	.	.	.	.	.
600	.	.	.	.	.	.	.	.	.	.	.	.	.	.	.	.	.	.	.
700	.	.	.	.	.	.	.	.	.	.	.	.	.	.	.	.	.	.	.
800	.	.	.	.	.	.	.	.	.	.	.	.	.	.	.	.	.	.	.
900	.	.	.	.	.	.	.	.	.	.	.	.	.	.	.	.	.	.	.
1000	.	.	.	.	.	.	.	.	.	.	.	.	.	.	.	.	.	.	.

Table 8.7. Power table for repeated measures ANOVA; $r = 0.40$, pattern M, 4 groups at alpha = 0.05

n	Hypothesized ES																		
	0.20	0.30	0.35	0.40	0.45	0.50	0.55	0.60	0.65	0.70	0.75	0.80	1.00	1.25	1.50	1.75	2.00	2.50	3.00
5	6	7	8	9	10	11	12	13	15	17	19	21	31	46	63	77	88	98	.
6	6	7	8	9	11	12	14	16	18	20	23	26	39	57	75	87	95	.	.
7	6	8	9	10	12	14	16	19	21	24	27	31	46	67	83	93	98	.	.
8	6	8	10	11	13	16	18	21	25	28	32	36	53	75	89	97	99	.	.
9	6	9	11	13	15	18	21	24	28	32	36	41	60	81	93	98	.	.	.
10	7	9	11	14	16	19	23	27	31	36	41	46	66	86	96	99	.	.	.
11	7	10	12	15	18	21	25	30	35	40	45	50	71	90	98	.	.	.	.
12	7	11	13	16	19	23	28	33	38	43	49	55	76	93	99	.	.	.	.
13	7	11	14	17	21	25	30	36	41	47	53	59	80	95	99	.	.	.	.
14	8	12	15	18	23	27	33	38	44	51	57	63	83	96	.	.	.	.	.
15	8	12	16	20	24	29	35	41	48	54	60	67	86	97	.	.	.	.	.
20	9	16	20	26	32	39	47	54	62	69	75	81	95	.	.	.	.	.	.
25	10	19	25	32	40	49	57	65	73	80	85	90	98	.	.	.	.	.	.
30	12	23	30	39	48	57	66	75	82	87	92	95	.	.	.	.	.	.	.
35	13	26	35	45	55	65	74	82	88	92	96	98	.	.	.	.	.	.	.
40	15	30	40	51	61	72	80	87	92	96	98	99	.	.	.	.	.	.	.
45	16	34	45	56	67	77	85	91	95	97	99	99	.	.	.	.	.	.	.
50	18	37	49	61	72	82	89	94	97	99	99	.	.	.	.	.	.	.	.
55	19	41	53	66	77	86	92	96	98	99	.	.	.	.	.	.	.	.	.
60	21	44	57	70	81	89	94	97	99	.	.	.	.	.	.	.	.	.	.
65	22	47	61	74	84	91	96	98	99	.	.	.	.	.	.	.	.	.	.
70	24	51	65	77	87	93	97	99	.	.	.	.	.	.	.	.	.	.	.
75	26	54	68	81	89	95	98	99	.	.	.	.	.	.	.	.	.	.	.
80	27	57	71	83	91	96	99	.	.	.	.	.	.	.	.	.	.	.	.
90	30	62	77	88	94	98	99	.	.	.	.	.	.	.	.	.	.	.	.
100	34	67	82	91	96	99	.	.	.	.	.	.	.	.	.	.	.	.	.
110	37	72	85	94	98	99	.	.	.	.	.	.	.	.	.	.	.	.	.
120	40	76	89	96	99	.	.	.	.	.	.	.	.	.	.	.	.	.	.
130	43	80	91	97	99	.	.	.	.	.	.	.	.	.	.	.	.	.	.
140	46	83	93	98	99	.	.	.	.	.	.	.	.	.	.	.	.	.	.
150	49	86	95	99	.	.	.	.	.	.	.	.	.	.	.	.	.	.	.
175	56	91	97	99	.	.	.	.	.	.	.	.	.	.	.	.	.	.	.
200	62	94	99	.	.	.	.	.	.	.	.	.	.	.	.	.	.	.	.
225	68	97	99	.	.	.	.	.	.	.	.	.	.	.	.	.	.	.	.
250	73	98	.	.	.	.	.	.	.	.	.	.	.	.	.	.	.	.	.
300	81	99	.	.	.	.	.	.	.	.	.	.	.	.	.	.	.	.	.
350	87	.	.	.	.	.	.	.	.	.	.	.	.	.	.	.	.	.	.
400	91	.	.	.	.	.	.	.	.	.	.	.	.	.	.	.	.	.	.
450	94	.	.	.	.	.	.	.	.	.	.	.	.	.	.	.	.	.	.
500	96	.	.	.	.	.	.	.	.	.	.	.	.	.	.	.	.	.	.
600	99	.	.	.	.	.	.	.	.	.	.	.	.	.	.	.	.	.	.
700	99	.	.	.	.	.	.	.	.	.	.	.	.	.	.	.	.	.	.
800	.	.	.	.	.	.	.	.	.	.	.	.	.	.	.	.	.	.	.
900	.	.	.	.	.	.	.	.	.	.	.	.	.	.	.	.	.	.	.
1000	.	.	.	.	.	.	.	.	.	.	.	.	.	.	.	.	.	.	.

Table 8.8. Power table for repeated measures ANOVA; $r = 0.60$, pattern M, 4 groups at alpha $= 0.05$

n	Hypothesized ES																		
	0.20	0.30	0.35	0.40	0.45	0.50	0.55	0.60	0.65	0.70	0.75	0.80	1.00	1.25	1.50	1.75	2.00	2.50	3.00
5	6	8	9	10	12	14	16	18	21	23	27	30	45	65	81	92	97	.	.
6	7	9	10	12	14	16	19	22	26	29	33	37	55	76	91	97	99	.	.
7	7	9	11	14	16	19	23	26	31	35	40	45	65	85	95	99	.	.	.
8	7	10	13	15	18	22	26	31	36	41	46	52	73	91	98	.	.	.	.
9	7	11	14	17	21	25	30	35	41	46	52	58	79	94	99	.	.	.	.
10	8	12	15	19	23	28	33	39	45	52	58	64	84	97	.	.	.	.	.
11	8	13	17	21	25	31	37	43	50	57	63	69	88	98	.	.	.	.	.
12	8	14	18	23	28	34	40	47	54	61	68	74	91	99	.	.	.	.	.
13	9	15	19	24	30	37	44	51	59	66	72	78	94	99	.	.	.	.	.
14	9	16	21	26	33	40	47	55	62	70	76	82	96	.	.	.	.	.	.
15	10	17	22	28	35	43	51	59	66	73	79	85	97	.	.	.	.	.	.
20	12	22	29	38	47	56	65	73	81	87	91	94	99	.	.	.	.	.	.
25	14	28	37	47	57	67	76	84	90	94	96	98	.	.	.	.	.	.	.
30	16	33	44	55	67	76	85	91	95	97	99	99	.	.	.	.	.	.	.
35	18	38	51	63	74	83	90	95	97	99	.	.	.	.	.	.	.	.	.
40	21	44	57	70	80	88	94	97	99	.	.	.	.	.	.	.	.	.	.
45	23	49	63	75	85	92	96	98	99	.	.	.	.	.	.	.	.	.	.
50	25	53	68	80	89	95	98	99	.	.	.	.	.	.	.	.	.	.	.
55	28	58	72	84	92	97	99	.	.	.	.	.	.	.	.	.	.	.	.
60	30	62	77	87	94	98	99	.	.	.	.	.	.	.	.	.	.	.	.
65	33	66	80	90	96	99	.	.	.	.	.	.	.	.	.	.	.	.	.
70	35	70	83	92	97	99	.	.	.	.	.	.	.	.	.	.	.	.	.
75	37	73	86	94	98	99	.	.	.	.	.	.	.	.	.	.	.	.	.
80	40	76	88	95	99	.	.	.	.	.	.	.	.	.	.	.	.	.	.
90	44	81	92	97	99	.	.	.	.	.	.	.	.	.	.	.	.	.	.
100	49	85	95	98	.	.	.	.	.	.	.	.	.	.	.	.	.	.	.
110	53	89	96	99	.	.	.	.	.	.	.	.	.	.	.	.	.	.	.
120	57	92	98	.	.	.	.	.	.	.	.	.	.	.	.	.	.	.	.
130	61	94	98	.	.	.	.	.	.	.	.	.	.	.	.	.	.	.	.
140	64	95	99	.	.	.	.	.	.	.	.	.	.	.	.	.	.	.	.
150	68	97	99	.	.	.	.	.	.	.	.	.	.	.	.	.	.	.	.
175	75	98	.	.	.	.	.	.	.	.	.	.	.	.	.	.	.	.	.
200	81	99	.	.	.	.	.	.	.	.	.	.	.	.	.	.	.	.	.
225	86	.	.	.	.	.	.	.	.	.	.	.	.	.	.	.	.	.	.
250	89	.	.	.	.	.	.	.	.	.	.	.	.	.	.	.	.	.	.
300	94	.	.	.	.	.	.	.	.	.	.	.	.	.	.	.	.	.	.
350	97	.	.	.	.	.	.	.	.	.	.	.	.	.	.	.	.	.	.
400	99	.	.	.	.	.	.	.	.	.	.	.	.	.	.	.	.	.	.
450	99	.	.	.	.	.	.	.	.	.	.	.	.	.	.	.	.	.	.
500	.	.	.	.	.	.	.	.	.	.	.	.	.	.	.	.	.	.	.
600	.	.	.	.	.	.	.	.	.	.	.	.	.	.	.	.	.	.	.
700	.	.	.	.	.	.	.	.	.	.	.	.	.	.	.	.	.	.	.
800	.	.	.	.	.	.	.	.	.	.	.	.	.	.	.	.	.	.	.
900	.	.	.	.	.	.	.	.	.	.	.	.	.	.	.	.	.	.	.
1000	.	.	.	.	.	.	.	.	.	.	.	.	.	.	.	.	.	.	.

Table 8.9. Power table for repeated measures ANOVA; $r = 0.40$, pattern H, 4 groups at alpha = 0.05

n	Hypothesized ES																	
	0.20	0.30	0.35	0.40	0.45	0.50	0.55	0.60	0.65	0.70	0.75	0.80	1.00	1.25	1.50	1.75	2.00	2.50	3.00
5	7	9	10	12	13	16	18	21	24	28	31	35	52	73	88	96	99	.	.
6	7	9	11	13	16	19	22	26	30	35	39	44	64	84	95	99	.	.	.
7	7	10	13	15	19	22	27	31	36	41	47	52	73	91	98	.	.	.	.
8	8	12	14	18	21	26	31	36	42	48	54	60	81	95	99	.	.	.	.
9	8	13	16	20	24	30	35	41	48	54	61	67	87	97	.	.	.	.	.
10	8	14	18	22	27	33	39	46	53	60	67	73	91	99	.	.	.	.	.
11	9	15	19	24	30	37	44	51	58	65	72	78	94	99	.	.	.	.	.
12	9	16	21	27	33	40	48	55	63	70	76	82	96	.	.	.	.	.	.
13	10	17	23	29	36	44	52	60	67	74	80	86	97	.	.	.	.	.	.
14	10	19	24	31	39	47	55	64	71	78	84	89	98	.	.	.	.	.	.
15	11	20	26	33	42	50	59	67	75	81	87	91	99	.	.	.	.	.	.
20	13	26	35	45	55	65	74	82	88	92	95	97	.	.	.	.	.	.	.
25	16	33	43	55	66	76	84	90	94	97	99	99	.	.	.	.	.	.	.
30	19	39	52	64	75	84	91	95	98	99	.	.	.	.	.	.	.	.	.
35	21	45	59	72	82	90	95	98	99	.	.	.	.	.	.	.	.	.	.
40	24	51	66	78	88	94	97	99	.	.	.	.	.	.	.	.	.	.	.
45	27	57	71	83	91	96	99	.	.	.	.	.	.	.	.	.	.	.	.
50	30	62	76	87	94	98	99	.	.	.	.	.	.	.	.	.	.	.	.
55	33	66	81	91	96	99	.	.	.	.	.	.	.	.	.	.	.	.	.
60	36	71	84	93	97	99	.	.	.	.	.	.	.	.	.	.	.	.	.
65	39	75	87	95	98	.	.	.	.	.	.	.	.	.	.	.	.	.	.
70	41	78	90	96	99	.	.	.	.	.	.	.	.	.	.	.	.	.	.
75	44	81	92	97	99	.	.	.	.	.	.	.	.	.	.	.	.	.	.
80	47	84	94	98	.	.	.	.	.	.	.	.	.	.	.	.	.	.	.
90	52	88	96	99	.	.	.	.	.	.	.	.	.	.	.	.	.	.	.
100	57	92	98	.	.	.	.	.	.	.	.	.	.	.	.	.	.	.	.
110	61	94	99	.	.	.	.	.	.	.	.	.	.	.	.	.	.	.	.
120	66	96	99	.	.	.	.	.	.	.	.	.	.	.	.	.	.	.	.
130	70	97	.	.	.	.	.	.	.	.	.	.	.	.	.	.	.	.	.
140	73	98	.	.	.	.	.	.	.	.	.	.	.	.	.	.	.	.	.
150	76	99	.	.	.	.	.	.	.	.	.	.	.	.	.	.	.	.	.
175	83	99	.	.	.	.	.	.	.	.	.	.	.	.	.	.	.	.	.
200	88	.	.	.	.	.	.	.	.	.	.	.	.	.	.	.	.	.	.
225	92	.	.	.	.	.	.	.	.	.	.	.	.	.	.	.	.	.	.
250	94	.	.	.	.	.	.	.	.	.	.	.	.	.	.	.	.	.	.
300	97	.	.	.	.	.	.	.	.	.	.	.	.	.	.	.	.	.	.
350	99	.	.	.	.	.	.	.	.	.	.	.	.	.	.	.	.	.	.
400	.	.	.	.	.	.	.	.	.	.	.	.	.	.	.	.	.	.	.
450	.	.	.	.	.	.	.	.	.	.	.	.	.	.	.	.	.	.	.
500	.	.	.	.	.	.	.	.	.	.	.	.	.	.	.	.	.	.	.
600	.	.	.	.	.	.	.	.	.	.	.	.	.	.	.	.	.	.	.
700	.	.	.	.	.	.	.	.	.	.	.	.	.	.	.	.	.	.	.
800	.	.	.	.	.	.	.	.	.	.	.	.	.	.	.	.	.	.	.
900	.	.	.	.	.	.	.	.	.	.	.	.	.	.	.	.	.	.	.
1000	.	.	.	.	.	.	.	.	.	.	.	.	.	.	.	.	.	.	.

Table 8.10. Power table for repeated measures ANOVA; $r = 0.60$, pattern H, 4 groups at alpha = 0.05

n	Hypothesized ES																		
	0.20	0.30	0.35	0.40	0.45	0.50	0.55	0.60	0.65	0.70	0.75	0.80	1.00	1.25	1.50	1.75	2.00	2.50	3.00
5	7	10	13	15	18	22	26	30	35	40	45	51	71	90	98	.	.	.	.
6	8	12	15	18	22	27	32	38	44	50	56	62	83	96	99	.	.	.	.
7	8	14	17	22	27	32	39	45	52	59	65	72	90	98	.	.	.	.	.
8	9	15	20	25	31	38	45	52	60	67	73	79	94	99	.	.	.	.	.
9	10	17	22	28	35	43	51	59	66	73	80	85	97	.	.	.	.	.	.
10	10	19	25	32	40	48	56	65	72	79	85	89	98	.	.	.	.	.	.
11	11	21	28	35	44	53	62	70	77	84	89	93	99	.	.	.	.	.	.
12	12	23	30	39	48	57	66	75	82	87	92	95	.	.	.	.	.	.	.
13	13	25	33	42	52	62	71	79	85	90	94	97	.	.	.	.	.	.	.
14	13	27	35	45	56	66	75	82	88	93	96	98	.	.	.	.	.	.	.
15	14	29	38	48	59	69	78	85	91	95	97	98	.	.	.	.	.	.	.
20	18	38	50	63	74	83	90	95	97	99	.	.	.	.	.	.	.	.	.
25	23	47	61	74	84	91	96	98	99	.	.	.	.	.	.	.	.	.	.
30	27	56	71	83	91	96	98	99	.	.	.	.	.	.	.	.	.	.	.
35	31	64	78	89	95	98	99	.	.	.	.	.	.	.	.	.	.	.	.
40	35	70	84	93	97	99	.	.	.	.	.	.	.	.	.	.	.	.	.
45	40	76	88	95	99	.	.	.	.	.	.	.	.	.	.	.	.	.	.
50	44	81	92	97	99	.	.	.	.	.	.	.	.	.	.	.	.	.	.
55	48	85	94	98	.	.	.	.	.	.	.	.	.	.	.	.	.	.	.
60	52	88	96	99	.	.	.	.	.	.	.	.	.	.	.	.	.	.	.
65	55	91	97	99	.	.	.	.	.	.	.	.	.	.	.	.	.	.	.
70	59	93	98	.	.	.	.	.	.	.	.	.	.	.	.	.	.	.	.
75	62	94	99	.	.	.	.	.	.	.	.	.	.	.	.	.	.	.	.
80	65	96	99	.	.	.	.	.	.	.	.	.	.	.	.	.	.	.	.
90	71	98	.	.	.	.	.	.	.	.	.	.	.	.	.	.	.	.	.
100	76	99	.	.	.	.	.	.	.	.	.	.	.	.	.	.	.	.	.
110	80	99	.	.	.	.	.	.	.	.	.	.	.	.	.	.	.	.	.
120	84	.	.	.	.	.	.	.	.	.	.	.	.	.	.	.	.	.	.
130	87	.	.	.	.	.	.	.	.	.	.	.	.	.	.	.	.	.	.
140	90	.	.	.	.	.	.	.	.	.	.	.	.	.	.	.	.	.	.
150	92	.	.	.	.	.	.	.	.	.	.	.	.	.	.	.	.	.	.
175	95	.	.	.	.	.	.	.	.	.	.	.	.	.	.	.	.	.	.
200	97	.	.	.	.	.	.	.	.	.	.	.	.	.	.	.	.	.	.
225	99	.	.	.	.	.	.	.	.	.	.	.	.	.	.	.	.	.	.
250	99	.	.	.	.	.	.	.	.	.	.	.	.	.	.	.	.	.	.
300	.	.	.	.	.	.	.	.	.	.	.	.	.	.	.	.	.	.	.
350	.	.	.	.	.	.	.	.	.	.	.	.	.	.	.	.	.	.	.
400	.	.	.	.	.	.	.	.	.	.	.	.	.	.	.	.	.	.	.
450	.	.	.	.	.	.	.	.	.	.	.	.	.	.	.	.	.	.	.
500	.	.	.	.	.	.	.	.	.	.	.	.	.	.	.	.	.	.	.
600	.	.	.	.	.	.	.	.	.	.	.	.	.	.	.	.	.	.	.
700	.	.	.	.	.	.	.	.	.	.	.	.	.	.	.	.	.	.	.
800	.	.	.	.	.	.	.	.	.	.	.	.	.	.	.	.	.	.	.
900	.	.	.	.	.	.	.	.	.	.	.	.	.	.	.	.	.	.	.
1000	.	.	.	.	.	.	.	.	.	.	.	.	.	.	.	.	.	.	.

Table 8.11. Power table for repeated measures ANOVA; $r = 0.40$, pattern L, 5 groups at alpha = 0.05

n	Hypothesized ES																		
	0.20	0.30	0.35	0.40	0.45	0.50	0.55	0.60	0.65	0.70	0.75	0.80	1.00	1.25	1.50	1.75	2.00	2.50	3.00
5	6	6	7	8	8	9	10	11	13	14	16	17	26	39	55	69	82	96	99
6	6	7	7	8	9	11	12	13	15	17	19	21	32	49	66	81	91	99	.
7	6	7	8	9	10	12	14	15	18	20	23	25	39	58	76	88	96	.	.
8	6	7	9	10	11	13	15	18	20	23	26	30	45	66	83	93	98	.	.
9	6	8	9	11	12	15	17	20	23	26	30	34	51	73	88	96	99	.	.
10	6	8	10	12	14	16	19	22	26	29	33	38	57	78	92	98	.	.	.
11	6	9	10	12	15	18	21	24	28	33	37	42	62	83	95	99	.	.	.
12	6	9	11	13	16	19	23	27	31	36	41	46	67	87	97	99	.	.	.
13	7	10	12	14	17	21	25	29	34	39	44	50	71	90	98	.	.	.	.
14	7	10	12	15	18	22	27	31	36	42	48	53	75	92	99	.	.	.	.
15	7	11	13	16	20	24	29	34	39	45	51	57	79	94	99	.	.	.	.
20	8	13	17	21	26	32	38	45	52	59	66	72	91	99	.	.	.	.	.
25	9	16	20	26	33	40	48	56	63	71	77	83	96	.	.	.	.	.	.
30	10	18	24	31	39	48	57	65	73	80	85	90	99	.	.	.	.	.	.
35	11	21	28	37	46	55	64	73	80	86	91	94	.	.	.	.	.	.	.
40	12	24	32	42	52	62	71	79	86	91	95	97	.	.	.	.	.	.	.
45	13	27	36	47	57	68	77	84	90	94	97	98	.	.	.	.	.	.	.
50	15	30	40	51	63	73	82	88	93	96	98	99	.	.	.	.	.	.	.
55	16	33	44	56	67	78	86	92	95	98	99	.	.	.	.	.	.	.	.
60	17	36	48	60	72	82	89	94	97	99	99	.	.	.	.	.	.	.	.
65	18	39	52	64	76	85	91	96	98	99	.	.	.	.	.	.	.	.	.
70	19	42	55	68	79	88	93	97	99	.	.	.	.	.	.	.	.	.	.
75	21	44	58	71	82	90	95	98	99	.	.	.	.	.	.	.	.	.	.
80	22	47	62	75	85	92	96	98	99	.	.	.	.	.	.	.	.	.	.
90	25	53	67	80	89	95	98	99	.	.	.	.	.	.	.	.	.	.	.
100	27	58	73	85	92	97	99	.	.	.	.	.	.	.	.	.	.	.	.
110	30	62	77	88	95	98	99	.	.	.	.	.	.	.	.	.	.	.	.
120	32	67	81	91	96	99	.	.	.	.	.	.	.	.	.	.	.	.	.
130	35	71	84	93	98	99	.	.	.	.	.	.	.	.	.	.	.	.	.
140	38	74	87	95	98	.	.	.	.	.	.	.	.	.	.	.	.	.	.
150	40	77	90	96	99	.	.	.	.	.	.	.	.	.	.	.	.	.	.
175	46	84	94	98	.	.	.	.	.	.	.	.	.	.	.	.	.	.	.
200	52	89	97	99	.	.	.	.	.	.	.	.	.	.	.	.	.	.	.
225	58	93	98	.	.	.	.	.	.	.	.	.	.	.	.	.	.	.	.
250	63	95	99	.	.	.	.	.	.	.	.	.	.	.	.	.	.	.	.
300	72	98	.	.	.	.	.	.	.	.	.	.	.	.	.	.	.	.	.
350	79	99	.	.	.	.	.	.	.	.	.	.	.	.	.	.	.	.	.
400	85	.	.	.	.	.	.	.	.	.	.	.	.	.	.	.	.	.	.
450	89	.	.	.	.	.	.	.	.	.	.	.	.	.	.	.	.	.	.
500	92	.	.	.	.	.	.	.	.	.	.	.	.	.	.	.	.	.	.
600	96	.	.	.	.	.	.	.	.	.	.	.	.	.	.	.	.	.	.
700	98	.	.	.	.	.	.	.	.	.	.	.	.	.	.	.	.	.	.
800	99	.	.	.	.	.	.	.	.	.	.	.	.	.	.	.	.	.	.
900	.	.	.	.	.	.	.	.	.	.	.	.	.	.	.	.	.	.	.
1000	.	.	.	.	.	.	.	.	.	.	.	.	.	.	.	.	.	.	.

Table 8.12. Power table for repeated measures ANOVA; $r = 0.60$, pattern L, 5 groups at alpha = 0.05

n	Hypothesized ES																		
	0.20	0.30	0.35	0.40	0.45	0.50	0.55	0.60	0.65	0.70	0.75	0.80	1.00	1.25	1.50	1.75	2.00	2.50	3.00
5	6	7	8	9	10	12	13	15	17	20	22	25	38	56	74	87	95	.	.
6	6	8	9	10	12	14	16	18	21	24	27	31	47	68	85	95	99	.	.
7	6	8	10	12	14	16	19	22	25	29	33	37	56	78	92	98	.	.	.
8	6	9	11	13	15	18	21	25	29	34	38	43	64	85	95	99	.	.	.
9	7	10	12	14	17	20	24	29	33	38	44	49	71	90	98	.	.	.	.
10	7	10	13	16	19	23	27	32	37	43	49	55	76	93	99	.	.	.	.
11	7	11	14	17	21	25	30	36	42	48	54	60	81	96	99	.	.	.	.
12	8	12	15	18	23	28	33	39	45	52	58	65	85	97	.	.	.	.	.
13	8	13	16	20	25	30	36	43	49	56	63	69	89	98	.	.	.	.	.
14	8	13	17	21	27	33	39	46	53	60	67	73	91	99	.	.	.	.	.
15	8	14	18	23	29	35	42	49	57	64	71	77	93	99	.	.	.	.	.
20	10	18	24	31	38	47	55	64	72	79	85	89	98	.	.	.	.	.	.
25	12	22	30	39	48	58	67	75	83	88	93	96	.	.	.	.	.	.	.
30	13	27	36	46	57	67	76	84	90	94	97	98	.	.	.	.	.	.	.
35	15	31	42	53	65	75	83	90	94	97	99	99	.	.	.	.	.	.	.
40	17	36	47	60	71	81	89	94	97	99	99	.	.	.	.	.	.	.	.
45	19	40	53	66	77	86	92	96	98	99	.	.	.	.	.	.	.	.	.
50	21	44	58	71	82	90	95	98	99	.	.	.	.	.	.	.	.	.	.
55	22	48	63	76	86	93	97	99	.	.	.	.	.	.	.	.	.	.	.
60	24	52	67	80	89	95	98	99	.	.	.	.	.	.	.	.	.	.	.
65	26	56	71	83	92	96	99	.	.	.	.	.	.	.	.	.	.	.	.
70	28	60	75	86	94	97	99	.	.	.	.	.	.	.	.	.	.	.	.
75	30	63	78	89	95	98	99	.	.	.	.	.	.	.	.	.	.	.	.
80	32	66	81	91	96	99	.	.	.	.	.	.	.	.	.	.	.	.	.
90	36	72	86	94	98	99	.	.	.	.	.	.	.	.	.	.	.	.	.
100	40	77	90	96	99	.	.	.	.	.	.	.	.	.	.	.	.	.	.
110	44	81	92	98	99	.	.	.	.	.	.	.	.	.	.	.	.	.	.
120	47	85	95	99	.	.	.	.	.	.	.	.	.	.	.	.	.	.	.
130	51	88	96	99	.	.	.	.	.	.	.	.	.	.	.	.	.	.	.
140	54	91	97	99	.	.	.	.	.	.	.	.	.	.	.	.	.	.	.
150	58	93	98	.	.	.	.	.	.	.	.	.	.	.	.	.	.	.	.
175	65	96	99	.	.	.	.	.	.	.	.	.	.	.	.	.	.	.	.
200	72	98	.	.	.	.	.	.	.	.	.	.	.	.	.	.	.	.	.
225	78	99	.	.	.	.	.	.	.	.	.	.	.	.	.	.	.	.	.
250	82	99	.	.	.	.	.	.	.	.	.	.	.	.	.	.	.	.	.
300	89	.	.	.	.	.	.	.	.	.	.	.	.	.	.	.	.	.	.
350	94	.	.	.	.	.	.	.	.	.	.	.	.	.	.	.	.	.	.
400	96	.	.	.	.	.	.	.	.	.	.	.	.	.	.	.	.	.	.
450	98	.	.	.	.	.	.	.	.	.	.	.	.	.	.	.	.	.	.
500	99	.	.	.	.	.	.	.	.	.	.	.	.	.	.	.	.	.	.
600	.	.	.	.	.	.	.	.	.	.	.	.	.	.	.	.	.	.	.
700	.	.	.	.	.	.	.	.	.	.	.	.	.	.	.	.	.	.	.
800	.	.	.	.	.	.	.	.	.	.	.	.	.	.	.	.	.	.	.
900	.	.	.	.	.	.	.	.	.	.	.	.	.	.	.	.	.	.	.
1000	.	.	.	.	.	.	.	.	.	.	.	.	.	.	.	.	.	.	.

Table 8.13. Power table for repeated measures ANOVA; $r = 0.40$, pattern M, 5 groups at alpha = 0.05

n	Hypothesized ES																		
	0.20	0.30	0.35	0.40	0.45	0.50	0.55	0.60	0.65	0.70	0.75	0.80	1.00	1.25	1.50	1.75	2.00	2.50	3.00
5	6	7	8	8	9	11	12	13	15	17	19	21	32	48	65	80	90	99	.
6	6	7	8	9	11	12	14	16	18	21	23	26	40	59	77	89	96	.	.
7	6	8	9	10	12	14	16	19	21	24	28	31	48	69	85	95	99	.	.
8	6	8	10	11	13	16	18	21	25	28	32	36	55	77	91	98	.	.	.
9	6	9	10	12	15	17	21	24	28	32	37	42	62	83	95	99	.	.	.
10	7	9	11	13	16	19	23	27	31	36	41	47	68	88	97	99	.	.	.
11	7	10	12	15	18	21	25	30	35	40	46	51	73	91	98	.	.	.	.
12	7	10	13	16	19	23	28	33	38	44	50	56	78	94	99	.	.	.	.
13	7	11	14	17	21	25	30	36	42	48	54	60	82	96	99	.	.	.	.
14	7	12	15	18	23	27	33	39	45	51	58	64	85	97	.	.	.	.	.
15	8	12	16	19	24	29	35	42	48	55	62	68	88	98	.	.	.	.	.
20	9	16	20	26	32	40	47	55	63	70	77	82	96	.	.	.	.	.	.
25	10	19	25	32	41	49	58	67	74	81	87	91	99	.	.	.	.	.	.
30	12	23	30	39	48	58	67	76	83	89	93	96	.	.	.	.	.	.	.
35	13	26	35	45	56	66	75	83	89	93	96	98	.	.	.	.	.	.	.
40	14	30	40	51	62	73	81	88	93	96	98	99	.	.	.	.	.	.	.
45	16	34	45	57	68	78	86	92	96	98	99	.	.	.	.	.	.	.	.
50	18	37	49	62	74	83	90	95	97	99	.	.	.	.	.	.	.	.	.
55	19	41	54	67	78	87	93	97	99	99	.	.	.	.	.	.	.	.	.
60	21	44	58	71	82	90	95	98	99	.	.	.	.	.	.	.	.	.	.
65	22	48	62	75	85	92	97	99	.	.	.	.	.	.	.	.	.	.	.
70	24	51	66	79	88	94	98	99	.	.	.	.	.	.	.	.	.	.	.
75	25	54	69	82	90	96	98	99	.	.	.	.	.	.	.	.	.	.	.
80	27	57	72	84	92	97	99	.	.	.	.	.	.	.	.	.	.	.	.
90	30	63	78	89	95	98	99	.	.	.	.	.	.	.	.	.	.	.	.
100	34	68	83	92	97	99	.	.	.	.	.	.	.	.	.	.	.	.	.
110	37	73	87	95	98	.	.	.	.	.	.	.	.	.	.	.	.	.	.
120	40	77	90	96	99	.	.	.	.	.	.	.	.	.	.	.	.	.	.
130	43	81	92	97	99	.	.	.	.	.	.	.	.	.	.	.	.	.	.
140	46	84	94	98	.	.	.	.	.	.	.	.	.	.	.	.	.	.	.
150	49	87	95	99	.	.	.	.	.	.	.	.	.	.	.	.	.	.	.
175	56	92	98	.	.	.	.	.	.	.	.	.	.	.	.	.	.	.	.
200	63	95	99	.	.	.	.	.	.	.	.	.	.	.	.	.	.	.	.
225	69	97	.	.	.	.	.	.	.	.	.	.	.	.	.	.	.	.	.
250	74	98	.	.	.	.	.	.	.	.	.	.	.	.	.	.	.	.	.
300	82	.	.	.	.	.	.	.	.	.	.	.	.	.	.	.	.	.	.
350	88	.	.	.	.	.	.	.	.	.	.	.	.	.	.	.	.	.	.
400	92	.	.	.	.	.	.	.	.	.	.	.	.	.	.	.	.	.	.
450	95	.	.	.	.	.	.	.	.	.	.	.	.	.	.	.	.	.	.
500	97	.	.	.	.	.	.	.	.	.	.	.	.	.	.	.	.	.	.
600	99	.	.	.	.	.	.	.	.	.	.	.	.	.	.	.	.	.	.
700	.	.	.	.	.	.	.	.	.	.	.	.	.	.	.	.	.	.	.
800	.	.	.	.	.	.	.	.	.	.	.	.	.	.	.	.	.	.	.
900	.	.	.	.	.	.	.	.	.	.	.	.	.	.	.	.	.	.	.
1000	.	.	.	.	.	.	.	.	.	.	.	.	.	.	.	.	.	.	.

Table 8.14. Power table for repeated measures ANOVA; $r = 0.60$, pattern M, 5 groups at alpha = 0.05

n	Hypothesized ES																		
	0.20	0.30	0.35	0.40	0.45	0.50	0.55	0.60	0.65	0.70	0.75	0.80	1.00	1.25	1.50	1.75	2.00	2.50	3.00
5	6	8	9	10	12	14	16	18	21	24	27	31	46	67	84	94	98	.	.
6	6	9	10	12	14	16	19	22	26	30	34	38	57	79	92	98	.	.	.
7	7	9	11	13	16	19	23	27	31	36	41	46	67	87	97	99	.	.	.
8	7	10	12	15	18	22	26	31	36	42	47	53	75	92	99	.	.	.	.
9	7	11	14	17	21	25	30	35	41	47	53	60	81	95	99	.	.	.	.
10	8	12	15	19	23	28	34	40	46	53	59	66	86	97	.	.	.	.	.
11	8	13	16	21	26	31	37	44	51	58	65	71	90	99	.	.	.	.	.
12	8	14	18	22	28	34	41	48	55	63	69	76	93	99	.	.	.	.	.
13	9	15	19	24	30	37	45	52	60	67	74	80	95	.	.	.	.	.	.
14	9	16	21	26	33	40	48	56	64	71	78	83	96	.	.	.	.	.	.
15	9	17	22	28	35	43	51	60	67	75	81	86	98	.	.	.	.	.	.
20	11	22	29	38	47	57	66	75	82	88	92	95	.	.	.	.	.	.	.
25	14	28	37	47	58	69	78	85	91	95	97	99	.	.	.	.	.	.	.
30	16	33	44	56	68	78	86	92	96	98	99	.	.	.	.	.	.	.	.
35	18	39	51	64	75	85	91	96	98	99	.	.	.	.	.	.	.	.	.
40	20	44	58	71	82	90	95	98	99	.	.	.	.	.	.	.	.	.	.
45	23	49	64	76	86	93	97	99	.	.	.	.	.	.	.	.	.	.	.
50	25	54	69	81	90	96	98	99	.	.	.	.	.	.	.	.	.	.	.
55	28	59	74	85	93	97	99	.	.	.	.	.	.	.	.	.	.	.	.
60	30	63	78	89	95	98	99	.	.	.	.	.	.	.	.	.	.	.	.
65	33	67	81	91	97	99	.	.	.	.	.	.	.	.	.	.	.	.	.
70	35	71	85	93	98	99	.	.	.	.	.	.	.	.	.	.	.	.	.
75	37	74	87	95	98	.	.	.	.	.	.	.	.	.	.	.	.	.	.
80	40	77	89	96	99	.	.	.	.	.	.	.	.	.	.	.	.	.	.
90	45	82	93	98	.	.	.	.	.	.	.	.	.	.	.	.	.	.	.
100	49	87	95	99	.	.	.	.	.	.	.	.	.	.	.	.	.	.	.
110	53	90	97	99	.	.	.	.	.	.	.	.	.	.	.	.	.	.	.
120	58	93	98	.	.	.	.	.	.	.	.	.	.	.	.	.	.	.	.
130	62	95	99	.	.	.	.	.	.	.	.	.	.	.	.	.	.	.	.
140	65	96	99	.	.	.	.	.	.	.	.	.	.	.	.	.	.	.	.
150	69	97	.	.	.	.	.	.	.	.	.	.	.	.	.	.	.	.	.
175	76	99	.	.	.	.	.	.	.	.	.	.	.	.	.	.	.	.	.
200	82	99	.	.	.	.	.	.	.	.	.	.	.	.	.	.	.	.	.
225	87	.	.	.	.	.	.	.	.	.	.	.	.	.	.	.	.	.	.
250	90	.	.	.	.	.	.	.	.	.	.	.	.	.	.	.	.	.	.
300	95	.	.	.	.	.	.	.	.	.	.	.	.	.	.	.	.	.	.
350	98	.	.	.	.	.	.	.	.	.	.	.	.	.	.	.	.	.	.
400	99	.	.	.	.	.	.	.	.	.	.	.	.	.	.	.	.	.	.
450	.	.	.	.	.	.	.	.	.	.	.	.	.	.	.	.	.	.	.
500	.	.	.	.	.	.	.	.	.	.	.	.	.	.	.	.	.	.	.
600	.	.	.	.	.	.	.	.	.	.	.	.	.	.	.	.	.	.	.
700	.	.	.	.	.	.	.	.	.	.	.	.	.	.	.	.	.	.	.
800	.	.	.	.	.	.	.	.	.	.	.	.	.	.	.	.	.	.	.
900	.	.	.	.	.	.	.	.	.	.	.	.	.	.	.	.	.	.	.
1000	.	.	.	.	.	.	.	.	.	.	.	.	.	.	.	.	.	.	.

Table 8.15. Power table for repeated measures ANOVA; $r = 0.40$, pattern H, 5 groups at alpha = 0.05

n	Hypothesized ES																	
	0.20	0.30	0.35	0.40	0.45	0.50	0.55	0.60	0.65	0.70	0.75	0.80	1.00	1.25	1.50	1.75	2.00	2.50	3.00
5	7	9	10	12	14	17	19	23	26	30	34	39	58	79	93	98	.	.	.
6	7	10	12	14	17	20	24	28	33	38	43	48	69	89	97	.	.	.	.
7	7	11	13	16	20	24	29	34	39	45	51	57	79	94	99	.	.	.	.
8	8	12	15	19	23	28	33	39	46	52	59	65	85	97	.	.	.	.	.
9	8	13	17	21	26	32	38	45	52	59	65	72	90	99	.	.	.	.	.
10	9	14	18	23	29	36	43	50	57	65	71	78	94	99	.	.	.	.	.
11	9	16	20	26	32	39	47	55	63	70	77	82	96	.	.	.	.	.	.
12	9	17	22	28	35	43	51	60	67	75	81	86	98	.	.	.	.	.	.
13	10	18	24	31	39	47	56	64	72	79	85	89	98	.	.	.	.	.	.
14	10	20	26	33	42	51	59	68	76	82	88	92	99	.	.	.	.	.	.
15	11	21	28	36	45	54	63	72	79	85	90	94	99	.	.	.	.	.	.
20	14	28	37	48	59	69	78	86	91	95	97	99	.	.	.	.	.	.	.
25	17	35	47	59	70	80	88	93	97	98	99	.	.	.	.	.	.	.	.
30	20	42	55	68	79	88	94	97	99	.	.	.	.	.	.	.	.	.	.
35	23	48	63	76	86	93	97	99	.	.	.	.	.	.	.	.	.	.	.
40	26	55	70	82	91	96	98	99	.	.	.	.	.	.	.	.	.	.	.
45	29	61	76	87	94	98	99	.	.	.	.	.	.	.	.	.	.	.	.
50	32	66	80	91	96	99	.	.	.	.	.	.	.	.	.	.	.	.	.
55	35	71	85	93	98	99	.	.	.	.	.	.	.	.	.	.	.	.	.
60	38	75	88	95	99	.	.	.	.	.	.	.	.	.	.	.	.	.	.
65	41	79	91	97	99	.	.	.	.	.	.	.	.	.	.	.	.	.	.
70	44	82	93	98	99	.	.	.	.	.	.	.	.	.	.	.	.	.	.
75	47	85	94	98	.	.	.	.	.	.	.	.	.	.	.	.	.	.	.
80	50	87	96	99	.	.	.	.	.	.	.	.	.	.	.	.	.	.	.
90	56	91	98	.	.	.	.	.	.	.	.	.	.	.	.	.	.	.	.
100	61	94	99	.	.	.	.	.	.	.	.	.	.	.	.	.	.	.	.
110	65	96	99	.	.	.	.	.	.	.	.	.	.	.	.	.	.	.	.
120	70	97	.	.	.	.	.	.	.	.	.	.	.	.	.	.	.	.	.
130	74	98	.	.	.	.	.	.	.	.	.	.	.	.	.	.	.	.	.
140	77	99	.	.	.	.	.	.	.	.	.	.	.	.	.	.	.	.	.
150	80	99	.	.	.	.	.	.	.	.	.	.	.	.	.	.	.	.	.
175	87	.	.	.	.	.	.	.	.	.	.	.	.	.	.	.	.	.	.
200	91	.	.	.	.	.	.	.	.	.	.	.	.	.	.	.	.	.	.
225	94	.	.	.	.	.	.	.	.	.	.	.	.	.	.	.	.	.	.
250	96	.	.	.	.	.	.	.	.	.	.	.	.	.	.	.	.	.	.
300	99	.	.	.	.	.	.	.	.	.	.	.	.	.	.	.	.	.	.
350	99	.	.	.	.	.	.	.	.	.	.	.	.	.	.	.	.	.	.
400	.	.	.	.	.	.	.	.	.	.	.	.	.	.	.	.	.	.	.
450	.	.	.	.	.	.	.	.	.	.	.	.	.	.	.	.	.	.	.
500	.	.	.	.	.	.	.	.	.	.	.	.	.	.	.	.	.	.	.
600	.	.	.	.	.	.	.	.	.	.	.	.	.	.	.	.	.	.	.
700	.	.	.	.	.	.	.	.	.	.	.	.	.	.	.	.	.	.	.
800	.	.	.	.	.	.	.	.	.	.	.	.	.	.	.	.	.	.	.
900	.	.	.	.	.	.	.	.	.	.	.	.	.	.	.	.	.	.	.
1000	.	.	.	.	.	.	.	.	.	.	.	.	.	.	.	.	.	.	.

Table 8.16. Power table for repeated measures ANOVA; $r = 0.60$, pattern H, 5 groups at alpha = 0.05

n	Hypothesized ES																		
	0.20	0.30	0.35	0.40	0.45	0.50	0.55	0.60	0.65	0.70	0.75	0.80	1.00	1.25	1.50	1.75	2.00	2.50	3.00
5	7	11	13	16	19	23	28	33	38	44	50	56	77	94	99	.	.	.	.
6	8	12	16	19	24	29	35	41	48	54	61	67	87	98	.	.	.	.	.
7	9	14	18	23	29	35	42	49	57	64	71	77	93	99	.	.	.	.	.
8	9	16	21	27	33	41	49	57	64	72	78	84	97	.	.	.	.	.	.
9	10	18	24	30	38	46	55	63	71	78	84	89	98	.	.	.	.	.	.
10	11	20	27	34	43	52	61	69	77	84	89	93	99	.	.	.	.	.	.
11	12	22	29	38	47	57	66	75	82	88	92	95	.	.	.	.	.	.	.
12	12	24	32	42	52	62	71	79	86	91	95	97	.	.	.	.	.	.	.
13	13	26	35	45	56	66	75	83	89	93	96	98	.	.	.	.	.	.	.
14	14	28	38	49	60	70	79	86	92	95	98	99	.	.	.	.	.	.	.
15	15	30	41	52	63	74	82	89	94	97	98	99	.	.	.	.	.	.	.
20	19	41	54	67	78	87	93	97	99	99	.	.	.	.	.	.	.	.	.
25	24	51	66	78	88	94	98	99	.	.	.	.	.	.	.	.	.	.	.
30	28	60	75	86	94	98	99	.	.	.	.	.	.	.	.	.	.	.	.
35	33	68	82	92	97	99	.	.	.	.	.	.	.	.	.	.	.	.	.
40	38	74	88	95	98	.	.	.	.	.	.	.	.	.	.	.	.	.	.
45	42	80	91	97	99	.	.	.	.	.	.	.	.	.	.	.	.	.	.
50	47	85	94	98	.	.	.	.	.	.	.	.	.	.	.	.	.	.	.
55	51	88	96	99	.	.	.	.	.	.	.	.	.	.	.	.	.	.	.
60	55	91	98	.	.	.	.	.	.	.	.	.	.	.	.	.	.	.	.
65	59	93	98	.	.	.	.	.	.	.	.	.	.	.	.	.	.	.	.
70	63	95	99	.	.	.	.	.	.	.	.	.	.	.	.	.	.	.	.
75	66	96	99	.	.	.	.	.	.	.	.	.	.	.	.	.	.	.	.
80	69	97	.	.	.	.	.	.	.	.	.	.	.	.	.	.	.	.	.
90	75	99	.	.	.	.	.	.	.	.	.	.	.	.	.	.	.	.	.
100	80	99	.	.	.	.	.	.	.	.	.	.	.	.	.	.	.	.	.
110	84	.	.	.	.	.	.	.	.	.	.	.	.	.	.	.	.	.	.
120	88	.	.	.	.	.	.	.	.	.	.	.	.	.	.	.	.	.	.
130	90	.	.	.	.	.	.	.	.	.	.	.	.	.	.	.	.	.	.
140	92	.	.	.	.	.	.	.	.	.	.	.	.	.	.	.	.	.	.
150	94	.	.	.	.	.	.	.	.	.	.	.	.	.	.	.	.	.	.
175	97	.	.	.	.	.	.	.	.	.	.	.	.	.	.	.	.	.	.
200	99	.	.	.	.	.	.	.	.	.	.	.	.	.	.	.	.	.	.
225	99	.	.	.	.	.	.	.	.	.	.	.	.	.	.	.	.	.	.
250	.	.	.	.	.	.	.	.	.	.	.	.	.	.	.	.	.	.	.
300	.	.	.	.	.	.	.	.	.	.	.	.	.	.	.	.	.	.	.
350	.	.	.	.	.	.	.	.	.	.	.	.	.	.	.	.	.	.	.
400	.	.	.	.	.	.	.	.	.	.	.	.	.	.	.	.	.	.	.
450	.	.	.	.	.	.	.	.	.	.	.	.	.	.	.	.	.	.	.
500	.	.	.	.	.	.	.	.	.	.	.	.	.	.	.	.	.	.	.
600	.	.	.	.	.	.	.	.	.	.	.	.	.	.	.	.	.	.	.
700	.	.	.	.	.	.	.	.	.	.	.	.	.	.	.	.	.	.	.
800	.	.	.	.	.	.	.	.	.	.	.	.	.	.	.	.	.	.	.
900	.	.	.	.	.	.	.	.	.	.	.	.	.	.	.	.	.	.	.
1000	.	.	.	.	.	.	.	.	.	.	.	.	.	.	.	.	.	.	.

Table 8.17. Sample size table for repeated measures ANOVA F-ratio; alpha = 0.05, r = 0.40

Chart A. 3 groups, pattern L/M

Power	Hypothesized ES																	
	0.20	0.30	0.35	0.40	0.45	0.50	0.55	0.60	0.65	0.70	0.75	0.80	1.00	1.25	1.50	1.75	2.00	2.50	3.00
0.80	287	129	95	73	58	48	40	34	29	25	22	20	13	9	7	6	5	4	4
0.90	377	169	125	96	76	62	52	44	38	33	29	26	17	12	9	7	6	5	4

Chart B. 3 groups, pattern H

Power	Hypothesized ES																	
	0.20	0.30	0.35	0.40	0.45	0.50	0.55	0.60	0.65	0.70	0.75	0.80	1.00	1.25	1.50	1.75	2.00	2.50	3.00
0.80	216	97	72	55	44	36	30	26	22	20	17	15	11	8	6	5	4	4	3
0.90	283	127	94	72	58	47	39	33	29	25	22	20	13	9	7	6	5	4	4

Chart C. 4 groups, pattern L

Power	Hypothesized ES																	
	0.20	0.30	0.35	0.40	0.45	0.50	0.55	0.60	0.65	0.70	0.75	0.80	1.00	1.25	1.50	1.75	2.00	2.50	3.00
0.80	325	146	107	83	66	54	45	38	32	28	25	22	15	10	8	6	5	4	3
0.90	424	189	140	107	85	69	58	49	42	36	32	28	19	13	9	7	6	5	5

Chart D. 4 groups, pattern M

Power	Hypothesized ES																	
	0.20	0.30	0.35	0.40	0.45	0.50	0.55	0.60	0.65	0.70	0.75	0.80	1.00	1.25	1.50	1.75	2.00	2.50	3.00
0.80	293	131	97	75	59	48	40	34	29	26	23	20	14	9	7	6	5	4	3
0.90	382	171	126	97	77	63	52	44	38	33	29	26	17	12	9	7	6	4	4

Chart E. 4 groups, pattern H

Power	Hypothesized ES																	
	0.20	0.30	0.35	0.40	0.45	0.50	0.55	0.60	0.65	0.70	0.75	0.80	1.00	1.25	1.50	1.75	2.00	2.50	3.00
0.80	164	74	55	42	34	28	23	20	17	15	13	12	8	6	5	4	4	3	3
0.90	213	96	71	55	44	36	30	25	22	19	17	15	10	7	6	5	4	3	3

Table 8.17. (*cont.*)

Chart F. 5 groups, pattern L																			
Power	Hypothesized ES																		
	0.20	0.30	0.35	0.40	0.45	0.50	0.55	0.60	0.65	0.70	0.75	0.80	1.00	1.25	1.50	1.75	2.00	2.50	3.00
0.80	356	159	118	90	72	58	49	41	35	31	27	24	16	11	8	6	5	4	3
0.90	462	206	152	117	93	75	63	53	45	39	34	30	20	14	10	8	6	5	4

Chart G. 5 groups, pattern M																			
Power	Hypothesized ES																		
	0.20	0.30	0.35	0.40	0.45	0.50	0.55	0.60	0.65	0.70	0.75	0.80	1.00	1.25	1.50	1.75	2.00	2.50	3.00
0.80	286	128	94	73	58	47	39	33	29	25	22	19	13	9	7	5	5	4	3
0.90	370	165	122	94	74	61	50	43	37	32	28	25	16	11	8	7	5	4	3

Chart H. 5 groups, pattern H																			
Power	Hypothesized ES																		
	0.20	0.30	0.35	0.40	0.45	0.50	0.55	0.60	0.65	0.70	0.75	0.80	1.00	1.25	1.50	1.75	2.00	2.50	3.00
0.80	150	67	50	39	31	25	21	18	16	14	12	11	8	6	4	4	3	3	3
0.90	193	87	64	50	40	32	27	23	20	17	15	14	9	7	5	4	4	3	3

Table 8.18. Sample size table for repeated measures ANOVA F-ratio; alpha = 0.05, r = 0.60

Chart A. 3 groups, pattern L/M

Power	Hypothesized ES																		
	0.20	0.30	0.35	0.40	0.45	0.50	0.55	0.60	0.65	0.70	0.75	0.80	1.00	1.25	1.50	1.75	2.00	2.50	3.00
0.80	192	86	64	49	40	32	27	23	20	18	16	14	10	7	6	5	4	3	3
0.90	252	113	84	65	51	42	35	30	26	22	20	18	12	9	7	5	5	4	3

Chart B. 3 groups, pattern H

Power	Hypothesized ES																		
	0.20	0.30	0.35	0.40	0.45	0.50	0.55	0.60	0.65	0.70	0.75	0.80	1.00	1.25	1.50	1.75	2.00	2.50	3.00
0.80	144	65	49	38	30	25	21	18	16	14	12	11	8	6	5	4	4	3	3
0.90	190	85	63	49	39	32	27	23	20	17	15	14	10	7	6	5	4	3	3

Chart C. 4 groups, pattern L

Power	Hypothesized ES																		
	0.20	0.30	0.35	0.40	0.45	0.50	0.55	0.60	0.65	0.70	0.75	0.80	1.00	1.25	1.50	1.75	2.00	2.50	3.00
0.80	217	98	72	56	44	36	30	26	22	19	17	15	11	7	6	5	4	3	3
0.90	283	127	94	72	57	47	39	33	28	25	22	19	13	9	7	6	5	4	3

Chart D. 4 groups, pattern M

Power	Hypothesized ES																		
	0.20	0.30	0.35	0.40	0.45	0.50	0.55	0.60	0.65	0.70	0.75	0.80	1.00	1.25	1.50	1.75	2.00	2.50	3.00
0.80	196	88	65	50	40	33	28	23	20	18	16	14	10	7	5	4	4	3	3
0.90	255	114	85	65	52	42	35	30	26	23	20	18	12	8	6	5	4	4	3

Chart E. 4 groups, pattern H

Power	Hypothesized ES																		
	0.20	0.30	0.35	0.40	0.45	0.50	0.55	0.60	0.65	0.70	0.75	0.80	1.00	1.25	1.50	1.75	2.00	2.50	3.00
0.80	110	50	37	29	23	19	16	14	12	11	10	9	6	5	4	3	3	3	2
0.90	142	64	48	37	30	24	20	17	15	13	12	11	8	6	4	4	3	3	3

Table 8.18. (*cont.*)

Chart F. 5 groups, pattern L

Power	Hypothesized ES																		
	0.20	0.30	0.35	0.40	0.45	0.50	0.55	0.60	0.65	0.70	0.75	0.80	1.00	1.25	1.50	1.75	2.00	2.50	3.00
0.80	238	107	79	61	48	40	33	28	24	21	19	17	11	8	6	5	4	3	3
0.90	308	138	102	78	62	51	42	36	31	27	24	21	14	10	7	6	5	4	3

Chart G. 5 groups, pattern M

Power	Hypothesized ES																		
	0.20	0.30	0.35	0.40	0.45	0.50	0.55	0.60	0.65	0.70	0.75	0.80	1.00	1.25	1.50	1.75	2.00	2.50	3.00
0.80	191	86	64	49	39	32	27	23	20	17	15	14	9	7	5	4	4	3	3
0.90	247	111	82	63	50	41	34	29	25	22	19	17	12	8	6	5	4	3	3

Chart H. 5 groups, pattern H

Power	Hypothesized ES																		
	0.20	0.30	0.35	0.40	0.45	0.50	0.55	0.60	0.65	0.70	0.75	0.80	1.00	1.25	1.50	1.75	2.00	2.50	3.00
0.80	100	46	34	26	21	18	15	13	11	10	9	8	6	4	4	3	3	3	2
0.90	129	58	43	34	27	22	19	16	14	12	11	10	7	5	4	4	3	3	2

Table 8.19. Sample size table for repeated measures ANOVA F-ratio; alpha = 0.01, r = 0.40

Chart A. 3 groups, pattern L/M

Power	Hypothesized ES																	
	0.20	0.30	0.35	0.40	0.45	0.50	0.55	0.60	0.65	0.70	0.75	0.80	1.00	1.25	1.50	1.75	2.00	2.50	3.00
0.80	415	186	138	106	84	69	57	49	42	37	32	29	19	13	10	8	7	6	5
0.90	522	233	172	133	105	86	71	60	52	45	40	35	24	16	12	10	8	6	5

Chart B. 3 groups, pattern H

Power	Hypothesized ES																		
	0.20	0.30	0.35	0.40	0.45	0.50	0.55	0.60	0.65	0.70	0.75	0.80	1.00	1.25	1.50	1.75	2.00	2.50	3.00
0.80	312	140	104	80	64	52	44	37	32	28	25	22	15	11	8	7	6	5	4
0.90	392	176	130	100	80	65	54	46	40	35	31	27	18	13	10	8	7	5	5

Chart C. 4 groups, pattern L

Power	Hypothesized ES																		
	0.20	0.30	0.35	0.40	0.45	0.50	0.55	0.60	0.65	0.70	0.75	0.80	1.00	1.25	1.50	1.75	2.00	2.50	3.00
0.80	463	207	153	118	93	76	63	54	46	40	35	31	21	14	11	8	7	5	5
0.90	577	258	190	146	116	94	78	66	57	49	43	38	25	17	13	10	8	6	5

Chart D. 4 groups, pattern M

Power	Hypothesized ES																		
	0.20	0.30	0.35	0.40	0.45	0.50	0.55	0.60	0.65	0.70	0.75	0.80	1.00	1.25	1.50	1.75	2.00	2.50	3.00
0.80	417	187	138	106	84	69	57	48	42	36	32	28	19	13	10	8	7	5	4
0.90	519	232	171	132	105	85	71	60	51	45	39	35	23	16	12	9	8	6	5

Chart E. 4 groups, pattern H

Power	Hypothesized ES																		
	0.20	0.30	0.35	0.40	0.45	0.50	0.55	0.60	0.65	0.70	0.75	0.80	1.00	1.25	1.50	1.75	2.00	2.50	3.00
0.80	233	105	78	60	48	39	33	28	24	21	19	17	12	8	7	5	5	4	3
0.90	290	130	96	74	59	48	40	34	30	26	23	20	14	10	8	6	5	4	4

Table 8.19. (*cont.*)

Chart F. 5 groups, pattern L																			
Power	Hypothesized ES																		
	0.20	0.30	0.35	0.40	0.45	0.50	0.55	0.60	0.65	0.70	0.75	0.80	1.00	1.25	1.50	1.75	2.00	2.50	3.00
0.80	502	224	165	127	101	82	68	58	49	43	38	33	22	15	11	9	7	5	4
0.90	622	278	205	157	125	101	84	71	61	53	46	41	27	18	13	10	8	6	5

Chart G. 5 groups, pattern M																			
Power	Hypothesized ES																		
	0.20	0.30	0.35	0.40	0.45	0.50	0.55	0.60	0.65	0.70	0.75	0.80	1.00	1.25	1.50	1.75	2.00	2.50	3.00
0.80	402	180	133	102	81	66	55	47	40	35	31	27	18	12	9	7	6	5	4
0.90	498	223	164	126	100	81	68	57	49	43	37	33	22	15	11	9	7	5	4

Chart H. 5 groups, pattern H																			
Power	Hypothesized ES																		
	0.20	0.30	0.35	0.40	0.45	0.50	0.55	0.60	0.65	0.70	0.75	0.80	1.00	1.25	1.50	1.75	2.00	2.50	3.00
0.80	210	95	70	54	43	35	30	25	22	19	17	15	11	8	6	5	4	4	3
0.90	260	117	86	67	53	43	36	31	27	23	21	18	13	9	7	6	5	4	3

Table 8.20. Sample size table for repeated measures ANOVA F-ratio; alpha = 0.01, r = 0.60

Chart A. 3 groups, pattern L/M																			
Power	Hypothesized ES																		
	0.20	0.30	0.35	0.40	0.45	0.50	0.55	0.60	0.65	0.70	0.75	0.80	1.00	1.25	1.50	1.75	2.00	2.50	3.00
0.80	278	125	93	72	57	47	39	33	29	25	22	20	14	10	8	6	6	5	4
0.90	349	157	116	89	71	58	49	41	36	31	27	24	17	12	9	7	6	5	4

Chart B. 3 groups, pattern H																			
Power	Hypothesized ES																		
	0.20	0.30	0.35	0.40	0.45	0.50	0.55	0.60	0.65	0.70	0.75	0.80	1.00	1.25	1.50	1.75	2.00	2.50	3.00
0.80	209	95	70	54	44	36	30	26	22	20	18	16	11	8	7	6	5	4	4
0.90	262	118	88	68	54	44	37	32	27	24	21	19	13	10	8	6	6	5	4

Chart C. 4 groups, pattern L																			
Power	Hypothesized ES																		
	0.20	0.30	0.35	0.40	0.45	0.50	0.55	0.60	0.65	0.70	0.75	0.80	1.00	1.25	1.50	1.75	2.00	2.50	3.00
0.80	309	139	103	79	63	52	43	37	31	27	24	22	15	10	8	6	6	4	4
0.90	385	173	127	98	78	64	53	45	39	34	30	26	18	12	9	7	6	5	4

Chart D. 4 groups, pattern M																			
Power	Hypothesized ES																		
	0.20	0.30	0.35	0.40	0.45	0.50	0.55	0.60	0.65	0.70	0.75	0.80	1.00	1.25	1.50	1.75	2.00	2.50	3.00
0.80	279	125	93	71	57	47	39	33	29	25	22	20	13	10	7	6	5	4	4
0.90	347	156	115	89	70	58	48	41	35	31	27	24	16	11	9	7	6	5	4

Chart E. 4 groups, pattern H																			
Power	Hypothesized ES																		
	0.20	0.30	0.35	0.40	0.45	0.50	0.55	0.60	0.65	0.70	0.75	0.80	1.00	1.25	1.50	1.75	2.00	2.50	3.00
0.80	156	71	53	41	33	27	23	19	17	15	13	12	9	6	5	4	4	3	3
0.90	194	87	65	50	40	33	28	24	21	18	16	14	10	7	6	5	4	4	3

Table 8.20. (*cont.*)

Chart F 5 groups, pattern L																			
Power	Hypothesized ES																		
	0.20	0.30	0.35	0.40	0.45	0.50	0.55	0.60	0.65	0.70	0.75	0.80	1.00	1.25	1.50	1.75	2.00	2.50	3.00
0.80	335	150	111	85	68	55	46	39	34	29	26	23	16	11	8	7	6	4	4
0.90	415	186	137	105	84	68	57	48	41	36	32	28	19	13	10	8	6	5	4

Chart G. 5 groups, pattern M																			
Power	Hypothesized ES																		
	0.20	0.30	0.35	0.40	0.45	0.50	0.55	0.60	0.65	0.70	0.75	0.80	1.00	1.25	1.50	1.75	2.00	2.50	3.00
0.80	269	121	89	69	55	45	37	32	27	24	21	19	13	9	7	6	5	4	3
0.90	333	149	110	85	67	55	46	39	33	29	26	23	15	11	8	7	6	4	4

Chart H. 5 groups, pattern H																			
Power	Hypothesized ES																		
	0.20	0.30	0.35	0.40	0.45	0.50	0.55	0.60	0.65	0.70	0.75	0.80	1.00	1.25	1.50	1.75	2.00	2.50	3.00
0.80	141	64	47	37	30	24	21	18	15	14	12	11	8	6	5	4	4	3	3
0.90	174	79	58	45	36	30	25	21	18	16	14	13	9	7	5	4	4	3	3

Table 8.21. Sample size table for repeated measures ANOVA F-ratio; alpha = 0.10, r = 0.40

Chart A. 3 groups, pattern L/M

Power	Hypothesized ES																		
	0.20	0.30	0.35	0.40	0.45	0.50	0.55	0.60	0.65	0.70	0.75	0.80	1.00	1.25	1.50	1.75	2.00	2.50	3.00
0.80	228	102	76	58	46	38	32	27	23	20	18	16	11	8	6	5	4	3	3
0.90	311	139	103	79	63	51	43	36	31	27	24	21	14	10	7	6	5	4	3

Chart B. 3 groups, pattern H

Power	Hypothesized ES																		
	0.20	0.30	0.35	0.40	0.45	0.50	0.55	0.60	0.65	0.70	0.75	0.80	1.00	1.25	1.50	1.75	2.00	2.50	3.00
0.80	172	77	57	44	35	29	24	21	18	16	14	12	9	6	5	4	4	3	3
0.90	234	105	77	60	47	39	32	27	24	21	18	16	11	8	6	5	4	3	3

Chart C. 4 groups, pattern L

Power	Hypothesized ES																		
	0.20	0.30	0.35	0.40	0.45	0.50	0.55	0.60	0.65	0.70	0.75	0.80	1.00	1.25	1.50	1.75	2.00	2.50	3.00
0.80	262	117	86	67	53	43	36	30	26	23	20	18	12	8	6	5	4	3	3
0.90	352	158	116	89	71	58	48	41	35	30	27	24	16	11	8	6	5	4	3

Chart D. 4 groups, pattern M

Power	Hypothesized ES																		
	0.20	0.30	0.35	0.40	0.45	0.50	0.55	0.60	0.65	0.70	0.75	0.80	1.00	1.25	1.50	1.75	2.00	2.50	3.00
0.80	236	106	78	60	48	39	33	28	24	21	18	16	11	8	6	5	4	3	3
0.90	317	142	105	81	64	52	43	37	31	27	24	21	14	10	7	6	5	4	3

Chart E. 4 groups, pattern H

Power	Hypothesized ES																		
	0.20	0.30	0.35	0.40	0.45	0.50	0.55	0.60	0.65	0.70	0.75	0.80	1.00	1.25	1.50	1.75	2.00	2.50	3.00
0.80	132	59	44	34	27	22	19	16	14	12	11	10	7	5	4	3	3	3	2
0.90	177	80	59	45	36	30	25	21	18	16	14	13	9	6	5	4	4	3	3

Table 8.21. *(cont.)*

Chart F. 5 groups, pattern L

Power	Hypothesized ES																		
	0.20	0.30	0.35	0.40	0.45	0.50	0.55	0.60	0.65	0.70	0.75	0.80	1.00	1.25	1.50	1.75	2.00	2.50	3.00
0.80	289	129	95	73	58	47	39	33	29	25	22	19	13	9	7	5	4	3	3
0.90	386	172	127	98	77	63	52	44	38	33	29	26	17	11	8	7	5	4	3

Chart G. 5 groups, pattern M

Power	Hypothesized ES																		
	0.20	0.30	0.35	0.40	0.45	0.50	0.55	0.60	0.65	0.70	0.75	0.80	1.00	1.25	1.50	1.75	2.00	2.50	3.00
0.80	231	104	77	59	47	38	32	27	23	20	18	16	11	7	6	5	4	3	3
0.90	309	138	102	78	62	51	42	36	31	27	23	21	14	9	7	6	5	4	3

Chart H. 5 groups, pattern H

Power	Hypothesized ES																		
	0.20	0.30	0.35	0.40	0.45	0.50	0.55	0.60	0.65	0.70	0.75	0.80	1.00	1.25	1.50	1.75	2.00	2.50	3.00
0.80	121	55	41	31	25	21	17	15	13	11	10	9	6	5	4	3	3	2	2
0.90	162	73	54	42	33	27	23	19	17	15	13	12	8	6	4	4	3	3	2

Table 8.22. Sample size table for repeated measures ANOVA F-ratio; alpha = 0.10, r = 0.60

Chart A. 3 groups, pattern L/M																			
Power	Hypothesized ES																		
	0.20	0.30	0.35	0.40	0.45	0.50	0.55	0.60	0.65	0.70	0.75	0.80	1.00	1.25	1.50	1.75	2.00	2.50	3.00
0.80	153	69	51	39	32	26	22	19	16	14	13	11	8	6	5	4	3	3	3
0.90	208	93	69	53	42	35	29	25	21	19	16	15	10	7	6	5	4	3	3

Chart B. 3 groups, pattern H																			
Power	Hypothesized ES																		
	0.20	0.30	0.35	0.40	0.45	0.50	0.55	0.60	0.65	0.70	0.75	0.80	1.00	1.25	1.50	1.75	2.00	2.50	3.00
0.80	115	52	39	30	24	20	17	14	12	11	10	9	6	5	4	3	3	3	3
0.90	156	70	52	40	32	26	22	19	16	14	13	11	8	6	5	4	3	3	3

Chart C. 4 groups, pattern L																			
Power	Hypothesized ES																		
	0.20	0.30	0.35	0.40	0.45	0.50	0.55	0.60	0.65	0.70	0.75	0.80	1.00	1.25	1.50	1.75	2.00	2.50	3.00
0.80	175	79	58	45	36	29	25	21	18	16	14	12	9	6	5	4	3	3	3
0.90	235	106	78	60	48	39	33	28	24	21	18	16	11	8	6	5	4	3	3

Chart D. 4 groups, pattern M																			
Power	Hypothesized ES																		
	0.20	0.30	0.35	0.40	0.45	0.50	0.55	0.60	0.65	0.70	0.75	0.80	1.00	1.25	1.50	1.75	2.00	2.50	3.00
0.80	158	71	53	41	32	27	22	19	16	14	13	11	8	6	4	4	3	3	3
0.90	212	95	70	54	43	35	29	25	22	19	17	15	10	7	5	4	4	3	3

Chart E. 4 groups, pattern H																			
Power	Hypothesized ES																		
	0.20	0.30	0.35	0.40	0.45	0.50	0.55	0.60	0.65	0.70	0.75	0.80	1.00	1.25	1.50	1.75	2.00	2.50	3.00
0.80	88	40	30	23	19	15	13	11	10	9	8	7	5	4	3	3	3	2	2
0.90	119	54	40	31	25	20	17	15	13	11	10	9	6	5	4	3	3	3	2

Table 8.22. (*cont.*)

Chart F. 5 groups, pattern L																			
Power	Hypothesized ES																		
	0.20	0.30	0.35	0.40	0.45	0.50	0.55	0.60	0.65	0.70	0.75	0.80	1.00	1.25	1.50	1.75	2.00	2.50	3.00
0.80	193	87	64	49	39	32	27	23	20	17	15	13	9	6	5	4	4	3	3
0.90	258	115	85	66	52	43	35	30	26	22	20	18	12	8	6	5	4	3	3
Chart G. 5 groups, pattern M																			
Power	Hypothesized ES																		
	0.20	0.30	0.35	0.40	0.45	0.50	0.55	0.60	0.65	0.70	0.75	0.80	1.00	1.25	1.50	1.75	2.00	2.50	3.00
0.80	155	70	52	40	32	26	22	19	16	14	12	11	8	6	4	4	3	3	2
0.90	207	93	68	53	42	34	29	24	21	18	16	14	10	7	5	4	4	3	3
Chart H. 5 groups, pattern H																			
Power	Hypothesized ES																		
	0.20	0.30	0.35	0.40	0.45	0.50	0.55	0.60	0.65	0.70	0.75	0.80	1.00	1.25	1.50	1.75	2.00	2.50	3.00
0.80	81	37	28	21	17	14	12	10	9	8	7	7	5	4	3	3	3	2	2
0.90	108	49	36	28	23	19	16	13	12	10	9	8	6	4	4	3	3	2	2

Table 8.23. Power table for repeated measures ANOVA; Tukey HSD MCP, $r = 0.40$, 3 groups at alpha = 0.05

n	Hypothesized ES																		
	0.20	0.30	0.35	0.40	0.45	0.50	0.55	0.60	0.65	0.70	0.75	0.80	1.00	1.25	1.50	1.75	2.00	2.50	3.00
5	2	3	4	5	6	7	8	9	11	12	14	16	25	40	57	72	84	97	.
6	3	4	5	6	7	8	10	12	14	16	18	21	33	52	70	84	93	99	.
7	3	4	5	7	8	10	12	14	17	20	23	26	41	62	80	91	97	.	.
8	3	5	6	8	10	12	14	17	20	23	27	31	49	71	87	96	99	.	.
9	3	5	7	9	11	14	16	20	23	27	32	36	56	78	92	98	.	.	.
10	3	6	8	10	12	15	19	23	27	31	36	41	62	83	95	99	.	.	.
11	4	7	9	11	14	17	21	25	30	35	40	46	68	88	97	99	.	.	.
12	4	7	9	12	15	19	23	28	33	39	45	51	73	91	98	.	.	.	.
13	4	8	10	13	17	21	26	31	37	43	49	55	77	94	99	.	.	.	.
14	4	8	11	15	19	23	28	34	40	46	53	59	81	95	99	.	.	.	.
15	5	9	12	16	20	25	31	37	43	50	57	63	84	97	.	.	.	.	.
20	6	12	17	22	28	35	43	50	58	66	73	79	94	99	.	.	.	.	.
25	7	15	21	28	36	45	54	62	70	77	84	88	98	.	.	.	.	.	.
30	9	19	26	35	44	54	63	72	80	86	91	94	99	.	.	.	.	.	.
35	10	22	31	41	51	62	71	80	86	91	95	97	.	.	.	.	.	.	.
40	11	26	36	47	58	69	78	85	91	95	97	99	.	.	.	.	.	.	.
45	13	30	41	53	64	75	83	90	94	97	99	99	.	.	.	.	.	.	.
50	14	33	45	58	70	80	88	93	96	98	99	.	.	.	.	.	.	.	.
55	16	37	50	63	75	84	91	95	98	99	.	.	.	.	.	.	.	.	.
60	17	40	54	67	79	87	93	97	99	99	.	.	.	.	.	.	.	.	.
65	19	44	58	71	82	90	95	98	99	.	.	.	.	.	.	.	.	.	.
70	20	47	62	75	85	92	97	99	99	.	.	.	.	.	.	.	.	.	.
75	22	50	65	78	88	94	98	99	.	.	.	.	.	.	.	.	.	.	.
80	23	53	69	81	90	96	98	99	.	.	.	.	.	.	.	.	.	.	.
90	27	59	75	86	94	97	99	.	.	.	.	.	.	.	.	.	.	.	.
100	30	65	80	90	96	99	.	.	.	.	.	.	.	.	.	.	.	.	.
110	33	69	84	93	97	99	.	.	.	.	.	.	.	.	.	.	.	.	.
120	36	74	87	95	98	.	.	.	.	.	.	.	.	.	.	.	.	.	.
130	39	78	90	96	99	.	.	.	.	.	.	.	.	.	.	.	.	.	.
140	42	81	92	97	99	.	.	.	.	.	.	.	.	.	.	.	.	.	.
150	45	84	94	98	.	.	.	.	.	.	.	.	.	.	.	.	.	.	.
175	52	90	97	99	.	.	.	.	.	.	.	.	.	.	.	.	.	.	.
200	59	94	98	.	.	.	.	.	.	.	.	.	.	.	.	.	.	.	.
225	65	96	99	.	.	.	.	.	.	.	.	.	.	.	.	.	.	.	.
250	70	98	.	.	.	.	.	.	.	.	.	.	.	.	.	.	.	.	.
300	79	99	.	.	.	.	.	.	.	.	.	.	.	.	.	.	.	.	.
350	86	.	.	.	.	.	.	.	.	.	.	.	.	.	.	.	.	.	.
400	90	.	.	.	.	.	.	.	.	.	.	.	.	.	.	.	.	.	.
450	94	.	.	.	.	.	.	.	.	.	.	.	.	.	.	.	.	.	.
500	96	.	.	.	.	.	.	.	.	.	.	.	.	.	.	.	.	.	.
600	98	.	.	.	.	.	.	.	.	.	.	.	.	.	.	.	.	.	.
700	99	.	.	.	.	.	.	.	.	.	.	.	.	.	.	.	.	.	.
800	.	.	.	.	.	.	.	.	.	.	.	.	.	.	.	.	.	.	.
900	.	.	.	.	.	.	.	.	.	.	.	.	.	.	.	.	.	.	.
1000	.	.	.	.	.	.	.	.	.	.	.	.	.	.	.	.	.	.	.

Table 8.24. Power table for repeated measures ANOVA; Tukey HSD MCP, $r = 0.60$, 3 groups at alpha = 0.05

n	Hypothesized ES																		
	0.20	0.30	0.35	0.40	0.45	0.50	0.55	0.60	0.65	0.70	0.75	0.80	1.00	1.25	1.50	1.75	2.00	2.50	3.00
5	3	4	5	7	8	10	11	13	16	18	21	24	39	59	77	89	96	.	.
6	3	5	6	8	10	12	15	17	21	24	28	32	50	72	88	96	99	.	.
7	3	6	8	10	12	15	18	22	26	30	35	40	60	82	94	99	.	.	.
8	4	7	9	11	14	18	22	26	31	36	41	47	69	88	97	.	.	.	.
9	4	8	10	13	16	21	25	30	36	42	48	54	76	93	99	.	.	.	.
10	4	8	11	15	19	24	29	35	41	47	54	60	82	96	99	.	.	.	.
11	5	9	13	17	21	27	32	39	46	52	59	66	86	97	.	.	.	.	.
12	5	10	14	18	24	30	36	43	50	57	64	71	90	99	.	.	.	.	.
13	6	11	15	20	26	33	40	47	55	62	69	75	93	99	.	.	.	.	.
14	6	12	17	22	28	36	43	51	59	66	73	79	95	.	.	.	.	.	.
15	6	13	18	24	31	38	46	55	63	70	77	83	96	.	.	.	.	.	.
20	8	18	25	34	43	52	62	71	78	85	90	93	99	.	.	.	.	.	.
25	10	24	33	43	54	64	74	82	88	93	96	98	.	.	.	.	.	.	.
30	12	29	40	52	63	74	83	89	94	97	98	99	.	.	.	.	.	.	.
35	15	34	47	60	72	81	89	94	97	99	99	.	.	.	.	.	.	.	.
40	17	40	53	67	78	87	93	97	98	99	.	.	.	.	.	.	.	.	.
45	19	45	59	73	84	91	96	98	99	.	.	.	.	.	.	.	.	.	.
50	22	50	65	78	88	94	97	99	.	.	.	.	.	.	.	.	.	.	.
55	24	54	70	82	91	96	98	99	.	.	.	.	.	.	.	.	.	.	.
60	26	59	74	86	93	97	99	.	.	.	.	.	.	.	.	.	.	.	.
65	29	63	78	89	95	98	99	.	.	.	.	.	.	.	.	.	.	.	.
70	31	67	81	91	97	99	.	.	.	.	.	.	.	.	.	.	.	.	.
75	33	70	84	93	98	99	.	.	.	.	.	.	.	.	.	.	.	.	.
80	36	74	87	95	98	.	.	.	.	.	.	.	.	.	.	.	.	.	.
90	41	79	91	97	99	.	.	.	.	.	.	.	.	.	.	.	.	.	.
100	45	84	94	98	.	.	.	.	.	.	.	.	.	.	.	.	.	.	.
110	49	88	96	99	.	.	.	.	.	.	.	.	.	.	.	.	.	.	.
120	54	90	97	99	.	.	.	.	.	.	.	.	.	.	.	.	.	.	.
130	58	93	98	.	.	.	.	.	.	.	.	.	.	.	.	.	.	.	.
140	61	95	99	.	.	.	.	.	.	.	.	.	.	.	.	.	.	.	.
150	65	96	99	.	.	.	.	.	.	.	.	.	.	.	.	.	.	.	.
175	73	98	.	.	.	.	.	.	.	.	.	.	.	.	.	.	.	.	.
200	79	99	.	.	.	.	.	.	.	.	.	.	.	.	.	.	.	.	.
225	84	.	.	.	.	.	.	.	.	.	.	.	.	.	.	.	.	.	.
250	88	.	.	.	.	.	.	.	.	.	.	.	.	.	.	.	.	.	.
300	94	.	.	.	.	.	.	.	.	.	.	.	.	.	.	.	.	.	.
350	97	.	.	.	.	.	.	.	.	.	.	.	.	.	.	.	.	.	.
400	98	.	.	.	.	.	.	.	.	.	.	.	.	.	.	.	.	.	.
450	99	.	.	.	.	.	.	.	.	.	.	.	.	.	.	.	.	.	.
500	.	.	.	.	.	.	.	.	.	.	.	.	.	.	.	.	.	.	.
600	.	.	.	.	.	.	.	.	.	.	.	.	.	.	.	.	.	.	.
700	.	.	.	.	.	.	.	.	.	.	.	.	.	.	.	.	.	.	.
800	.	.	.	.	.	.	.	.	.	.	.	.	.	.	.	.	.	.	.
900	.	.	.	.	.	.	.	.	.	.	.	.	.	.	.	.	.	.	.
1000	.	.	.	.	.	.	.	.	.	.	.	.	.	.	.	.	.	.	.

Table 8.25. Power table for repeated measures ANOVA; Tukey HSD MCP, $r = 0.40$, 4 groups at alpha $= 0.05$

n	Hypothesized ES																		
	0.20	0.30	0.35	0.40	0.45	0.50	0.55	0.60	0.65	0.70	0.75	0.80	1.00	1.25	1.50	1.75	2.00	2.50	3.00
5	1	2	3	3	4	5	6	7	8	9	11	13	21	36	53	70	83	97	.
6	2	3	3	4	5	6	7	9	10	12	14	17	28	47	66	82	92	99	.
7	2	3	4	5	6	7	9	11	13	15	18	21	35	57	76	90	97	.	.
8	2	3	4	5	7	8	10	13	15	18	22	25	43	66	84	94	99	.	.
9	2	4	5	6	8	10	12	15	18	22	26	30	49	73	90	97	99	.	.
10	2	4	5	7	9	11	14	17	21	25	30	34	56	79	93	98	.	.	.
11	2	4	6	8	10	13	16	20	24	29	34	39	61	84	96	99	.	.	.
12	2	5	6	9	11	14	18	22	27	32	38	43	67	88	97	.	.	.	.
13	3	5	7	9	12	16	20	25	30	36	42	48	72	91	98	.	.	.	.
14	3	6	8	10	14	18	22	27	33	39	45	52	76	94	99	.	.	.	.
15	3	6	8	11	15	19	24	30	36	42	49	56	80	95	99	.	.	.	.
20	4	8	12	16	22	28	35	42	50	58	66	73	92	99	.	.	.	.	.
25	5	11	16	22	29	37	45	54	63	71	78	84	97	.	.	.	.	.	.
30	6	14	20	27	36	45	55	65	73	81	87	91	99	.	.	.	.	.	.
35	7	16	24	33	43	53	64	73	81	87	92	95	.	.	.	.	.	.	.
40	8	19	28	38	50	61	71	80	87	92	96	98	.	.	.	.	.	.	.
45	9	23	33	44	56	68	78	86	91	95	98	99	.	.	.	.	.	.	.
50	10	26	37	49	62	73	83	90	94	97	99	99	.	.	.	.	.	.	.
55	11	29	41	54	67	78	87	93	96	98	99	.	.	.	.	.	.	.	.
60	12	32	45	59	72	82	90	95	98	99	.	.	.	.	.	.	.	.	.
65	13	35	49	64	76	86	93	96	99	99	.	.	.	.	.	.	.	.	.
70	15	38	53	68	80	89	94	98	99	.	.	.	.	.	.	.	.	.	.
75	16	41	57	72	83	91	96	98	99	.	.	.	.	.	.	.	.	.	.
80	17	45	61	75	86	93	97	99	.	.	.	.	.	.	.	.	.	.	.
90	20	51	67	81	90	96	98	.	.	.	.	.	.	.	.	.	.	.	.
100	23	56	73	86	94	98	99	.	.	.	.	.	.	.	.	.	.	.	.
110	25	61	78	89	96	99	.	.	.	.	.	.	.	.	.	.	.	.	.
120	28	66	82	92	97	99	.	.	.	.	.	.	.	.	.	.	.	.	.
130	31	71	85	94	98	.	.	.	.	.	.	.	.	.	.	.	.	.	.
140	34	74	88	96	99	.	.	.	.	.	.	.	.	.	.	.	.	.	.
150	37	78	91	97	99	.	.	.	.	.	.	.	.	.	.	.	.	.	.
175	44	85	95	99	.	.	.	.	.	.	.	.	.	.	.	.	.	.	.
200	50	90	97	.	.	.	.	.	.	.	.	.	.	.	.	.	.	.	.
225	56	94	99	.	.	.	.	.	.	.	.	.	.	.	.	.	.	.	.
250	62	96	99	.	.	.	.	.	.	.	.	.	.	.	.	.	.	.	.
300	72	98	.	.	.	.	.	.	.	.	.	.	.	.	.	.	.	.	.
350	80	99	.	.	.	.	.	.	.	.	.	.	.	.	.	.	.	.	.
400	86	.	.	.	.	.	.	.	.	.	.	.	.	.	.	.	.	.	.
450	90	.	.	.	.	.	.	.	.	.	.	.	.	.	.	.	.	.	.
500	93	.	.	.	.	.	.	.	.	.	.	.	.	.	.	.	.	.	.
600	97	.	.	.	.	.	.	.	.	.	.	.	.	.	.	.	.	.	.
700	99	.	.	.	.	.	.	.	.	.	.	.	.	.	.	.	.	.	.
800	.	.	.	.	.	.	.	.	.	.	.	.	.	.	.	.	.	.	.
900	.	.	.	.	.	.	.	.	.	.	.	.	.	.	.	.	.	.	.
1000	.	.	.	.	.	.	.	.	.	.	.	.	.	.	.	.	.	.	.

Table 8.26. Power table for repeated measures ANOVA; Tukey HSD MCP, $r = 0.60$, 4 groups at alpha = 0.05

n	Hypothesized ES																		
	0.20	0.30	0.35	0.40	0.45	0.50	0.55	0.60	0.65	0.70	0.75	0.80	1.00	1.25	1.50	1.75	2.00	2.50	3.00
5	2	3	4	5	6	7	9	10	13	15	17	20	34	55	75	89	96	.	.
6	2	3	4	6	7	9	11	14	16	20	23	27	45	68	86	95	99	.	.
7	2	4	5	7	9	11	14	17	21	25	29	34	55	78	93	98	.	.	.
8	2	5	6	8	11	13	17	21	25	30	35	41	63	86	96	99	.	.	.
9	3	5	7	9	12	16	20	24	30	35	41	47	71	91	98	.	.	.	.
10	3	6	8	11	14	18	23	28	34	40	47	53	77	94	99	.	.	.	.
11	3	6	9	12	16	21	26	32	39	45	52	59	82	96	.	.	.	.	.
12	3	7	10	14	18	23	29	36	43	50	58	65	87	98	.	.	.	.	.
13	4	8	11	15	20	26	32	40	47	55	62	69	90	99	.	.	.	.	.
14	4	9	12	17	22	29	36	43	51	59	67	74	92	99	.	.	.	.	.
15	4	9	13	18	24	31	39	47	55	63	71	78	94	.	.	.	.	.	.
20	6	13	19	27	35	44	54	63	72	80	86	91	99	.	.	.	.	.	.
25	7	18	26	35	46	56	67	76	84	89	94	96	.	.	.	.	.	.	.
30	9	22	32	43	55	67	77	85	91	95	97	99	.	.	.	.	.	.	.
35	10	27	39	51	64	75	84	91	95	98	99	.	.	.	.	.	.	.	.
40	12	32	45	59	71	82	90	95	98	99	.	.	.	.	.	.	.	.	.
45	14	36	51	65	78	87	93	97	99	.	.	.	.	.	.	.	.	.	.
50	16	41	57	71	83	91	96	98	99	.	.	.	.	.	.	.	.	.	.
55	18	46	62	76	87	94	97	99	.	.	.	.	.	.	.	.	.	.	.
60	20	50	67	81	90	96	98	.	.	.	.	.	.	.	.	.	.	.	.
65	22	54	71	84	93	97	99	.	.	.	.	.	.	.	.	.	.	.	.
70	24	58	75	87	95	98	99	.	.	.	.	.	.	.	.	.	.	.	.
75	26	62	79	90	96	99	.	.	.	.	.	.	.	.	.	.	.	.	.
80	28	66	82	92	97	99	.	.	.	.	.	.	.	.	.	.	.	.	.
90	32	72	87	95	99	.	.	.	.	.	.	.	.	.	.	.	.	.	.
100	36	78	91	97	99	.	.	.	.	.	.	.	.	.	.	.	.	.	.
110	41	82	93	98	.	.	.	.	.	.	.	.	.	.	.	.	.	.	.
120	45	86	96	99	.	.	.	.	.	.	.	.	.	.	.	.	.	.	.
130	49	89	97	99	.	.	.	.	.	.	.	.	.	.	.	.	.	.	.
140	53	92	98	.	.	.	.	.	.	.	.	.	.	.	.	.	.	.	.
150	56	94	99	.	.	.	.	.	.	.	.	.	.	.	.	.	.	.	.
175	65	97	.	.	.	.	.	.	.	.	.	.	.	.	.	.	.	.	.
200	72	98	.	.	.	.	.	.	.	.	.	.	.	.	.	.	.	.	.
225	78	99	.	.	.	.	.	.	.	.	.	.	.	.	.	.	.	.	.
250	83	.	.	.	.	.	.	.	.	.	.	.	.	.	.	.	.	.	.
300	90	.	.	.	.	.	.	.	.	.	.	.	.	.	.	.	.	.	.
350	95	.	.	.	.	.	.	.	.	.	.	.	.	.	.	.	.	.	.
400	97	.	.	.	.	.	.	.	.	.	.	.	.	.	.	.	.	.	.
450	98	.	.	.	.	.	.	.	.	.	.	.	.	.	.	.	.	.	.
500	99	.	.	.	.	.	.	.	.	.	.	.	.	.	.	.	.	.	.
600	.	.	.	.	.	.	.	.	.	.	.	.	.	.	.	.	.	.	.
700	.	.	.	.	.	.	.	.	.	.	.	.	.	.	.	.	.	.	.
800	.	.	.	.	.	.	.	.	.	.	.	.	.	.	.	.	.	.	.
900	.	.	.	.	.	.	.	.	.	.	.	.	.	.	.	.	.	.	.
1000	.	.	.	.	.	.	.	.	.	.	.	.	.	.	.	.	.	.	.

Table 8.27. Power table for repeated measures ANOVA; Tukey HSD MCP, $r = 0.40$, 5 groups at alpha = 0.05

n	Hypothesized ES																		
	0.20	0.30	0.35	0.40	0.45	0.50	0.55	0.60	0.65	0.70	0.75	0.80	1.00	1.25	1.50	1.75	2.00	2.50	3.00
5	1	2	2	2	3	4	4	5	6	8	9	10	18	33	50	67	81	96	.
6	1	2	2	3	4	5	6	7	8	10	12	14	25	43	63	80	91	99	.
7	1	2	3	3	4	5	7	8	10	12	15	18	31	53	73	88	96	.	.
8	1	2	3	4	5	7	8	10	12	15	18	21	38	62	81	93	98	.	.
9	1	3	3	5	6	8	10	12	15	18	22	26	44	69	87	96	99	.	.
10	1	3	4	5	7	9	11	14	17	21	25	30	51	76	92	98	.	.	.
11	2	3	4	6	8	10	13	16	20	24	29	34	56	81	95	99	.	.	.
12	2	3	5	7	9	11	15	18	23	27	33	38	62	86	97	99	.	.	.
13	2	4	5	7	10	13	16	20	25	31	36	42	67	89	98	.	.	.	.
14	2	4	6	8	11	14	18	23	28	34	40	46	72	92	99	.	.	.	.
15	2	4	6	9	12	15	20	25	31	37	44	50	76	94	99	.	.	.	.
20	3	6	9	13	18	23	30	37	45	52	60	68	90	99	.	.	.	.	.
25	3	8	12	17	24	31	40	48	57	66	74	80	96	.	.	.	.	.	.
30	4	11	16	22	30	39	49	59	68	76	83	89	99	.	.	.	.	.	.
35	5	13	19	28	37	47	58	68	77	84	90	94	.	.	.	.	.	.	.
40	6	15	23	33	44	55	66	76	84	90	94	97	.	.	.	.	.	.	.
45	6	18	27	38	50	62	73	82	89	93	97	98	.	.	.	.	.	.	.
50	7	21	31	43	56	68	79	87	92	96	98	99	.	.	.	.	.	.	.
55	8	24	35	48	62	74	83	90	95	98	99	.	.	.	.	.	.	.	.
60	9	27	39	53	67	78	87	93	97	99	99	.	.	.	.	.	.	.	.
65	10	30	43	58	71	82	90	95	98	99	.	.	.	.	.	.	.	.	.
70	11	33	47	62	75	86	93	97	99	.	.	.	.	.	.	.	.	.	.
75	12	36	51	66	79	89	94	98	99	.	.	.	.	.	.	.	.	.	.
80	13	38	55	70	82	91	96	98	99	.	.	.	.	.	.	.	.	.	.
90	16	44	61	76	87	94	98	99	.	.	.	.	.	.	.	.	.	.	.
100	18	50	67	82	91	97	99	.	.	.	.	.	.	.	.	.	.	.	.
110	21	55	73	86	94	98	99	.	.	.	.	.	.	.	.	.	.	.	.
120	23	60	78	90	96	99	.	.	.	.	.	.	.	.	.	.	.	.	.
130	26	65	82	92	97	99	.	.	.	.	.	.	.	.	.	.	.	.	.
140	28	69	85	94	98	.	.	.	.	.	.	.	.	.	.	.	.	.	.
150	31	73	88	96	99	.	.	.	.	.	.	.	.	.	.	.	.	.	.
175	37	81	93	98	.	.	.	.	.	.	.	.	.	.	.	.	.	.	.
200	44	87	96	99	.	.	.	.	.	.	.	.	.	.	.	.	.	.	.
225	50	91	98	.	.	.	.	.	.	.	.	.	.	.	.	.	.	.	.
250	56	94	99	.	.	.	.	.	.	.	.	.	.	.	.	.	.	.	.
300	67	98	.	.	.	.	.	.	.	.	.	.	.	.	.	.	.	.	.
350	75	99	.	.	.	.	.	.	.	.	.	.	.	.	.	.	.	.	.
400	82	.	.	.	.	.	.	.	.	.	.	.	.	.	.	.	.	.	.
450	87	.	.	.	.	.	.	.	.	.	.	.	.	.	.	.	.	.	.
500	91	.	.	.	.	.	.	.	.	.	.	.	.	.	.	.	.	.	.
600	96	.	.	.	.	.	.	.	.	.	.	.	.	.	.	.	.	.	.
700	98	.	.	.	.	.	.	.	.	.	.	.	.	.	.	.	.	.	.
800	99	.	.	.	.	.	.	.	.	.	.	.	.	.	.	.	.	.	.
900	.	.	.	.	.	.	.	.	.	.	.	.	.	.	.	.	.	.	.
1000	.	.	.	.	.	.	.	.	.	.	.	.	.	.	.	.	.	.	.

Table 8.28. Power table for repeated measures ANOVA; Tukey HSD MCP, $r = 0.60$, 5 groups at alpha = 0.05

n	Hypothesized ES																		
	0.20	0.30	0.35	0.40	0.45	0.50	0.55	0.60	0.65	0.70	0.75	0.80	1.00	1.25	1.50	1.75	2.00	2.50	3.00
5	1	2	3	3	4	6	7	8	10	12	15	17	31	52	73	88	96	.	.
6	1	2	3	4	6	7	9	11	14	17	20	23	41	65	84	95	99	.	.
7	2	3	4	5	7	9	11	14	17	21	25	30	50	76	91	98	.	.	.
8	2	3	5	6	8	11	14	17	21	26	31	36	59	83	96	99	.	.	.
9	2	4	5	7	10	13	16	20	25	31	36	42	67	89	98	.	.	.	.
10	2	4	6	8	11	15	19	24	29	35	42	48	74	93	99	.	.	.	.
11	2	5	7	10	13	17	22	27	34	40	47	54	79	95	.	.	.	.	.
12	2	5	8	11	15	19	25	31	38	45	52	60	84	97	.	.	.	.	.
13	3	6	9	12	16	22	28	34	42	50	57	65	87	98	.	.	.	.	.
14	3	7	9	13	18	24	31	38	46	54	62	69	90	99	.	.	.	.	.
15	3	7	10	15	20	26	34	42	50	58	66	73	93	99	.	.	.	.	.
20	4	10	15	22	30	39	48	58	67	75	82	88	98	.	.	.	.	.	.
25	5	14	21	30	40	51	61	71	80	87	92	95	.	.	.	.	.	.	.
30	6	18	27	38	49	61	72	81	88	93	96	98	.	.	.	.	.	.	.
35	8	22	33	45	58	70	81	88	93	97	98	99	.	.	.	.	.	.	.
40	9	26	39	53	66	78	87	93	97	98	99	.	.	.	.	.	.	.	.
45	11	31	45	60	73	84	91	96	98	99	.	.	.	.	.	.	.	.	.
50	12	35	51	66	79	88	94	98	99	.	.	.	.	.	.	.	.	.	.
55	14	40	56	71	83	92	96	99	.	.	.	.	.	.	.	.	.	.	.
60	16	44	61	76	87	94	98	99	.	.	.	.	.	.	.	.	.	.	.
65	17	48	66	80	90	96	99	.	.	.	.	.	.	.	.	.	.	.	.
70	19	52	70	84	93	97	99	.	.	.	.	.	.	.	.	.	.	.	.
75	21	56	74	87	95	98	99	.	.	.	.	.	.	.	.	.	.	.	.
80	23	60	77	89	96	99	.	.	.	.	.	.	.	.	.	.	.	.	.
90	27	67	83	93	98	99	.	.	.	.	.	.	.	.	.	.	.	.	.
100	31	73	88	96	99	.	.	.	.	.	.	.	.	.	.	.	.	.	.
110	35	78	91	97	99	.	.	.	.	.	.	.	.	.	.	.	.	.	.
120	39	82	94	98	.	.	.	.	.	.	.	.	.	.	.	.	.	.	.
130	43	86	96	99	.	.	.	.	.	.	.	.	.	.	.	.	.	.	.
140	46	89	97	99	.	.	.	.	.	.	.	.	.	.	.	.	.	.	.
150	50	91	98	.	.	.	.	.	.	.	.	.	.	.	.	.	.	.	.
175	59	96	99	.	.	.	.	.	.	.	.	.	.	.	.	.	.	.	.
200	67	98	.	.	.	.	.	.	.	.	.	.	.	.	.	.	.	.	.
225	73	99	.	.	.	.	.	.	.	.	.	.	.	.	.	.	.	.	.
250	79	99	.	.	.	.	.	.	.	.	.	.	.	.	.	.	.	.	.
300	87	.	.	.	.	.	.	.	.	.	.	.	.	.	.	.	.	.	.
350	93	.	.	.	.	.	.	.	.	.	.	.	.	.	.	.	.	.	.
400	96	.	.	.	.	.	.	.	.	.	.	.	.	.	.	.	.	.	.
450	98	.	.	.	.	.	.	.	.	.	.	.	.	.	.	.	.	.	.
500	99	.	.	.	.	.	.	.	.	.	.	.	.	.	.	.	.	.	.
600	.	.	.	.	.	.	.	.	.	.	.	.	.	.	.	.	.	.	.
700	.	.	.	.	.	.	.	.	.	.	.	.	.	.	.	.	.	.	.
800	.	.	.	.	.	.	.	.	.	.	.	.	.	.	.	.	.	.	.
900	.	.	.	.	.	.	.	.	.	.	.	.	.	.	.	.	.	.	.
1000	.	.	.	.	.	.	.	.	.	.	.	.	.	.	.	.	.	.	.

Table 8.29. Sample size table for one-way repeated measures ANOVA MCP; $p = 0.05$, $r = 0.40$

Chart A. 3 groups, $r = 0.40$ at alpha = 0.05; NK no intervening groups																			
Power	Hypothesized ES																		
	0.20	0.30	0.35	0.40	0.45	0.50	0.55	0.60	0.65	0.70	0.75	0.80	1.00	1.25	1.50	1.75	2.00	2.50	3.00
0.80	237	106	79	61	48	39	33	28	24	21	18	16	11	8	6	5	4	4	3
0.90	317	142	105	80	64	52	43	37	31	27	24	21	14	10	7	6	5	4	3

Chart B. 3 groups, $r = 0.40$ at alpha = 0.05; NK 1 intervening group, Tukey 3 groups																			
Power	Hypothesized ES																		
	0.20	0.30	0.35	0.40	0.45	0.50	0.55	0.60	0.65	0.70	0.75	0.80	1.00	1.25	1.50	1.75	2.00	2.50	3.00
0.80	306	137	101	78	62	51	42	36	31	27	24	21	14	10	8	6	5	4	4
0.90	396	177	131	101	80	65	54	46	39	34	30	27	18	12	9	7	6	5	4

Chart C. 4 groups, $r = 0.40$ at alpha = 0.05; NK no intervening groups																			
Power	Hypothesized ES																		
	0.20	0.30	0.35	0.40	0.45	0.50	0.55	0.60	0.65	0.70	0.75	0.80	1.00	1.25	1.50	1.75	2.00	2.50	3.00
0.80	237	106	78	60	48	39	32	27	24	21	18	15	11	7	6	5	4	3	3
0.90	316	141	104	80	64	52	43	36	31	27	24	21	14	9	7	6	5	4	3

Chart D. 4 groups, $r = 0.40$ at alpha = 0.05; NK 1 intervening group																			
Power	Hypothesized ES																		
	0.20	0.30	0.35	0.40	0.45	0.50	0.55	0.60	0.65	0.70	0.75	0.80	1.00	1.25	1.50	1.75	2.00	2.50	3.00
0.80	306	137	101	78	62	50	42	35	30	26	23	21	14	9	7	6	5	4	3
0.90	396	177	130	100	79	65	54	45	39	34	30	26	17	12	9	7	6	4	4

Chart E. 4 groups, $r = 0.40$ at alpha = 0.05; NK 2 intervening groups, Tukey 4 groups																			
Power	Hypothesized ES																		
	0.20	0.30	0.35	0.40	0.45	0.50	0.55	0.60	0.65	0.70	0.75	0.80	1.00	1.25	1.50	1.75	2.00	2.50	3.00
0.80	351	157	116	89	71	57	48	40	35	30	26	23	16	11	8	6	5	4	3
0.90	446	199	147	113	89	73	61	51	44	38	33	29	19	13	10	8	6	5	4

Chart F. 5 groups, $r = 0.40$ at alpha = 0.05; NK no intervening groups																			
Power	Hypothesized ES																		
	0.20	0.30	0.35	0.40	0.45	0.50	0.55	0.60	0.65	0.70	0.75	0.80	1.00	1.25	1.50	1.75	2.00	2.50	3.00
0.80	237	106	78	60	48	39	32	27	23	20	18	16	11	7	5	4	4	3	3
0.90	316	141	104	80	63	52	43	36	31	27	23	21	14	9	7	5	4	3	3

Table 8.29. (*cont.*)

Chart G. 5 groups, $r = 0.40$ at alpha = 0.05; NK 1 intervening groups																			
Power	Hypothesized ES																		
	0.20	0.30	0.35	0.40	0.45	0.50	0.55	0.60	0.65	0.70	0.75	0.80	1.00	1.25	1.50	1.75	2.00	2.50	3.00
0.80	306	137	101	77	61	50	41	35	30	26	23	20	13	9	7	5	4	3	3
0.90	396	176	130	100	79	64	53	45	39	33	29	26	17	11	8	6	5	4	3

Chart H. 5 groups, $r = 0.40$ at alpha = 0.05; NK 2 intervening groups																			
Power	Hypothesized ES																		
	0.20	0.30	0.35	0.40	0.45	0.50	0.55	0.60	0.65	0.70	0.75	0.80	1.00	1.25	1.50	1.75	2.00	2.50	3.00
0.80	350	156	115	89	70	57	47	40	34	30	26	23	15	10	8	6	5	4	3
0.90	446	199	147	113	89	73	60	51	43	38	33	29	19	13	9	7	6	4	4

Chart I. 5 groups, $r = 0.40$ at alpha = 0.05; NK 3 intervening groups, Tukey 5 groups																			
Power	Hypothesized ES																		
	0.20	0.30	0.35	0.40	0.45	0.50	0.55	0.60	0.65	0.70	0.75	0.80	1.00	1.25	1.50	1.75	2.00	2.50	3.00
0.80	384	171	126	97	77	63	52	44	38	33	29	25	17	11	8	7	5	4	3
0.90	484	216	159	122	97	79	65	55	47	41	36	32	21	14	10	8	6	5	4

Table 8.30. Sample size table for one-way repeated measures ANOVA MCP; $p = 0.05$, $r = 0.60$

Chart A. 3 groups, $r = 0.60$ at alpha = 0.05; NK no intervening groups

Power	Hypothesized ES																		
	0.20	0.30	0.35	0.40	0.45	0.50	0.55	0.60	0.65	0.70	0.75	0.80	1.00	1.25	1.50	1.75	2.00	2.50	3.00
0.80	159	71	53	41	33	27	22	19	17	15	13	12	8	6	5	4	4	3	3
0.90	212	95	70	54	43	35	29	25	22	19	17	15	10	7	6	5	4	3	3

Chart B. 3 groups, $r = 0.60$ at alpha = 0.05; NK 1 intervening group, Tukey 3 groups

Power	Hypothesized ES																		
	0.20	0.30	0.35	0.40	0.45	0.50	0.55	0.60	0.65	0.70	0.75	0.80	1.00	1.25	1.50	1.75	2.00	2.50	3.00
0.80	205	92	68	53	42	34	29	25	21	19	16	15	10	7	6	5	4	4	3
0.90	265	119	88	68	54	44	37	31	27	23	21	18	13	9	7	6	5	4	3

Chart C. 4 groups, $r = 0.60$ at alpha = 0.05; NK no intervening groups

Power	Hypothesized ES																		
	0.20	0.30	0.35	0.40	0.45	0.50	0.55	0.60	0.65	0.70	0.75	0.80	1.00	1.25	1.50	1.75	2.00	2.50	3.00
0.80	158	71	53	41	32	26	22	19	16	14	12	11	8	5	4	4	3	3	2
0.90	211	95	70	54	43	35	29	25	21	18	16	14	10	7	5	4	4	3	3

Chart D. 4 groups, $r = 0.60$ at alpha = 0.05; NK 1 intervening group

Power	Hypothesized ES																		
	0.20	0.30	0.35	0.40	0.45	0.50	0.55	0.60	0.65	0.70	0.75	0.80	1.00	1.25	1.50	1.75	2.00	2.50	3.00
0.80	206	92	68	52	42	34	28	24	21	18	16	14	10	7	5	4	4	3	3
0.90	264	118	87	67	53	44	36	31	26	23	20	18	12	8	6	5	4	3	3

Chart E. 4 groups, $r = 0.60$ at alpha = 0.05; NK 2 intervening groups, Tukey 4 groups

Power	Hypothesized ES																		
	0.20	0.30	0.35	0.40	0.45	0.50	0.55	0.60	0.65	0.70	0.75	0.80	1.00	1.25	1.50	1.75	2.00	2.50	3.00
0.80	234	105	78	60	48	39	32	27	24	21	18	16	11	8	6	5	4	3	3
0.90	298	133	98	76	60	49	41	35	30	26	23	20	14	9	7	6	5	4	3

Chart F. 5 groups, $r = 0.60$ at alpha = 0.05; NK no intervening groups

Power	Hypothesized ES																		
	0.20	0.30	0.35	0.40	0.45	0.50	0.55	0.60	0.65	0.70	0.75	0.80	1.00	1.25	1.50	1.75	2.00	2.50	3.00
0.80	158	71	52	40	32	26	22	19	16	14	12	11	7	5	4	3	3	3	2
0.90	211	94	70	54	43	35	29	24	21	18	16	14	10	7	5	4	3	3	2

Table 8.30. (*cont.*)

Chart G. 5 groups, $r = 0.60$ at alpha = 0.05; NK 1 intervening group																			
Power	Hypothesized ES																		
	0.20	0.30	0.35	0.40	0.45	0.50	0.55	0.60	0.65	0.70	0.75	0.80	1.00	1.25	1.50	1.75	2.00	2.50	3.00
0.80	204	91	68	52	41	34	28	24	20	18	16	14	9	7	5	4	4	3	3
0.90	264	118	87	67	53	43	36	30	26	23	20	18	12	8	6	5	4	3	3

Chart H. 5 groups, $r = 0.60$ at alpha = 0.05; NK 2 intervening groups																			
Power	Hypothesized ES																		
	0.20	0.30	0.35	0.40	0.45	0.50	0.55	0.60	0.65	0.70	0.75	0.80	1.00	1.25	1.50	1.75	2.00	2.50	3.00
0.80	234	105	77	60	47	39	32	27	23	20	18	16	11	7	6	5	4	3	3
0.90	298	133	98	75	60	49	41	34	29	26	22	20	13	9	7	5	4	4	3

Chart I. 5 groups, $r = 0.60$ at alpha = 0.05; NK 3 intervening groups, Tukey 5 groups																			
Power	Hypothesized ES																		
	0.20	0.30	0.35	0.40	0.45	0.50	0.55	0.60	0.65	0.70	0.75	0.80	1.00	1.25	1.50	1.75	2.00	2.50	3.00
0.80	256	115	85	65	53	42	35	30	26	22	20	17	12	8	6	5	4	3	3
0.90	323	144	106	82	65	53	44	37	32	28	24	22	14	10	7	6	5	4	3

Table 8.31. Sample size table for one-way repeated measures ANOVA MCP; $p = 0.01$, $r = 0.40$

Chart A. 3 groups, $r = 0.40$ at alpha = 0.01; NK no intervening groups

Power	Hypothesized ES																		
	0.20	0.30	0.35	0.40	0.45	0.50	0.55	0.60	0.65	0.70	0.75	0.80	1.00	1.25	1.50	1.75	2.00	2.50	3.00
0.80	353	158	117	90	72	58	49	41	36	31	27	24	16	11	9	7	6	5	4
0.90	449	201	148	114	91	74	61	52	45	39	34	30	20	14	10	8	7	5	5

Chart B. 3 groups, $r = 0.40$ at alpha = 0.01; NK 1 intervening group, Tukey 3 groups

Power	Hypothesized ES																		
	0.20	0.30	0.35	0.40	0.45	0.50	0.55	0.60	0.65	0.70	0.75	0.80	1.00	1.25	1.50	1.75	2.00	2.50	3.00
0.80	426	191	141	109	86	70	59	50	43	37	33	29	20	14	10	8	7	6	5
0.90	531	237	175	135	107	87	73	61	53	46	40	36	24	16	12	10	8	6	5

Chart C. 4 groups, $r = 0.40$ at alpha = 0.01; NK no intervening groups

Power	Hypothesized ES																		
	0.20	0.30	0.35	0.40	0.45	0.50	0.55	0.60	0.65	0.70	0.75	0.80	1.00	1.25	1.50	1.75	2.00	2.50	3.00
0.80	352	157	116	89	71	58	48	41	35	30	27	24	16	11	8	6	5	4	4
0.90	448	200	147	113	90	73	61	51	44	38	33	30	20	13	10	8	6	5	4

Chart D. 4 groups, $r = 0.40$ at alpha = 0.01; NK 1 intervening group

Power	Hypothesized ES																		
	0.20	0.30	0.35	0.40	0.45	0.50	0.55	0.60	0.65	0.70	0.75	0.80	1.00	1.25	1.50	1.75	2.00	2.50	3.00
0.80	425	190	140	108	86	70	58	49	42	37	32	28	19	13	10	8	6	5	4
0.90	530	237	174	134	106	86	72	61	52	45	40	35	23	16	11	9	7	6	5

Chart E. 4 groups, $r = 0.40$ at alpha = 0.01; NK 2 intervening groups, Tukey 4 groups

Power	Hypothesized ES																		
	0.20	0.30	0.35	0.40	0.45	0.50	0.55	0.60	0.65	0.70	0.75	0.80	1.00	1.25	1.50	1.75	2.00	2.50	3.00
0.80	471	211	155	119	95	77	64	54	47	40	36	32	21	14	11	8	7	5	4
0.90	582	260	191	147	117	95	79	67	57	49	43	38	25	17	13	10	8	6	5

Chart F. 5 groups, $r = 0.40$ at alpha = 0.01; NK no intervening groups

Power	Hypothesized ES																		
	0.20	0.30	0.35	0.40	0.45	0.50	0.55	0.60	0.65	0.70	0.75	0.80	1.00	1.25	1.50	1.75	2.00	2.50	3.00
0.80	352	157	116	89	71	58	48	40	35	30	26	23	15	10	8	6	5	4	3
0.90	448	200	147	113	90	73	60	51	44	38	33	29	19	13	9	7	6	5	4

Table 8.31. (*cont.*)

Chart G. 5 groups, $r = 0.40$ at alpha = 0.01; NK 1 intervening group																			
Power	Hypothesized ES																		
	0.20	0.30	0.35	0.40	0.45	0.50	0.55	0.60	0.65	0.70	0.75	0.80	1.00	1.25	1.50	1.75	2.00	2.50	3.00
0.80	425	190	140	107	85	69	58	49	42	36	32	28	19	13	9	7	6	5	4
0.90	530	236	174	134	106	86	71	60	52	45	39	35	23	15	11	9	7	5	4

Chart H. 5 groups, $r = 0.40$ at alpha = 0.01; NK 2 intervening groups																			
Power	Hypothesized ES																		
	0.20	0.30	0.35	0.40	0.45	0.50	0.55	0.60	0.65	0.70	0.75	0.80	1.00	1.25	1.50	1.75	2.00	2.50	3.00
0.80	471	210	155	119	94	77	64	54	46	40	35	31	21	14	10	8	7	5	4
0.90	581	259	191	147	116	94	78	66	57	49	43	38	25	17	12	9	8	6	5

Chart I. 5 groups, $r = 0.40$ at alpha = 0.01; NK 3 intervening groups, Tukey 5 groups																			
Power	Hypothesized ES																		
	0.20	0.30	0.35	0.40	0.45	0.50	0.55	0.60	0.65	0.70	0.75	0.80	1.00	1.25	1.50	1.75	2.00	2.50	3.00
0.80	505	226	166	128	101	82	68	58	50	43	38	33	22	15	11	9	7	5	4
0.90	619	276	203	156	124	101	83	70	60	52	46	40	27	18	13	10	8	6	5

Table 8.32. Sample size table for one-way repeated measures ANOVA MCP; $p=0.01$, $r=0.60$

Chart A. 3 groups, $r=0.60$ at alpha = 0.01; NK no intervening groups

Power	Hypothesized ES																		
	0.20	0.30	0.35	0.40	0.45	0.50	0.55	0.60	0.65	0.70	0.75	0.80	1.00	1.25	1.50	1.75	2.00	2.50	3.00
0.80	236	106	79	61	49	40	33	28	25	21	19	17	12	8	7	6	5	4	4
0.90	300	135	100	77	61	50	42	35	31	27	24	21	14	10	8	6	6	4	4

Chart B. 3 groups, $r=0.60$ at alpha = 0.01; NK 1 intervening group, Tukey 3 groups

Power	Hypothesized ES																		
	0.20	0.30	0.35	0.40	0.45	0.50	0.55	0.60	0.65	0.70	0.75	0.80	1.00	1.25	1.50	1.75	2.00	2.50	3.00
0.80	285	128	95	73	58	48	40	34	29	26	23	20	14	10	8	7	6	5	4
0.90	355	159	118	91	72	59	49	42	36	32	28	25	17	12	9	7	6	5	4

Chart C. 4 groups, $r=0.60$ at alpha = 0.01; NK no intervening groups

Power	Hypothesized ES																		
	0.20	0.30	0.35	0.40	0.45	0.50	0.55	0.60	0.65	0.70	0.75	0.80	1.00	1.25	1.50	1.75	2.00	2.50	3.00
0.80	235	106	78	60	48	39	33	28	24	21	18	16	11	8	6	5	4	4	3
0.90	299	134	99	76	61	49	41	35	30	26	23	20	14	9	7	6	5	4	3

Chart D. 4 groups, $r=0.60$ at alpha = 0.01; NK 1 intervening group

Power	Hypothesized ES																		
	0.20	0.30	0.35	0.40	0.45	0.50	0.55	0.60	0.65	0.70	0.75	0.80	1.00	1.25	1.50	1.75	2.00	2.50	3.00
0.80	284	127	94	73	58	47	39	33	29	25	22	20	13	9	7	6	5	4	4
0.90	354	158	117	90	72	58	49	41	35	31	27	24	16	11	8	7	6	4	4

Chart E. 4 groups, $r=0.60$ at alpha = 0.01; NK 2 intervening groups, Tukey 4 groups

Power	Hypothesized ES																		
	0.20	0.30	0.35	0.40	0.45	0.50	0.55	0.60	0.65	0.70	0.75	0.80	1.00	1.25	1.50	1.75	2.00	2.50	3.00
0.80	315	141	104	80	64	52	44	37	32	28	24	22	15	10	8	6	5	4	4
0.90	388	174	128	99	78	64	53	45	39	34	30	26	18	12	9	7	6	5	4

Chart F. 5 groups, $r=0.60$ at alpha = 0.01; NK no intervening groups

Power	Hypothesized ES																		
	0.20	0.30	0.35	0.40	0.45	0.50	0.55	0.60	0.65	0.70	0.75	0.80	1.00	1.25	1.50	1.75	2.00	2.50	3.00
0.80	235	105	78	60	48	39	32	27	24	21	18	16	11	8	6	5	4	3	3
0.90	299	134	99	76	60	49	41	35	30	26	23	20	13	9	7	5	5	4	3

Table 8.32. (*cont.*)

Chart G. 5 groups, $r = 0.60$ at alpha $= 0.01$; NK 1 intervening group																			
Power	Hypothesized ES																		
	0.20	0.30	0.35	0.40	0.45	0.50	0.55	0.60	0.65	0.70	0.75	0.80	1.00	1.25	1.50	1.75	2.00	2.50	3.00
0.80	284	127	94	72	57	47	39	33	28	25	22	19	13	9	7	5	5	4	3
0.90	354	158	117	90	71	58	48	41	35	30	27	24	16	11	8	6	5	4	3

Chart H. 5 groups, $r = 0.60$ at alpha $= 0.01$; NK 2 intervening groups																			
Power	Hypothesized ES																		
	0.20	0.30	0.35	0.40	0.45	0.50	0.55	0.60	0.65	0.70	0.75	0.80	1.00	1.25	1.50	1.75	2.00	2.50	3.00
0.80	315	141	104	80	64	52	43	37	31	27	24	21	14	10	7	6	5	4	3
0.90	388	173	128	98	78	64	53	45	38	33	29	26	17	12	9	7	6	4	4

Chart I. 5 groups, $r = 0.60$ at alpha $= 0.01$; NK 3 intervening groups, Tukey 5 groups																			
Power	Hypothesized ES																		
	0.20	0.30	0.35	0.40	0.45	0.50	0.55	0.60	0.65	0.70	0.75	0.80	1.00	1.25	1.50	1.75	2.00	2.50	3.00
0.80	337	151	111	86	68	56	46	39	34	29	26	23	15	11	8	6	5	4	4
0.90	413	185	136	105	83	68	56	48	41	35	31	28	18	12	9	7	6	5	4

Table 8.33. Sample size table for one-way repeated measures ANOVA MCP; $p = 0.10$, $r = 0.40$

Chart A. 3 groups, $r = 0.40$ at alpha = 0.10; NK no intervening groups

Power	Hypothesized ES																		
	0.20	0.30	0.35	0.40	0.45	0.50	0.55	0.60	0.65	0.70	0.75	0.80	1.00	1.25	1.50	1.75	2.00	2.50	3.00
0.80	187	84	62	48	38	31	26	22	19	17	15	13	9	6	5	4	4	3	3
0.90	258	116	85	66	52	42	35	30	26	22	20	17	12	8	6	5	4	3	3

Chart B. 3 groups, $r = 0.40$ at alpha = 0.10; NK 1 intervening group, Tukey 3 groups

Power	Hypothesized ES																		
	0.20	0.30	0.35	0.40	0.45	0.50	0.55	0.60	0.65	0.70	0.75	0.80	1.00	1.25	1.50	1.75	2.00	2.50	3.00
0.80	253	113	84	64	51	42	35	30	25	22	20	17	12	8	6	5	4	4	3
0.90	335	150	110	85	67	55	46	39	33	29	25	22	15	10	8	6	5	4	3

Chart C. 4 groups, $r = 0.40$ at alpha = 0.10; NK no intervening groups

Power	Hypothesized ES																		
	0.20	0.30	0.35	0.40	0.45	0.50	0.55	0.60	0.65	0.70	0.75	0.80	1.00	1.25	1.50	1.75	2.00	2.50	3.00
0.80	187	84	62	47	38	31	26	22	19	16	14	13	9	6	5	4	3	3	2
0.90	258	115	85	65	52	42	35	30	25	22	19	17	11	8	6	5	4	3	3

Chart D. 4 groups, $r = 0.40$ at alpha = 0.10; NK 1 intervening group

Power	Hypothesized ES																		
	0.20	0.30	0.35	0.40	0.45	0.50	0.55	0.60	0.65	0.70	0.75	0.80	1.00	1.25	1.50	1.75	2.00	2.50	3.00
0.80	252	113	83	64	51	41	34	29	25	22	19	17	11	8	6	5	4	3	3
0.90	335	149	119	85	67	55	45	38	33	28	25	22	15	10	7	6	5	4	3

Chart E. 4 groups, $r = 0.40$ at alpha = 0.10; NK 2 intervening groups, Tukey 4 groups

Power	Hypothesized ES																		
	0.20	0.30	0.35	0.40	0.45	0.50	0.55	0.60	0.65	0.70	0.75	0.80	1.00	1.25	1.50	1.75	2.00	2.50	3.00
0.80	296	132	98	75	60	48	40	34	29	25	22	20	13	9	7	5	5	4	3
0.90	384	172	126	97	77	63	52	44	38	33	29	25	17	11	8	6	5	4	3

Chart F. 5 groups, $r = 0.40$ at alpha = 0.10; NK no intervening groups

Power	Hypothesized ES																		
	0.20	0.30	0.35	0.40	0.45	0.50	0.55	0.60	0.65	0.70	0.75	0.80	1.00	1.25	1.50	1.75	2.00	2.50	3.00
0.80	186	83	61	47	38	31	25	22	19	16	14	13	8	6	4	4	3	2	2
0.90	258	115	85	65	52	42	35	29	25	22	19	17	11	8	6	4	4	3	2

Table 8.33. (*cont.*)

Chart G. 5 groups $r = 0.40$ at alpha = 0.10; NK 1 intervening group																			
Power	Hypothesized ES																		
	0.20	0.30	0.35	0.40	0.45	0.50	0.55	0.60	0.65	0.70	0.75	0.80	1.00	1.25	1.50	1.75	2.00	2.50	3.00
0.80	252	113	83	64	51	41	34	29	25	22	19	17	11	8	6	4	4	3	3
0.90	334	149	110	84	67	54	45	38	33	28	25	22	14	10	7	6	5	3	3

Chart H. 5 groups $r = 0.40$ at alpha = 0.10; NK 2 intervening groups																			
Power	Hypothesized ES																		
	0.20	0.30	0.35	0.40	0.45	0.50	0.55	0.60	0.65	0.70	0.75	0.80	1.00	1.25	1.50	1.75	2.00	2.50	3.00
0.80	296	132	97	75	59	48	40	34	29	25	22	20	13	9	6	5	4	3	3
0.90	384	171	126	97	77	62	52	44	37	32	28	25	17	11	8	6	5	4	3

Chart I. 5 groups $r = 0.40$ at alpha = 0.10; NK 3 intervening groups, Tukey 5 groups																			
Power	Hypothesized ES																		
	0.20	0.30	0.35	0.40	0.45	0.50	0.55	0.60	0.65	0.70	0.75	0.80	1.00	1.25	1.50	1.75	2.00	2.50	3.00
0.80	328	147	108	83	66	54	44	38	32	28	24	22	14	10	7	6	5	4	3
0.90	421	188	138	106	84	68	57	48	41	36	31	27	18	12	9	7	6	4	3

Table 8.34. Sample size table for one-way repeated measures ANOVA MCP; $p = 0.10$, $r = 0.60$

Chart A. 3 groups, $r = 0.60$ at alpha = 0.10; NK no intervening groups																			
Power	Hypothesized ES																		
	0.20	0.30	0.35	0.40	0.45	0.50	0.55	0.60	0.65	0.70	0.75	0.80	1.00	1.25	1.50	1.75	2.00	2.50	3.00
0.80	125	56	42	32	26	21	18	15	13	12	10	9	6	5	4	3	3	3	2
0.90	173	77	57	44	35	29	24	20	18	15	14	12	8	6	5	4	3	3	3

Chart B. 3 groups, $r = 0.60$ at alpha = 0.10; NK 1 intervening group, Tukey 3 groups																			
Power	Hypothesized ES																		
	0.20	0.30	0.35	0.40	0.45	0.50	0.55	0.60	0.65	0.70	0.75	0.80	1.00	1.25	1.50	1.75	2.00	2.50	3.00
0.80	169	76	56	44	35	28	24	20	18	15	14	12	8	6	5	4	4	3	3
0.90	224	100	74	57	46	37	31	26	23	20	17	16	10	7	6	5	4	3	3

Chart C. 4 groups, $r = 0.60$ at alpha = 0.10; NK no intervening groups																			
Power	Hypothesized ES																		
	0.20	0.30	0.35	0.40	0.45	0.50	0.55	0.60	0.65	0.70	0.75	0.80	1.00	1.25	1.50	1.75	2.00	2.50	3.00
0.80	125	56	41	32	26	21	17	15	13	11	10	9	6	4	4	3	3	2	2
0.90	172	77	57	44	35	28	24	20	17	15	13	12	8	6	4	4	3	3	2

Chart D. 4 groups, $r = 0.60$ at alpha = 0.10; NK 1 intervening group																			
Power	Hypothesized ES																		
	0.20	0.30	0.35	0.40	0.45	0.50	0.55	0.60	0.65	0.70	0.75	0.80	1.00	1.25	1.50	1.75	2.00	2.50	3.00
0.80	169	76	56	43	34	28	23	20	17	15	13	12	8	6	4	4	3	3	2
0.90	224	100	74	57	45	37	31	26	22	19	17	15	10	7	5	4	4	3	3

Chart E. 4 groups, $r = 0.60$ at alpha = 0.10; NK 2 intervening groups, Tukey 4 groups																			
Power	Hypothesized ES																		
	0.20	0.30	0.35	0.40	0.45	0.50	0.55	0.60	0.65	0.70	0.75	0.80	1.00	1.25	1.50	1.75	2.00	2.50	3.00
0.80	198	89	65	50	40	33	27	23	20	17	15	14	9	7	5	4	4	3	3
0.90	257	115	85	65	52	42	35	30	26	22	20	17	12	8	6	5	4	3	3

Chart F. 5 groups, $r = 0.60$ at alpha = 0.10; NK no intervening groups																			
Power	Hypothesized ES																		
	0.20	0.30	0.35	0.40	0.45	0.50	0.55	0.60	0.65	0.70	0.75	0.80	1.00	1.25	1.50	1.75	2.00	2.50	3.00
0.80	125	56	41	32	25	21	17	15	13	11	10	9	6	4	3	3	3	2	2
0.90	172	77	57	44	35	28	24	20	17	15	13	12	8	5	4	3	3	2	2

Table 8.34. (*cont.*)

Chart G. 5 groups, $r = 0.60$ at alpha = 0.10; NK 1 intervening group																			
Power	Hypothesized ES																		
	0.20	0.30	0.35	0.40	0.45	0.50	0.55	0.60	0.65	0.70	0.75	0.80	1.00	1.25	1.50	1.75	2.00	2.50	3.00
0.80	169	76	56	43	34	28	23	20	17	15	13	12	8	5	4	3	3	3	2
0.90	223	100	74	57	45	37	30	26	22	19	17	15	10	7	5	4	4	3	2

Chart H. 5 groups, $r = 0.60$ at alpha = 0.10; NK 2 intervening groups																			
Power	Hypothesized ES																		
	0.20	0.30	0.35	0.40	0.45	0.50	0.55	0.60	0.65	0.70	0.75	0.80	1.00	1.25	1.50	1.75	2.00	2.50	3.00
0.80	197	88	65	50	40	33	27	23	20	17	15	13	9	6	5	4	3	3	2
0.90	256	115	85	65	52	42	35	30	25	22	19	17	11	8	6	5	4	3	3

Chart I. 5 groups, $r = 0.60$ at alpha = 0.10; NK 3 intervening groups, Tukey 5 groups																			
Power	Hypothesized ES																		
	0.20	0.30	0.35	0.40	0.45	0.50	0.55	0.60	0.65	0.70	0.75	0.80	1.00	1.25	1.50	1.75	2.00	2.50	3.00
0.80	219	98	72	56	44	36	30	25	22	19	17	15	10	7	5	4	4	3	3
0.90	281	126	93	71	57	46	38	32	28	24	21	19	12	8	6	5	4	3	3

9 Interaction effects for factorial analysis of variance

Purpose of the statistic

To this point we have considered only differences among groups that can be conceptualized as representing a single independent variable. There are many occasions, however, when the investigator is interested in ascertaining the *joint* or *differential* effects that two or more independent variables exert on a dependent variable.

The two most common of these scenarios involve (a) testing whether or not an intervention is differentially more effective for one group (e.g., males vs. females, severely ill vs. less severely ill patients with the same diagnosis) than another and (b) testing whether or not subjects receiving the intervention change across assessment intervals more during the course of the study than do those in the control group.

In most cases the first scenario involves a between subjects design (i.e., different people are in all of the groups) while the second is usually a mixed design (i.e., although different individuals are contained in the treatment groups *everyone* in these groups is measured two or more times, such as at baseline and at the end of the study). In this chapter we will present power and sample size tables for two-factor interactions involving both between subject (including ANCOVA) and mixed designs.

Regardless of the type of design employed, however, an interaction tests a completely different hypothesis than occurs when the two independent variables are considered separately. In fact, all two-factor designs provide the opportunity to test a total of three hypotheses: (a) two for each of the independent variables (called main effects) which are basically identical to all of the hypotheses that have been discussed to this point in Chapters 4 through 8 and (b) one for the interaction between them. Let us illustrate these hypotheses and the method by which their all important effect sizes are calculated.

Experimental interactions

Let us assume that an intervention is designed that is expected to result in a greater reduction in pain, when compared to an appropriate control, among males than females. Assuming that no baseline assessments nor additional independent variables were employed, the most appropriate analysis for this study would be described as a 2 (intervention vs. control) × 2 (males vs. females) between subjects factorial ANOVA which would, in turn, yield three *F*-ratios addressing the following questions:

(1) Do males, irrespective of whether they received the intervention or not, differ from females with respect to self-reported pain at the end of the study?
(2) Do both males and females (considered together) who receive the intervention differ from males and females who do not with respect to self-reported pain measured at the end of the study?
(3) Is the intervention more (or less) effective for one gender than the other?

Now obviously these three questions would probably not be of equal interest, but all three of the *F*-ratios resulting from this study are independent of one another and all three would be tested via the 2 × 2 ANOVA. The interaction *F*, for example, could be significant even if males did not differ overall from females and if the intervention was not superior to the control for both genders combined. Figure 9.1 illustrates two of several scenarios reflecting this possibility.

Scenario A, which fortunately rarely occurs in actual clinical research, reflects a situation in which (a) the treatment main effect is zero (that is the grand mean of males and females receiving experimental treatment (E) is equal to the grand mean of both genders receiving control treatment (C)), (b) the gender main effect is zero (the grand mean of males receiving E and males receiving C is equal to the grand mean of females receiving both conditions), but (c) the interaction effect is statistically significant since males receiving E experience less pain than control males while females receiving E actually experience *more* pain than control females. Scenario B, which is much less rare although not common, reflects a situation in which (a) the grand mean for E would differ slightly from the grand mean of C (indicating that overall individuals receiving the intervention experienced less pain than those not receiving same), (b) the grand mean for males would be slightly less than the grand mean for females, but (c) the difference between males who receive E and those who receive C is considerably less than the difference between females who receive E and

Scenario A. No treatment or gender main effects but the intervention is effective for males and harmful for females.

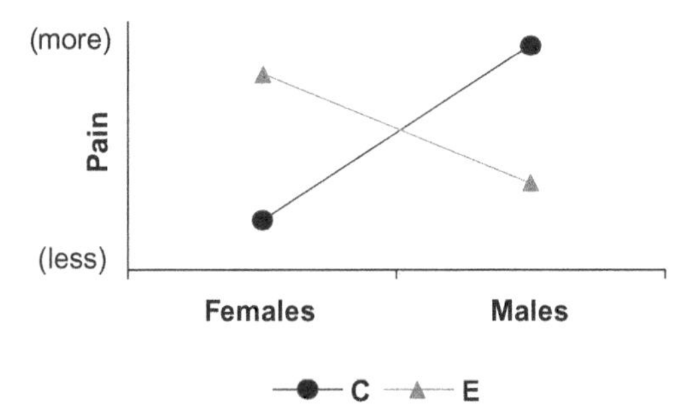

Scenario B. Small (but non-significant) treatment and gender main effects but the intervention is less effective for males than females.

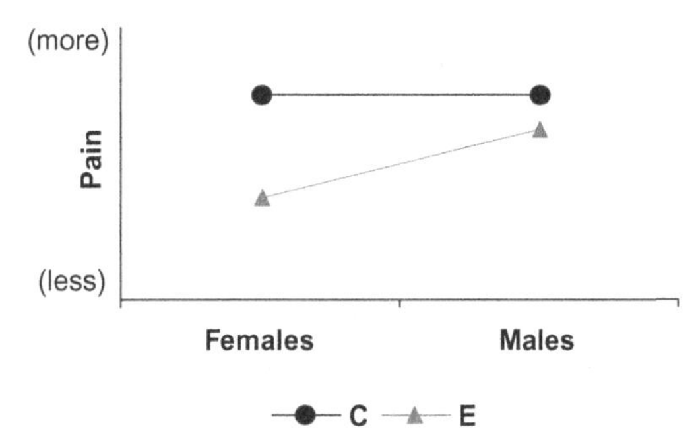

Figure 9.1. Two (of several) scenarios in which a statistically significant interaction may occur in the absence of any main effects.

those who receive C (which is defined as the interaction between the two variables: treatments and gender). Interaction *F*-ratios, then, test differences between completely different sets of means than do main effect *F*-ratios (which only look at one independent variable irrespective of any other(s)).

The purpose of this chapter, therefore, is to elucidate the estimation of power/required sample size for this third genre of research question, the interaction between two independent variables. (To calculate the power available for answering the first two questions the researcher could use the *t*-test tables in Chapter 4 after making the indicated *N*/group adjustment detailed in Chapter 10.) A between subjects ANOVA interaction term, then, is used when:

(1) at least two independent, non-continuous variables are present,
(2) the dependent variable is continuous, and

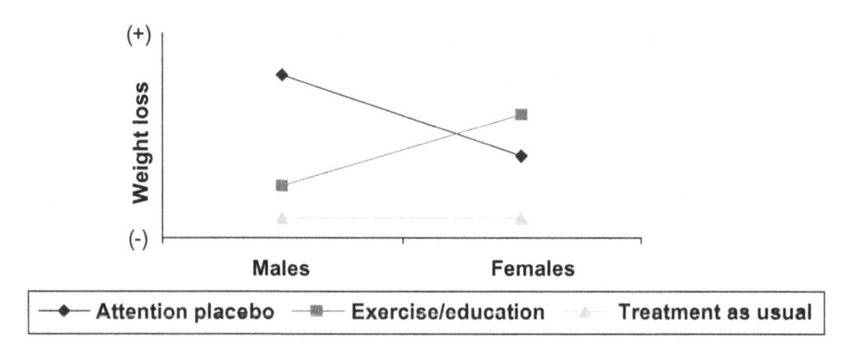

Figure 9.2. A complex (and unlikely) three-group interaction.

(3) the hypothesis being tested is expressed in terms of the differences between one or more levels of one independent variable being greater or less than the differences between one or more of the other levels of the same independent variable (see the third question above).

There is no theoretical limit on the number of independent variables that can be represented as factors in an ANOVA design, but in the vast majority of cases the interaction of interest is between two independent variables – hence we will only consider these two-way interactions in the present chapter. There is also no limit on the number of levels (i.e., groups) any given independent variable can have. Thus, in the hypothetical weight loss trial first presented in Chapter 6, a differential treatment effect for gender would be represented as a 3 (exercise/education intervention vs. attention placebo vs. treatment-as-usual) $\times$ 2 (male vs. female) interaction. On the surface, relationships such as this can appear relatively complicated, but usually even potentially complex interactions conceptually reduce to 2×2 interactions and, as we will demonstrate later, these 2×2 conceptualizations are quite convenient in facilitating estimation of the interaction ES values for a large number of designs. (It would be quite unusual, for example, for an investigator to hypothesize that (a) the exercise/education intervention would be less effective for males but more effective for females when compared to the attention placebo *while* (b) there would be no difference between males and females with respect to the amount of weight lost when they were receiving treatment-as-usual (see Figure 9.2).)

A much more likely hypothesized interaction in real world research would be something to the effect that (a) the weight loss intervention would be more effective for males than females but (b) there would be no concomitant differential effect for gender within the two control groups (see Figure

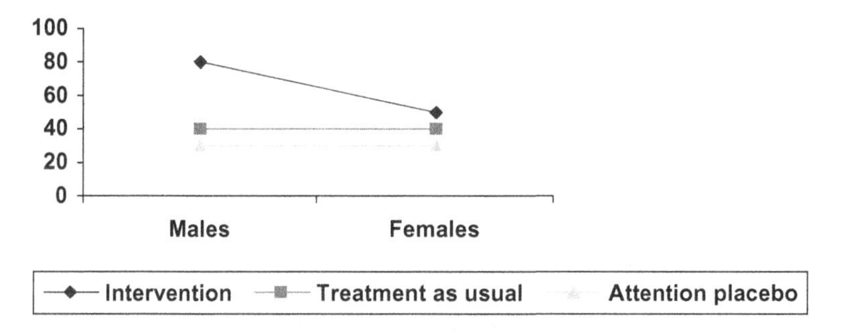

Figure 9.3. A more typical treatment × attribute interaction.

9.3). This hypothesis is conceptually quite similar to a 2 (male vs. females) × 2 (exercise/education intervention vs. the placebo + treatment-as-usual control groups combined), although the power of this interaction would be calculated based upon a two-degree rather than a one-degree of freedom model.

Regardless of the form the interaction is hypothesized to take, however, it is important to remember that one *integral* aspect of the power analytic process is that the ES *must* be hypothesized a priori. As difficult as this process is for main effects, it is an even more tenuous enterprise for interactions, especially anything more complex than those represented by 2 × 2 models. Let us therefore consider this more simple, and practical, model first since (a) they are much easier to conceptualize and, perhaps partly because of this, (b) most interaction hypotheses are conceived either in terms of a 2 × 2 framework (or are reducible thereto), and (c) because of this, we will conclude our discussion by recommending that all interaction ES values be estimated based upon this 2 × 2 model anyway.

A note on within subjects and ANCOVA interactions. Conceptually within subjects (including mixed designs containing both between and within subjects factors) and ANCOVA interactions are interpreted identically to their between subjects counterparts. Of equal importance, their interaction effect sizes are also computed identically, although different power tables are employed. Usually in within subjects or mixed designs, one main effect is represented as a time factor (e.g., baseline, EOT, first follow-up, second follow-up) and the other is a between subjects factor in which participants are randomly assigned to groups. (There are occasions, of course, when the same subjects are administered all treatments as well, producing a pure repeated measures design.) For an ANCOVA interaction, which can be used in either between or within subjects designs (see Chapter 10), the only practical effect that the covariate has from a power analytic perspective is that (a) more power is available for the interaction and (b) the patterns of means being tested are adjusted based upon pre-existing differences on the covariate.

Effect sizes for 2 × 2 interactions

In a 2 (E vs. C) × 2 (attribute A vs. attribute B) design, for a non-zero interaction to occur there must be a differential treatment effect between the two attributes. (In a 2 (E vs. C) × 2 (baseline vs. end-of-treatment) mixed design, for a non-zero interaction to occur one of the two groups must change more from baseline to EOT than the other.) From a between subject perspective, then, a statistically significant interaction implies that the intervention (or its control) must be relatively more or less effective for attribute A than it is for attribute B. (Alternatively, in the relatively rare scenario in which both factors represent experimental manipulations, an interaction implies an additive effect for the two treatment factors.) Given this, the possibilities by which this differential effect occurs can be conceptualized as reducing to one of the following: (a) there is a treatment effect (e.g., the intervention is superior to the control) for attribute A but none for attribute B, (b) there is a larger intervention effect for one attribute than another, or (c) there is a reverse effect whereby the intervention will have the opposite effect upon attribute B as it had on attribute A.

Let us illustrate one scenario (of several) for each of these possibilities employing gender as the attribute and a single experimental and control group as the treatment groups. As always, we recommend the use of standardized means (produced by dividing raw means by their standard deviations), which allows us to compute ES values quite directly by arbitrarily setting the lowest cell means to zero. Once this is done, the ES for a 2 × 2 interaction is nothing more than the difference between the two means for each level of one factor divided by the number of observations (or 2):

$$\text{Interaction ES} = \frac{[\text{ES difference A1 (or B1)} - \text{ES difference A2 (or B2)}]}{2}$$

As will be illustrated shortly, this very simple formula results in much smaller ES values than appears to be the case from simply examining the means involved. Formula 9.3 in the Technical appendix presents the etiology of this phenomenon more clearly, but basically it is due to the fact that the interaction ES must be independent of any residual effects due to either of the main effects considered in isolation. Let us illustrate this computational process via the three *general* forms (there are of course many specific variations that can emanate from these themes):

Chart 9.1. *Three general forms for a 2 (treatment) × 2 (attribute) interaction*
Scenario A. *Intervention effective for males and not effective for females*

	B1 Females	B2 Males	ES Difference
A1 Intervention	0.00	1.00	1.00
A2 Control	0.00	0.00	0.00
ES Difference	0.00	1.00	1.0 − 0 = 1.0[a]

Notes:
[a] Or 0 − 1.0 = −1.0 if column ES differences are subtracted from one another.
Interaction ES = [ES difference (A1 or B1) − ES difference (A2 or B2)]/2 = (1.0–0.0)/2 = 0.50

Scenario B. *Intervention effective for both genders, but more so for males than females*

	B1 Females	B2 Males	ES Difference
A1 Intervention	0.50	1.00	0.50
A2 Control	0.00	0.00	0.00
ES Difference	0.50	1.00	1 − 0.5 = 0.5[a]

Notes:
[a] Or 0.5 − 1.0 = −0.50 if column ES differences are subtracted from one another.
Interaction ES = [ES difference (A1 or B1) − ES difference (A2 or B2)]/2 = (0.5–0.0)/2 = 0.25

Scenario C. *Intervention effective for males but harmful for females*

	B1 Females	B2 Males	ES Difference
A1 Intervention	0.00	1.00	1.00
A2 Control	1.00	0.00	−1.00
ES Difference	−1.00	1.00	1.0 − (−1.0) = 2[a]

Notes:
[a] Or −1.0 − 1.0 = −2 if column ES differences are subtracted from one another.
Interaction ES = [ES difference (A1 or B1) − ES difference (A2 or B2)]/2 = (1.0 − (−1.0)/2) = 1.0

In 2×2 designs such as this, the meaning of the interaction is quite easy to grasp from these standardized means and the interaction ES is also quite easy to calculate. In scenario A, for example, the investigator is basically positing that the intervention will be effective *only* for males while the control will have no effect for either males or females. This is a very dramatic interaction and is not very likely to occur in an actual experiment. If the second factor were conceived as the baseline vs. the end of treatment (EOT) administrations of the outcome measure, however, this interaction could be quite realistic:

Chart 9.2. *Scenario A recast as a mixed design*

	B1 Baseline	B2 EOT	ES Difference
A1 Intervention	0.00	1.00	1.00
A2 Control	0.00	0.00	0.00
ES Difference	0.00	1.00	1.0 − 0 = 1.0

In either case, the calculation of the interaction ES (expressed in terms of *d*) involves nothing more than subtracting the difference between the experimental standardized means at the different levels of B (whether gender or time) and then subtracting these two differences from one another and dividing by 2. In other words, in Chart 9.2 the ES for the intervention alone between baseline and end of treatment is 1.00 (1.00 − 0.00) while the ES for the control across time is 0, hence the interaction ES is the difference between these two ES values divided by 2 or (1.00 − 0)/2 = 0.50.

In scenario B of Chart 9.1, the investigator is also hypothesizing that the treatment will be effective for males but relatively less effective for females. Here, however, the same pattern occurs in the control group but less dramatically, since a positive mean difference exists between both genders considered separately (i.e., with respect to the intervention vs. the control group), although it is higher for males than females, indicating that the treatments were differential with respect to gender.

Recast as a mixed design in Chart 9.3, scenario B translates to a much more realistic hypothesis since both the intervention and the control subjects improved over time, although the intervention group improved considerably more:

Chart 9.3. *Scenario B recast as a mixed design*

	B1 Baseline	B2 EOT	ES Difference
A1 Intervention	0.00	1.00	1.00
A2 Control	0.00	0.50	0.50
ES Difference	0.00	0.50	1.0 − 0.5 = 0.5

Again, the ES expressed in terms of *d* is obtained by subtracting the ES for the control group *alone* from the ES for the experimental group *alone* ($1.0 - 0.5 = 0.5$) and dividing by 2 (or the intervention vs. control ES at baseline (0) from the intervention vs. control ES at EOT (0.5) producing a negative 0.5 (the differences in signs between the row subtractions and the column subtractions are irrelevant in this situation)) which yields an interaction ES of 0.25.

Converting raw interaction means to standardized means. Prior to demonstrating the use of the 2×2 templates for calculating the power of an interaction, we will illustrate how raw cell means can be standardized. Basically all that is entailed, regardless of whether a between subjects or within subjects design is being employed, is for the investigator to write in the hypothesized means based upon the actual outcome measure to be employed as depicted in Chart 9.4. (Naturally the investigator will need to be familiar with the measure involved, both with respect to the likely mean scores to be obtained and their standard deviation. This information, however, is normally readily available from his/her preliminary studies or knowledge of the literature, thus let us assume that the measure in question has a score range from 10 to 50 in previous studies involving similar populations with a standard deviation of 11.5.)

Chart 9.4. *Positing a 2×2 interaction using raw hypothesized scores*

	Males	Females	Raw difference
Experimental	24.00	42.00	18.00
Control	11.50	16.10	4.60

This group of means is then converted to standard scores with a standard deviation of one by dividing each mean by the scale's standard deviation of 11.5, producing the following ES differences:

Chart 9.5. *Converting raw hypothesized scores to standard scores*

	Males	Females	ES Difference
Experimental	2.1	3.7	1.6
Control	1.0	1.4	0.4
ES Difference	1.1	2.3	1.2

Next, as illustrated in Chart 9.6, the smallest cell mean (males in the control group) is subtracted from each of the cells (including itself), which will permit the direct calculations of all the ES values relevant to this design.

Chart 9.6. *Ensuring at least one cell will have a standardized mean of zero*

	Males	Females	ES Difference
Experimental	2.1 − 1.0 = 1.1	3.7 − 1.0 = 2.7	2.7 − 1.1 = 1.6
Control	1.0 − 1.0 = 0.0	1.4 − 1.0 = 0.4	0.4 − 0.0 = 0.4
ES Difference	1.1 − 0.0 = 1.1	2.7 − 0.4 = 2.3	1.6 − 0.4 = 1.2

Thus the interaction ES can be calculated by simply finding the difference between the male vs. female ES values within each treatment and then dividing by 2 (which yields an interaction ES expressed in terms of d), or $(1.6 - 0.4)/2 = 1.2/2 = 0.6$. (Alternatively, employing column differences: $(1.1 - 2.3)/2 = -1.2/2 = -0.6$, which is the same as 0.6 since the sign is irrelevant except for interpretative purposes.)

Estimating the power for a between subject 2 × 2 interaction. At this point, all that is necessary for computing the power of a 2 × 2 between subject interaction, once the ES is estimated as described above, is to locate the intersection of the ES column and the N/cell row in Table 9.1, interpolating or rounding as desired. To facilitate this process for more complex studies, however, a preliminary template (9.1) is provided that will permit the computation of the interaction ES for both between subject ANOVA and ANCOVA designs and the five different interaction table sizes (i.e., 2 × 2, 2 × 3, 2 × 4, 3 × 3, and 3 × 4) covered in the present chapter, after which the investigator is instructed to access the template specific to his/her design.

Let us illustrate this process via the hypothesized interaction presented in Chart 9.6. Turning to the preliminary template (9.1), then, our investigator would follow the following steps:

Preliminary step 1. The 2 × 2 table shell would be selected and the hypothesized standardized means would be filled in as follows:

2×2	B1 M	B2 F
A1 E	1.1	2.7
A2 C	0.0	0.4

Preliminary step 2. Let us assume that 25 subjects per cell (total $N=100$) were available.

Preliminary step 3. The values from step 1 would basically be copied and the appropriate differences computed as follows:

	Males	Females	ES Difference
Experimental	1.1	2.7	1.6
Control	0.0	0.4	0.4
ES Difference	1.1	2.3	Diff of diff = 1.2

Preliminary step 4. The interaction ES would be computed by dividing the "differences of the effect size differences" by 2:

Interaction ES = 1.2/2 = 0.6

Preliminary step 5. This instructs the investigator to access Template 9.2 since this design does not employ a covariate.

Following the instructions in Template 9.2 yields the following results.

Power. Step 6. Since this is a 2×2 design and power is desired, Table 9.1 is accessed.

Step 7. Locating the intersection of the 0.60 ES column and the row N/cell = 25 produces an estimated power level of 0.84 for this particular hypothesized interaction.

Sample size. Alternatively, if the N/cell required to produce a power level of, say, 0.80 is desired, the first five steps would be identical except this desired value would be specified in step 2 instead of the N/cell. (Other than this, the steps involved in the preliminary template are identical to those employed for power.) Then, accessing Template 9.2:

Step 6. This instructs the investigator to access Table 9.26, Chart A.

Step 7. Locating the intersection of the 0.80 row and the 0.60 ES would indicate that an N/cell of 23 is needed to assure an 80% chance of achieving statistical significance for this particular hypothesized ES. Should values other than 0.80 and 0.90 be desired, step 7 of Template 9.2 instructs the investigator to again use Table 9.1 with the desired power level being located in the 0.60 ES column and the N/cell read to the left of the row in which this power level resides. Interpolation or estimation is used as desired. Thus if, for some reason, a power level of 0.60 were deemed sufficient, the closest value in the ES = 0.60 column would be 0.59 and the corresponding N/cell value would be 14.

Assuming that 0.80 was the desired power level and that an N/cell of 25 subjects (total $N = 100$) was the maximum sample size likely to be available to the investigator, the results of this analysis might be described as follows:

> A power analysis indicated that an N/cell of 25 subjects (total $N = 100$) would produce a power level of 0.84 for the 2 (treatment) × 2 (gender) between subjects interaction assuming an ES of 0.60 based upon the following pattern of means (standard deviation 11.5) observed in our pilot study (it is always assumed that a citation or description is provided to justify the summary statistics upon which a power/sample size analysis is based):

	Males	Females
Experimental	24.00	42.00
Control	11.50	16.10

(Note that, depending upon the audience, raw means and standard deviations may be more meaningful when describing the results of a power analysis, especially to other investigators who are familiar with the dependent variable metric as opposed to statisticians or methodologists.)

Alternatively, the interaction ES can simply be posited with reference to either the raw or standardized means sans a table (or a graph such as the ones presented in Figures 9.1 to 9.3 can be employed). Assuming that the necessary sample size were required to produce a desired power level of 0.80, the resulting analysis could be described as follows (if the targeted reader could be assumed to be conversant with the ES concept):

> A sample size analysis indicated that 23 participants per cell (or a total N of 92) would be required to yield an 80% chance of obtaining statistical significance for the 2 (intervention vs. control) × 2 (gender) between subjects interaction assuming a hypothesized interaction ES of 0.60 (Bausell & Li, 2002).

Within subject example of more complex two-factor interactions

As previously discussed, it is not at all uncommon to employ experimental designs with three or more levels of a treatment variable (e.g., an intervention and two different controls) or as many as four different time periods in a mixed design. Unfortunately interactions employing more than one degree of freedom are not as easily conceptualized as 2×2 designs nor are their effect sizes as simple to compute.

What we suggest, therefore, as witnessed by preliminary step 3 in Template 9.1, is that the interactions emanating from these more complicated designs be reduced by averaging across cells until the 2×2 interaction most closely associated with the actual hypothesized interaction is represented. In most cases this will be the largest possible ES resulting from the collapsing process. (If it is not, an investigator might be wise to redesign his/her study, since it is the largest possible interaction ES that will be tested inferentially.) This process makes these more complex interactions easier to conceptualize and it allows their ES values to be computed in the same metric (d) as employed in the previous chapters (i.e., by simply subtracting two ES differences from one another and dividing the resulting difference by 2 as discussed above). Admittedly this process introduces a degree of imprecision into the power analytic process since the resulting ES is not precisely the same as would be computed using Formula 9.3 in the Technical appendix. It is our opinion, however, that for most purposes the resulting estimate will be sufficiently accurate. Should more precision be desired, then Formula 9.3 can be employed to produce an exact hypothesized ES.

Collapsing a 2×3 interaction. Let us begin by illustrating this process via a study designed to assess the effects of an intervention across three time periods: (a) baseline (prior to implementation of the intervention), (b) end-of treatment (EOT, in which subjects are assessed after the completion of the intervention), and (c) a follow-up interval (measured at some pre-determined point in time after the EOT assessment to ascertain how long the intervention effect, if any, is capable of lasting). This thus results in a 2×3 mixed ANOVA model in which the first factor (intervention vs. control) is represented as a between subject factor and the time factor (baseline vs. EOT vs. follow-up) would serve as a within subject factor (or repeated measure), since the same subjects are represented within these cells.

In this type of design, then, it is the interaction (which assesses whether or not the two treatment groups change differentially across time) that is of primary scientific interest. As with power/sample size analyses for all interactions, the first step involves hypothesizing the pattern of cell means that will make up this interaction. To do this, we would begin by filling in the hypothesized pattern of standardized means into a 2×3 cell exactly as we did for the between subject 2×2 design. Let us suppose that this process resulted in the following hypothesized pattern of standardized means:

Chart 9.7. *A hypothetical 2×3 interaction*

	B1 Baseline	B2 EOT	B3 Follow-up
A1 Intervention	0.0	0.8	0.7
A2 Control	0.0	0.1	0.0

In words, what this pattern is hypothesizing is that the intervention will result in significantly greater changes from baseline to the end of the intervention interval and this superiority will largely be maintained during the entire follow-up time interval. The second step, assuming that power (and not sample size) were desired, would be to estimate how many subjects would be available in each cell, which we will assume to be equal to 25 (total $N=50$).

Step 3. Here the investigator will need to collapse the pattern of means in Chart 9.7 to the best 2×2 approximation possible. It is probably obvious how this would be accomplished, since A1 and A2 cannot be collapsed and overall the two follow-up measures (B2 and B3) are hypothesized to be considerably more similar to one another than B1 (baseline) will be to either. Sometimes the results are not quite this obvious, however, hence we suggest that the investigator always write the original interaction hypothesis in prose since this often holds a key to its subsequent collapsing. One way to express the interaction hypothesized in Chart 9.7 in prose would be:

> Subjects in the intervention group are hypothesized to improve more from baseline to EOT than are subjects in the control group. While subjects' scores are expected to decrease slightly over the follow-up period, the initial EOT superiority witnessed by the intervention group will be largely maintained.

Alternatively, this hypothesized interaction can be expressed more mechanistically by simply reiterating the relationships represented as though they have already occurred: (a) no difference exists between the two groups at baseline (which is reasonable since subjects were randomly assigned to the two groups), (b) the intervention is superior to the control by EOT (the control group is expected to change slightly, perhaps because of a placebo effect, but the intervention is hypothesized to change considerably more), and (c) by the follow-up interval both groups have regressed slightly but the intervention still appears to be superior to the control. (Note that the zero values hypothesized for both groups at baseline and for the control group after six months does not imply that the subjects have really achieved a zero score on whatever the outcome variable happens to be. This is simply due to our convention of making the table easier to read and interpret and is generated by first standardizing the means by dividing them by the standard deviation and then subtracting the lowest resulting standardized mean from each cell.)

Thus, either verbally or empirically, it would appear that cells B2 and B3 are the ones that should be combined. There are, however, three possible ways to collapse the data represented by a 2×3 design into a 2×2 table and these are depicted in Chart 9.8:

Chart 9.8. *Possible ways to collapse a 2×3 table to a 2×2 table*

Scenario A

	Baseline/EOT	Follow-up	ES Difference
Intervention	0.40	0.70	0.30
Control	0.05	0.00	−0.05
ES Difference	0.25	0.50	Diff − diff = 0.35

Interaction ES = (ES Difference − ES Difference)/2 = 0.35/2 = 0.175

Scenario B

	Baseline/Follow-up	EOT	ES Difference
Intervention	0.35	0.80	0.45
Control	0.00	0.10	0.10
ES Difference	0.35	0.70	Diff − diff = 0.35

Interaction ES = (ES Difference − ES Difference)/2 = 0.35/2 = 0.175

Scenario C

	Baseline	EOT/Follow-up	ES Difference
Intervention	0.00	0.75	0.75
Control	0.00	0.05	0.05
ES Difference	0.00	0.60	Diff − diff = 0.70

Interaction ES = (ES Difference − ES Difference)/2 = 0.70/2 = 0.35

While any of the three could theoretically be chosen as the collapsed 2×2 table from which to calculate the interaction ES, obviously scenario C best reflects the actual interaction posited. *Not coincidentally, this permutation also produces the largest possible interaction ES since it represents the hypothesis that generated the projected means in the first place.* (It is the largest collapsed ES that is actually tested by the interaction term, hence it is the largest possible 2×2 ES that should normally be employed in estimating the power of the overall interaction.)

Armed with this ES, the investigator is instructed to turn to Template 9.4 (after completing the preliminary Template 9.1), since the represented design involves one between and one within subjects factor. Here the investigator is asked (step 6) to estimate the most likely average correlation among the repeated measures (i.e., 0.40 or 0.60, for other values see Chapter 10). Assuming that 0.60 was chosen, step 7 instructs the investigator to turn to Table 9.19. Step 8 indicates that the power for this particular interaction would be found at the intersection of the *N*/cell row of 25 and the 0.35 ES column and found to be 0.85. Should the *N*/cell value to achieve a power level of exactly 0.80 be required for these parameters, Table 9.36 Chart B would indicate this value to be 22 by following the instructions in steps 7 and 8, which in turn would translate to a total *N* of 22×2=44, since the same subjects are represented in the B2 and B3 cells.

Collapsing a 2×4 table. In most cases, the more cells involved in a table, the more combinations of 2×2 tables by which these cells can be collapsed. With a 2×3 shell there are three possible 2×2 tables that can be constructed; for a 2×4 design there are six, for a 3×3 there are nine, and the possibilities increase correspondingly thereafter. This potential proliferation highlights the importance of constructing a definitive interaction hypothesis to guide the collapsing process, since there is almost always one 2×2 combination that "best" captures the investigator's actual expectations. (If there is not, then there is no real choice other than to employ Formula 9.3 in the Technical appendix to compute the ES, although the investigator should remember that he/she will need to *double* this value when employing one of the tables in the present chapter since their ES values are expressed in terms of *d* rather than *f*.)

To illustrate, let us extend the previous design to a 2×4 configuration by adding a fourth level to our original design. (For a between subjects design, the most likely scenario would be a trial in which the differential effects of two treatments upon four time periods were to be tested.)

Even though there are six possible ways to collapse a 2×4 table into a 2×2 representation (i.e., B1 vs. B2/B3/B4, B2 vs. B1/B3/B4, B3 vs. B1/B2/B4, B4 vs. B1/B2/B3, B1/B2 vs. B3/B4, and B1/B3 vs. B2/B4), let us consider the scenario in which originally the same configuration of means illustrated in Chart 9.8 were expected to accrue through the first follow-up, but the intervention superiority was expected to dissipate completely by the second follow-up assessment:

Chart 9.9. *A hypothetical 2 × 4 interaction*

	B1 Baseline	B2 EOT	B3 1st Follow-up	B4 2nd Follow-up
A1 Intervention	0.0	0.8	0.7	0.0
A2 Control	0.0	0.1	0.0	0.0

In words, this might be described as follows: Subjects in the intervention group are hypothesized to improve more from baseline to EOT than are subjects in the control group. While subjects' scores are expected to decrease only slightly over the first follow-up period, this process will accelerate by the second follow-up interval to the point that scores basically return to baseline values. Now admittedly this is a wordy description of the hypothesized interaction, but the more complex the interactions become, the more unwieldy their prose descriptions become. It is an accurate translation of the numbers the investigator is hypothesizing in Chart 9.9, however, and it suggests the best 2 × 2 representation of this interaction: namely that the EOT/first follow-up and the baseline/second follow-up intervals are more similar to one another than is true of any of the other possible combinations, and as usual this produces the largest 2 × 2 ES of the six possible combinations.

Combining cells B1 with B4 and B2 with B3, then, produces the following 2 × 2 representation:

Chart 9.10. *Chart 9.9 collapsed as a 2 × 2 interaction*

	Baseline/2nd Follow-up	EOT/1st Follow-up	ES Difference
Intervention	0.00	0.75	0.75
Control	0.00	0.05	0.05
ES Difference	0.00	0.70	Diff − diff = 0.70

Interaction ES = (ES Difference − ES Difference)/2 = 0.70/2 = 0.35

Once this step is completed, the computation of power is quite simple. If we assume that the *N*/cell (again we are referring to the number of observations, not the number of "unique" subjects) is 25 and the average correlation across repeated measures is 0.60, then the power to test this interaction would be interpolated to be approximately 0.91 via the use of Table 9.21.

Collapsing a 3 × 3 table. Although we cannot illustrate all of the possible interaction designs, we will provide one final example in which both factors involve more than two levels. Let us employ a between subjects

ANCOVA example for this interaction. Suppose, therefore, that a trial involved three treatments (one intervention, a treatment-as-usual control, and an attention placebo control) and three levels of an attribute for which the treatment was hypothesized to be differentially effective (e.g., adults 18 to 39, 40 to 65, and over 65 years of age) with pre-experimental values on the dependent variable serving as the covariate. Let us assume that the investigator hypothesized the following interaction: (a) the intervention is expected to be more effective than either of the two controls for the two younger age groups but only marginally so for subjects 65 years of age and older, (b) there will be very little difference between the two control groups although a slight placebo effect is expected, and (c) there should be little or no difference between the age groups with respect to whether or not they manifest a placebo effect.

This particular 3 × 3 interaction hypothesis therefore might be associated with the following hypothesized pattern of standardized means:

Chart 9.11. *A hypothetical 3 × 3 interaction*

	18–39 years	40–65 years	Over 65 years
Intervention	1.0	0.8	0.2
Attention control	0.2	0.2	0.2
Treat-as-usual	0.0	0.0	0.0

Collapsed, this table would produce the following 2 × 2 pattern:

Chart 9.12. *Chart 9.11 collapsed into a 2 × 2 interaction shell*

	18-65 years	Over 65 years	ES Difference
Intervention	0.90	0.20	0.70
Control	0.10	0.10	0.00
ES Difference	0.80	0.10	Diff − diff = 0.70

Interaction ES = (ES Difference − ES Difference)/2 = 0.70/2 = 0.35

Assuming the same *N*/cell of 25, step 5 in the preliminary template would send the investigator to Template 9.3, where, assuming a covariate to dependent variable correlation of 0.60 (step 6) were hypothesized, step 7 would in turn indicate that the power (0.74) for this interaction could be interpolated from Table 9.13 (or the necessary *N*/cell (29 for power of 0.80 and 37 for 0.90) for a desired power level could be obtained in Table 9.30 Chart D).

The results of this analysis might be communicated as follows:

> A power analysis indicated that an N/cell of 25 subjects (total $N=225$) would produce a power level of 0.74 for the 3 (intervention vs. attention placebo vs. treatment-as-usual) $\times$3 (19–39 vs. 40–65 vs. >65 years of age) between subjects ANCOVA assuming an ES of 0.35 and a covariate–dependent variable relationship of 0.60 based upon the pattern of standardized means hypothesized in Chart 9.11. (Some justification of this pattern should be tendered.) A sample size analysis indicated that 29 subjects per cell would be required to yield an 80% chance of this hypothesized interaction reaching statistical significance.

Should such a sample size (total *N* for the trial of 261) be judged as unrealistic, the investigator would need to explore seriously his/her other options, among which would be to (a) settle for testing the main effect hypothesis that the treatment was effective, (b) employ fewer cells (e.g., contrasting younger vs. over 65 subjects sans the middle group), (c) exclude over 65 years of age subjects altogether since the intervention was not hypothesized to be effective for this group. Assuming the latter option were chosen, the investigator would have a three condition, one-way ANCOVA design with the following hypothesized standardized means: intervention group (0.90) and attention placebo group (0.20) vs. treatment-as-usual (0.00). Accessing the relevant tables in Chapter 7 would indicate that an *N*/group of only 17 subjects (total $N=51$) would be necessary to obtain an 80% chance of achieving a statistically significant omnibus *F*-ratio and only 28 subjects per group (total $N=84$) to have an 80% chance of obtaining statistically significant pairwise comparisons between the intervention and the attention control group using the conservative Tukey HSD procedure – hence our advice in Chapter 2 concerning hypothesizing main effects rather than interactions when possible, since the latter come with a heavy price.

Summary

Interaction hypotheses involve the differential effects that two separate, categorical independent variables have upon one another. In experimental research they are normally written in terms of the various treatments (a) being more or less effective for different attributes of the participating subjects or (b) changing differentially across time (e.g., from baseline to EOT to follow-up).

While the steps in performing a power/sample size analysis for an interaction hypothesis are the same as those for main effects (i.e., specifying the hypothesized ES, *p*-value, and number of subjects/desired power), the process of hypothesizing an interaction ES is somewhat more complex than was the case for the ES values discussed in previous chapters. We have, however, provided templates and suggestions to facilitate this process including:

(1) selecting the table shell that reflects the configuration of cells involved (i.e., 2×2, 2×3, 2×4, 3×3, and 3×4),
(2) writing a detailed prose description of the expected interaction,
(3) filling the table shell in with standardized means that best reflect this hypothesis,
(4) collapsing the table shell to the 2×2 configuration that best represents the hypothesis (or that produces the most dramatic 2×2 interaction representation),
(5) computing the ES by subtracting the difference of differences of the treatment effect for each level of the second independent variable and dividing by two, and
(6) accessing the appropriate table indicated in the power/sample size templates provided.

Power estimates based upon interactions appear to be considerably smaller than those for overall between groups because their ES values must be corrected for any effects due to the main effects involved. For this reason it is especially important to conduct such analyses if an interaction is an integral component of the study being conducted.

Template 9.1. Generic preliminary two-factor ANOVA interaction template

Because of the number of permutations involved in estimating the power/sample size of a two-factor ANOVA, it is not possible to present separate templates for every possibility. Regardless of the design, however, all two-factor interactions involve the same preliminary steps, hence the following instructions:

Preliminary step 1. Choose one of the following table shells and fill in the standardized means that are expected to make up the hypothesized interaction. (It is recommended that actual variable names be written next to the letters representing the various cells and factor levels.)

2×2	B1	B2
A1		
A2		

2×3	B1	B2	B3
A1			
A2			

2×4	B1	B2	B3	B4
A1				
A2				

3×3	B1	B2	B3
A1			
A2			
A3			

3×4	B1	B2	B3	B4	B5
A1					
A2					
A3					

Preliminary step 2. For the estimation of power, estimate the "*N*/cell" available (recalling that it is always wise to model power parameters); for the estimation of sample size, fill in the desired power:

N/cell= or, desired power=

Preliminary step 3. Collapse the standardized means from step 1 above into the following 2×2 table. (For actual 2×2 analyses, simply copy the results from step 1.)

	B1	B2	ES Difference
A1	A1B1=	A1B2=	B1 − B2=
A2	A2B1=	A2B2=	B1 − B2=
ES Difference	A1 − A2=	A1 − A2=	Diff of diff=

Preliminary step 4. Calculate the interaction ES by dividing the differences of the ES differences by 2:

Interaction ES=diff of diff, e.g., [(B1 − B2 for A1) − (B1 − B2 for A2)]/2=

Preliminary step 5. If both factors involve different, unmatched subjects, go to Template 9.2. If both factors involve different subjects, but a covariate is used, go to Template 9.3. If one factor involves a repeated measure, go to Template 9.4.

Template 9.2. Interaction power and sample size for between subject designs

Step 6. Locate the appropriate table based upon the following chart:

Design	Power	Sample size (for powers 0.80 or 0.90) $p=0.05$	$p=0.01$	$p=0.10$
2×2	Table 9.1	Table 9.26 Chart A	Table 9.27 Chart A	Table 9.28 Chart A
2×3	Table 9.2	Table 9.26 Chart B	Table 9.27 Chart B	Table 9.28 Chart B
2×4	Table 9.3	Table 9.26 Chart C	Table 9.27 Chart C	Table 9.28 Chart C

3×3	Table 9.4	Table 9.26 Chart D	Table 9.27 Chart D	Table 9.28 Chart D
3×4	Table 9.5	Table 9.26 Chart E	Table 9.27 Chart E	Table 9.28 Chart E

Step 7. Locate power (interpolate as necessary) at the intersection of the ES column (step 4) and the *N*/cell row (step 2).

Power =

Or, to locate the *N*/cell necessary to achieve desired power:

(1) For desired power levels of 0.80 and 0.90, locate the *N*/cell at the intersection of the ES column and the desired power row for the indicated table detailed in step 6.
(2) for other power levels, use the power tables indicated in step 6 and locate the nearest value to the desired power level in the ES column (or interpolate as desired) and read the *N*/cell at the beginning of that row.

N/cell =

Template 9.3. Interaction power and sample size for ANCOVA designs

---

Step 6. Select 0.40 or 0.60 as the most likely value for the covariate–dependent variable correlation:

r =

Step 7. Locate the appropriate table based upon the following chart:
(A = Chart A; B = Chart B; C = Chart C; D = Chart D; E = Chart E).

Design	Power		Sample size					
			$p=0.05$		$p=0.01$		$p=0.10$	
	$r=0.40$	$r=0.60$	$r=0.40$	$r=0.60$	$r=0.40$	$r=0.60$	$r=0.40$	$r=0.60$
2×2	9.6	9.7	9.29 A	9.30 A	9.31 A	9.32 A	9.33 A	9.34 A
2×3	9.8	9.9	9.29 B	9.30 B	9.31 B	9.32 B	9.33 B	9.34 B
2×4	9.10	9.11	9.29 C	9.30 C	9.31 C	9.32 C	9.33 C	9.34 C
3×3	9.12	9.13	9.29 D	9.30 D	9.31 D	9.32 D	9.33 D	9.34 D
3×4	9.14	9.15	9.29 E	9.30 E	9.31 E	9.32 E	9.33 E	9.34 E

Step 8. Locate power (interpolate as necessary) at the intersection of the ES column (step 4) and the *N*/cell row (step 3), interpolating or rounding as desired.

Power =

Or, to locate the *N*/cell necessary to achieve desired power:

(3) For desired power levels of 0.80 and 0.90, locate the *N*/cell at the intersection of the ES column and the desired power row for the indicated table detailed in step 7.
(4) For other power levels, locate the nearest value to desired power level in the ES column (or interpolate as desired) and read the *N*/cell at the beginning of that row.

N/cell =

Template 9.4. Interaction power and sample size for mixed designs

Step 6. Select 0.40 or 0.60 as the most likely correlation for the within subject factor:

$r=$

Step 7. Locate the appropriate table based upon the following chart:
(A = Chart A; B = Chart B; C = Chart C; D = Chart D; E = Chart E).

Design	Power		Sample size					
			$p=0.05$		$p=0.01$		$p=0.10$	
	$r=0.40$	$r=0.60$	$r=0.40$	$r=0.60$	$r=0.40$	$r=0.60$	$r=0.40$	$r=0.60$
2×2	9.16	9.17	9.35 A	9.36 A	9.37 A	9.38 A	9.39 A	9.40 A
2×3	9.18	9.19	9.35 B	9.36 B	9.37 B	9.38 B	9.39 B	9.40 B
2×4	9.20	9.21	9.35 C	9.36 C	9.37 C	9.38 C	9.39 C	9.40 C
3×3	9.22	9.23	9.35 D	9.36 D	9.37 D	9.38 D	9.39 D	9.40 D
3×4	9.24	9.25	9.35 E	9.36 E	9.37 E	9.38 E	9.39 E	9.40 E

Step 8. Locate power (interpolate as necessary) at the intersection of the ES column (step 4) and the *N*/cell row (step 2), interpolating or rounding as desired.

Power =

Or, to locate the *N*/cell necessary to achieve desired power:

(5) For desired power levels of 0.80 and 0.90, locate the *N*/cell at the intersection of the ES column and the desired power row for the indicated table detailed in step 7.
(6) For other power levels, access the power table indicated in step 7 and locate the nearest value to the desired power level in the ES column (or interpolate as desired) and read the *N*/cell at the beginning of that row.

N/cell =

Table 9.1. Power table for between subjects analysis; 2 × 2 design at alpha = 0.05

n	Hypothesized ES																				
	0.10	0.15	0.20	0.30	0.35	0.40	0.45	0.50	0.55	0.60	0.65	0.70	0.75	0.80	1.00	1.25	1.50	1.75	2.00	2.50	3.00
3	5	5	6	7	8	8	9	11	12	13	15	16	18	20	29	42	55	69	80	93	99
4	5	5	6	8	9	10	12	13	16	18	20	23	25	28	41	58	74	86	93	99	.
5	5	5	6	9	10	12	14	17	19	22	26	29	33	36	52	71	85	94	98	.	.
6	5	5	6	10	12	14	17	20	23	27	31	35	39	44	62	80	92	98	99	.	.
7	5	6	7	11	13	16	19	23	27	32	36	41	46	51	70	87	96	99	.	.	.
8	5	6	7	12	15	18	22	26	31	36	41	47	52	57	76	92	98	.	.	.	.
9	5	6	8	13	16	20	25	30	35	40	46	52	57	63	82	95	99	.	.	.	.
10	5	6	8	14	18	22	27	33	39	45	51	57	62	68	86	97	99	.	.	.	.
11	5	7	9	15	20	24	30	36	42	48	55	61	67	73	89	98	.	.	.	.	.
12	5	7	9	16	21	27	33	39	46	52	59	65	71	77	92	99	.	.	.	.	.
13	5	7	10	18	23	29	35	42	49	56	63	69	75	80	94	99	.	.	.	.	.
14	5	7	10	19	24	31	37	45	52	59	66	72	78	83	95	.	.	.	.	.	.
15	6	8	11	20	26	33	40	47	55	62	69	76	81	86	97	.	.	.	.	.	.
20	6	9	13	26	34	42	51	60	68	75	82	87	91	94	99	.	.	.	.	.	.
25	7	11	16	32	41	51	61	70	78	84	90	93	96	98	.	.	.	.	.	.	.
30	7	12	19	37	48	59	69	78	85	90	94	97	98	99	.	.	.	.	.	.	.
35	8	13	21	43	54	66	76	84	90	94	97	98	99	.	.	.	.	.	.	.	.
40	9	15	24	48	60	72	81	88	93	97	98	99	.	.	.	.	.	.	.	.	.
45	9	16	27	52	65	77	86	92	96	98	99	.	.	.	.	.	.	.	.	.	.
50	10	18	29	57	70	81	89	94	97	99	.	.	.	.	.	.	.	.	.	.	.
55	11	19	32	61	74	84	92	96	98	99	.	.	.	.	.	.	.	.	.	.	.
60	11	21	34	65	78	87	94	97	99	.	.	.	.	.	.	.	.	.	.	.	.
65	12	22	37	68	81	90	95	98	99	.	.	.	.	.	.	.	.	.	.	.	.
70	12	24	39	71	84	92	97	99	.	.	.	.	.	.	.	.	.	.	.	.	.
75	13	25	41	74	86	94	97	99	.	.	.	.	.	.	.	.	.	.	.	.	.
80	14	27	44	77	88	95	98	99	.	.	.	.	.	.	.	.	.	.	.	.	.
90	15	30	48	82	92	97	99	.	.	.	.	.	.	.	.	.	.	.	.	.	.
100	16	33	52	86	94	98	99	.	.	.	.	.	.	.	.	.	.	.	.	.	.
110	18	35	56	89	96	99	.	.	.	.	.	.	.	.	.	.	.	.	.	.	.
120	19	38	60	91	97	99	.	.	.	.	.	.	.	.	.	.	.	.	.	.	.
130	20	41	63	93	98	.	.	.	.	.	.	.	.	.	.	.	.	.	.	.	.
140	22	43	67	95	99	.	.	.	.	.	.	.	.	.	.	.	.	.	.	.	.
150	23	46	70	96	99	.	.	.	.	.	.	.	.	.	.	.	.	.	.	.	.
175	26	52	76	98	.	.	.	.	.	.	.	.	.	.	.	.	.	.	.	.	.
200	30	57	82	99	.	.	.	.	.	.	.	.	.	.	.	.	.	.	.	.	.
225	33	62	86	99	.	.	.	.	.	.	.	.	.	.	.	.	.	.	.	.	.
250	36	67	89	.	.	.	.	.	.	.	.	.	.	.	.	.	.	.	.	.	.
300	42	75	94	.	.	.	.	.	.	.	.	.	.	.	.	.	.	.	.	.	.
350	47	81	96	.	.	.	.	.	.	.	.	.	.	.	.	.	.	.	.	.	.
400	53	86	98	.	.	.	.	.	.	.	.	.	.	.	.	.	.	.	.	.	.
450	58	90	99	.	.	.	.	.	.	.	.	.	.	.	.	.	.	.	.	.	.
500	62	92	99	.	.	.	.	.	.	.	.	.	.	.	.	.	.	.	.	.	.
600	70	96	.	.	.	.	.	.	.	.	.	.	.	.	.	.	.	.	.	.	.
700	76	98	.	.	.	.	.	.	.	.	.	.	.	.	.	.	.	.	.	.	.
800	82	99	.	.	.	.	.	.	.	.	.	.	.	.	.	.	.	.	.	.	.
900	86	99	.	.	.	.	.	.	.	.	.	.	.	.	.	.	.	.	.	.	.
1000	89	.	.	.	.	.	.	.	.	.	.	.	.	.	.	.	.	.	.	.	.

Table 9.2. Power table for between subjects analysis; 2 × 3 design at alpha = 0.05

n	Hypothesized ES																				
	0.10	0.15	0.20	0.30	0.35	0.40	0.45	0.50	0.55	0.60	0.65	0.70	0.75	0.80	1.00	1.25	1.50	1.75	2.00	2.50	3.00
3	5	5	6	7	8	9	10	11	12	14	16	17	19	22	32	47	63	77	87	97	.
4	5	5	6	8	9	11	12	14	16	19	22	25	28	31	46	66	82	92	97	.	.
5	5	6	6	9	11	13	15	18	21	24	28	32	36	40	59	79	92	98	.	.	.
6	5	6	7	10	12	15	18	21	25	30	34	39	44	49	69	88	97	99	.	.	.
7	5	6	7	11	14	17	21	25	30	35	40	46	52	57	77	93	99	.	.	.	.
8	5	6	8	13	16	20	24	29	34	40	46	52	58	64	84	96	99	.	.	.	.
9	5	7	8	14	18	22	27	33	39	45	52	58	65	70	89	98	.	.	.	.	.
10	5	7	9	15	19	24	30	36	43	50	57	64	70	76	92	99	.	.	.	.	.
11	6	7	9	16	21	27	33	40	47	54	62	69	75	80	95	99	.	.	.	.	.
12	6	7	10	18	23	29	36	43	51	59	66	73	79	84	96	.	.	.	.	.	.
13	6	8	10	19	25	32	39	47	55	63	70	77	83	87	98	.	.	.	.	.	.
14	6	8	11	20	27	34	42	50	58	66	74	80	86	90	98	.	.	.	.	.	.
15	6	8	12	22	29	36	45	53	62	70	77	83	88	92	99	.	.	.	.	.	.
20	7	10	14	28	38	47	58	67	76	83	89	93	96	98	.	.	.	.	.	.	.
25	7	11	17	35	46	57	68	78	85	91	95	97	99	99	.	.	.	.	.	.	.
30	8	13	20	42	54	66	77	85	91	95	98	99	.	.	.	.	.	.	.	.	.
35	8	14	23	48	61	73	83	91	95	98	99	.	.	.	.	.	.	.	.	.	.
40	9	16	26	54	67	79	88	94	97	99	.	.	.	.	.	.	.	.	.	.	.
45	10	18	29	59	73	84	92	96	99	.	.	.	.	.	.	.	.	.	.	.	.
50	11	19	32	64	78	88	94	98	99	.	.	.	.	.	.	.	.	.	.	.	.
55	11	21	35	68	82	91	96	99	.	.	.	.	.	.	.	.	.	.	.	.	.
60	12	23	38	72	85	93	97	99	.	.	.	.	.	.	.	.	.	.	.	.	.
65	13	24	41	76	88	95	98	.	.	.	.	.	.	.	.	.	.	.	.	.	.
70	13	26	44	79	90	96	99	.	.	.	.	.	.	.	.	.	.	.	.	.	.
75	14	28	46	82	92	97	99	.	.	.	.	.	.	.	.	.	.	.	.	.	.
80	15	29	49	85	94	98	.	.	.	.	.	.	.	.	.	.	.	.	.	.	.
90	16	33	54	89	96	99	.	.	.	.	.	.	.	.	.	.	.	.	.	.	.
100	18	36	59	92	98	.	.	.	.	.	.	.	.	.	.	.	.	.	.	.	.
110	19	39	63	94	99	.	.	.	.	.	.	.	.	.	.	.	.	.	.	.	.
120	21	43	67	96	99	.	.	.	.	.	.	.	.	.	.	.	.	.	.	.	.
130	22	46	71	97	.	.	.	.	.	.	.	.	.	.	.	.	.	.	.	.	.
140	24	49	74	98	.	.	.	.	.	.	.	.	.	.	.	.	.	.	.	.	.
150	25	52	77	99	.	.	.	.	.	.	.	.	.	.	.	.	.	.	.	.	.
175	29	58	84	99	.	.	.	.	.	.	.	.	.	.	.	.	.	.	.	.	.
200	33	65	89	.	.	.	.	.	.	.	.	.	.	.	.	.	.	.	.	.	.
225	36	70	92	.	.	.	.	.	.	.	.	.	.	.	.	.	.	.	.	.	.
250	40	75	95	.	.	.	.	.	.	.	.	.	.	.	.	.	.	.	.	.	.
300	47	82	97	.	.	.	.	.	.	.	.	.	.	.	.	.	.	.	.	.	.
350	53	88	99	.	.	.	.	.	.	.	.	.	.	.	.	.	.	.	.	.	.
400	59	92	.	.	.	.	.	.	.	.	.	.	.	.	.	.	.	.	.	.	.
450	65	95	.	.	.	.	.	.	.	.	.	.	.	.	.	.	.	.	.	.	.
500	70	97	.	.	.	.	.	.	.	.	.	.	.	.	.	.	.	.	.	.	.
600	78	99	.	.	.	.	.	.	.	.	.	.	.	.	.	.	.	.	.	.	.
700	84	99	.	.	.	.	.	.	.	.	.	.	.	.	.	.	.	.	.	.	.
800	89	.	.	.	.	.	.	.	.	.	.	.	.	.	.	.	.	.	.	.	.
900	92	.	.	.	.	.	.	.	.	.	.	.	.	.	.	.	.	.	.	.	.
1000	95	.	.	.	.	.	.	.	.	.	.	.	.	.	.	.	.	.	.	.	.

Table 9.3. Power table for between subjects analysis; 2 × 4 design at alpha = 0.05

n	Hypothesized ES																				
	0.10	0.15	0.20	0.30	0.35	0.40	0.45	0.50	0.55	0.60	0.65	0.70	0.75	0.80	1.00	1.25	1.50	1.75	2.00	2.50	3.00
3	5	5	6	7	8	9	10	12	13	15	17	19	21	24	36	53	70	84	93	99	.
4	5	5	6	8	10	11	13	15	18	21	24	27	31	35	52	73	88	96	99	.	.
5	5	6	7	9	11	14	16	19	23	27	31	36	40	46	66	86	96	99	.	.	.
6	5	6	7	11	13	16	20	24	28	33	38	44	50	55	76	93	99	.	.	.	.
7	5	6	8	12	15	19	23	28	33	39	45	52	58	64	84	97	.	.	.	.	.
8	5	6	8	13	17	21	27	32	38	45	52	59	65	71	90	98	.	.	.	.	.
9	5	7	9	15	19	24	30	37	44	51	58	65	72	78	94	99	.	.	.	.	.
10	6	7	9	16	21	27	34	41	48	56	64	71	77	83	96	.	.	.	.	.	.
11	6	7	10	18	23	30	37	45	53	61	69	76	82	87	98	.	.	.	.	.	.
12	6	8	10	19	25	32	40	49	57	66	73	80	86	90	99	.	.	.	.	.	.
13	6	8	11	21	27	35	44	53	61	70	77	84	89	93	99	.	.	.	.	.	.
14	6	8	12	22	30	38	47	56	65	74	81	87	91	94	99	.	.	.	.	.	.
15	6	9	12	24	32	41	50	60	69	77	84	89	93	96	.	.	.	.	.	.	.
20	7	10	15	32	42	53	64	74	83	89	94	97	98	99	.	.	.	.	.	.	.
25	7	12	19	39	52	64	75	84	91	95	98	99	.	.	.	.	.	.	.	.	.
30	8	14	22	47	61	73	84	91	96	98	99	.	.	.	.	.	.	.	.	.	.
35	9	16	26	54	68	80	89	95	98	99	.	.	.	.	.	.	.	.	.	.	.
40	10	17	29	60	75	86	93	97	99	.	.	.	.	.	.	.	.	.	.	.	.
45	10	19	33	66	80	90	96	99	.	.	.	.	.	.	.	.	.	.	.	.	.
50	11	21	36	71	84	93	97	99	.	.	.	.	.	.	.	.	.	.	.	.	.
55	12	23	39	75	88	95	98	.	.	.	.	.	.	.	.	.	.	.	.	.	.
60	13	25	43	79	91	97	99	.	.	.	.	.	.	.	.	.	.	.	.	.	.
65	13	27	46	83	93	98	99	.	.	.	.	.	.	.	.	.	.	.	.	.	.
70	14	29	49	86	95	99	.	.	.	.	.	.	.	.	.	.	.	.	.	.	.
75	15	31	52	88	96	99	.	.	.	.	.	.	.	.	.	.	.	.	.	.	.
80	16	33	55	90	97	99	.	.	.	.	.	.	.	.	.	.	.	.	.	.	.
90	18	37	61	94	98	.	.	.	.	.	.	.	.	.	.	.	.	.	.	.	.
100	19	40	66	96	99	.	.	.	.	.	.	.	.	.	.	.	.	.	.	.	.
110	21	44	70	97	.	.	.	.	.	.	.	.	.	.	.	.	.	.	.	.	.
120	23	48	74	98	.	.	.	.	.	.	.	.	.	.	.	.	.	.	.	.	.
130	24	51	78	99	.	.	.	.	.	.	.	.	.	.	.	.	.	.	.	.	.
140	26	55	81	99	.	.	.	.	.	.	.	.	.	.	.	.	.	.	.	.	.
150	28	58	84	.	.	.	.	.	.	.	.	.	.	.	.	.	.	.	.	.	.
175	32	65	90	.	.	.	.	.	.	.	.	.	.	.	.	.	.	.	.	.	.
200	36	72	93	.	.	.	.	.	.	.	.	.	.	.	.	.	.	.	.	.	.
225	41	77	96	.	.	.	.	.	.	.	.	.	.	.	.	.	.	.	.	.	.
250	45	82	97	.	.	.	.	.	.	.	.	.	.	.	.	.	.	.	.	.	.
300	53	89	99	.	.	.	.	.	.	.	.	.	.	.	.	.	.	.	.	.	.
350	60	93	.	.	.	.	.	.	.	.	.	.	.	.	.	.	.	.	.	.	.
400	66	96	.	.	.	.	.	.	.	.	.	.	.	.	.	.	.	.	.	.	.
450	72	98	.	.	.	.	.	.	.	.	.	.	.	.	.	.	.	.	.	.	.
500	77	99	.	.	.	.	.	.	.	.	.	.	.	.	.	.	.	.	.	.	.
600	84	.	.	.	.	.	.	.	.	.	.	.	.	.	.	.	.	.	.	.	.
700	90	.	.	.	.	.	.	.	.	.	.	.	.	.	.	.	.	.	.	.	.
800	94	.	.	.	.	.	.	.	.	.	.	.	.	.	.	.	.	.	.	.	.
900	96	.	.	.	.	.	.	.	.	.	.	.	.	.	.	.	.	.	.	.	.
1000	98	.	.	.	.	.	.	.	.	.	.	.	.	.	.	.	.	.	.	.	.

Table 9.4. Power table for between subjects analysis; 3 × 3 design at alpha = 0.05

n	Hypothesized ES																				
	0.10	0.15	0.20	0.30	0.35	0.40	0.45	0.50	0.55	0.60	0.65	0.70	0.75	0.80	1.00	1.25	1.50	1.75	2.00	2.50	3.00
3	5	5	6	7	8	9	10	11	13	15	17	19	21	24	36	54	72	86	94	99	.
4	5	5	6	8	9	11	13	15	18	20	24	27	31	35	53	74	89	97	99	.	.
5	5	6	7	9	11	13	16	19	23	27	31	36	41	46	67	87	97	99	.	.	.
6	5	6	7	10	13	16	19	23	28	33	38	44	50	56	78	94	99	.	.	.	.
7	5	6	8	12	15	18	23	28	33	39	46	52	59	65	85	97	.	.	.	.	.
8	5	6	8	13	17	21	26	32	39	45	52	59	66	73	91	99	.	.	.	.	.
9	5	7	9	15	19	24	30	37	44	51	59	66	73	79	94	.	.	.	.	.	.
10	6	7	9	16	21	27	33	41	49	57	65	72	78	84	97	.	.	.	.	.	.
11	6	7	10	17	23	30	37	45	54	62	70	77	83	88	98	.	.	.	.	.	.
12	6	8	10	19	25	32	41	49	58	67	74	81	87	91	99	.	.	.	.	.	.
13	6	8	11	21	27	35	44	53	62	71	78	85	90	93	99	.	.	.	.	.	.
14	6	8	11	22	29	38	47	57	66	75	82	88	92	95	.	.	.	.	.	.	.
15	6	8	12	24	32	41	51	60	70	78	85	90	94	97	.	.	.	.	.	.	.
20	7	10	15	32	42	54	65	75	84	90	94	97	99	99	.	.	.	.	.	.	.
25	7	12	18	39	52	65	76	86	92	96	98	99	.	.	.	.	.	.	.	.	.
30	8	13	22	47	61	74	85	92	96	98	99	.	.	.	.	.	.	.	.	.	.
35	9	15	25	54	69	82	90	96	98	99	.	.	.	.	.	.	.	.	.	.	.
40	9	17	29	61	76	87	94	98	99	.	.	.	.	.	.	.	.	.	.	.	.
45	10	19	32	67	81	91	96	99	.	.	.	.	.	.	.	.	.	.	.	.	.
50	11	21	36	72	86	94	98	99	.	.	.	.	.	.	.	.	.	.	.	.	.
55	12	23	39	76	89	96	99	.	.	.	.	.	.	.	.	.	.	.	.	.	.
60	12	25	43	80	92	97	99	.	.	.	.	.	.	.	.	.	.	.	.	.	.
65	13	27	46	84	94	98	.	.	.	.	.	.	.	.	.	.	.	.	.	.	.
70	14	29	49	87	96	99	.	.	.	.	.	.	.	.	.	.	.	.	.	.	.
75	15	31	53	89	97	99	.	.	.	.	.	.	.	.	.	.	.	.	.	.	.
80	16	33	56	91	98	.	.	.	.	.	.	.	.	.	.	.	.	.	.	.	.
90	17	37	61	94	99	.	.	.	.	.	.	.	.	.	.	.	.	.	.	.	.
100	19	41	67	96	99	.	.	.	.	.	.	.	.	.	.	.	.	.	.	.	.
110	21	44	71	98	.	.	.	.	.	.	.	.	.	.	.	.	.	.	.	.	.
120	22	48	75	99	.	.	.	.	.	.	.	.	.	.	.	.	.	.	.	.	.
130	24	52	79	99	.	.	.	.	.	.	.	.	.	.	.	.	.	.	.	.	.
140	26	55	82	.	.	.	.	.	.	.	.	.	.	.	.	.	.	.	.	.	.
150	28	58	85	.	.	.	.	.	.	.	.	.	.	.	.	.	.	.	.	.	.
175	32	66	91	.	.	.	.	.	.	.	.	.	.	.	.	.	.	.	.	.	.
200	36	73	94	.	.	.	.	.	.	.	.	.	.	.	.	.	.	.	.	.	.
225	41	78	97	.	.	.	.	.	.	.	.	.	.	.	.	.	.	.	.	.	.
250	45	83	98	.	.	.	.	.	.	.	.	.	.	.	.	.	.	.	.	.	.
300	53	90	99	.	.	.	.	.	.	.	.	.	.	.	.	.	.	.	.	.	.
350	60	94	.	.	.	.	.	.	.	.	.	.	.	.	.	.	.	.	.	.	.
400	67	97	.	.	.	.	.	.	.	.	.	.	.	.	.	.	.	.	.	.	.
450	73	98	.	.	.	.	.	.	.	.	.	.	.	.	.	.	.	.	.	.	.
500	78	99	.	.	.	.	.	.	.	.	.	.	.	.	.	.	.	.	.	.	.
600	85	.	.	.	.	.	.	.	.	.	.	.	.	.	.	.	.	.	.	.	.
700	91	.	.	.	.	.	.	.	.	.	.	.	.	.	.	.	.	.	.	.	.
800	94	.	.	.	.	.	.	.	.	.	.	.	.	.	.	.	.	.	.	.	.
900	97	.	.	.	.	.	.	.	.	.	.	.	.	.	.	.	.	.	.	.	.
1000	98	.	.	.	.	.	.	.	.	.	.	.	.	.	.	.	.	.	.	.	.

Table 9.5. Power table for between subjects analysis; 3 × 4 design at alpha = 0.05

n	Hypothesized ES																				
	0.10	0.15	0.20	0.30	0.35	0.40	0.45	0.50	0.55	0.60	0.65	0.70	0.75	0.80	1.00	1.25	1.50	1.75	2.00	2.50	3.00
3	5	5	6	7	8	9	11	12	14	16	18	21	23	27	41	62	80	92	97	.	.
4	5	6	6	8	10	12	14	16	19	23	26	30	35	40	60	82	94	99	.	.	.
5	5	6	7	10	12	14	17	21	25	30	35	40	46	52	75	93	99	.	.	.	.
6	5	6	7	11	14	17	21	26	31	37	43	50	57	63	85	97	.	.	.	.	.
7	5	6	8	13	16	20	25	31	37	44	52	59	66	72	91	99	.	.	.	.	.
8	5	7	8	14	18	23	29	36	44	51	59	67	74	80	95	.	.	.	.	.	.
9	6	7	9	16	21	27	33	41	49	58	66	73	80	86	98	.	.	.	.	.	.
10	6	7	10	17	23	30	38	46	55	64	72	79	85	90	99	.	.	.	.	.	.
11	6	7	10	19	25	33	42	51	60	69	77	84	89	93	99	.	.	.	.	.	.
12	6	8	11	21	28	36	46	56	65	74	82	88	92	95	.	.	.	.	.	.	.
13	6	8	11	23	30	40	50	60	70	78	85	91	94	97	.	.	.	.	.	.	.
14	6	8	12	24	33	43	53	64	74	82	88	93	96	98	.	.	.	.	.	.	.
15	6	9	13	26	35	46	57	68	77	85	91	95	97	99	.	.	.	.	.	.	.
20	7	11	16	35	48	60	72	82	90	95	98	99	.	.	.	.	.	.	.	.	.
25	8	12	20	44	59	72	83	91	96	98	99	.	.	.	.	.	.	.	.	.	.
30	8	14	24	53	68	81	90	96	98	.	.	.	.	.	.	.	.	.	.	.	.
35	9	16	28	61	76	88	95	98	99	.	.	.	.	.	.	.	.	.	.	.	.
40	10	19	32	68	83	92	97	99	.	.	.	.	.	.	.	.	.	.	.	.	.
45	11	21	36	74	87	95	99	.	.	.	.	.	.	.	.	.	.	.	.	.	.
50	12	23	40	79	91	97	99	.	.	.	.	.	.	.	.	.	.	.	.	.	.
55	12	25	44	83	94	98	.	.	.	.	.	.	.	.	.	.	.	.	.	.	.
60	13	27	48	87	96	99	.	.	.	.	.	.	.	.	.	.	.	.	.	.	.
65	14	30	52	90	97	99	.	.	.	.	.	.	.	.	.	.	.	.	.	.	.
70	15	32	56	92	98	.	.	.	.	.	.	.	.	.	.	.	.	.	.	.	.
75	16	34	59	94	99	.	.	.	.	.	.	.	.	.	.	.	.	.	.	.	.
80	17	37	62	95	99	.	.	.	.	.	.	.	.	.	.	.	.	.	.	.	.
90	19	41	68	97	.	.	.	.	.	.	.	.	.	.	.	.	.	.	.	.	.
100	21	46	74	99	.	.	.	.	.	.	.	.	.	.	.	.	.	.	.	.	.
110	23	50	78	99	.	.	.	.	.	.	.	.	.	.	.	.	.	.	.	.	.
120	25	54	82	.	.	.	.	.	.	.	.	.	.	.	.	.	.	.	.	.	.
130	27	58	86	.	.	.	.	.	.	.	.	.	.	.	.	.	.	.	.	.	.
140	29	62	89	.	.	.	.	.	.	.	.	.	.	.	.	.	.	.	.	.	.
150	31	65	91	.	.	.	.	.	.	.	.	.	.	.	.	.	.	.	.	.	.
175	36	73	95	.	.	.	.	.	.	.	.	.	.	.	.	.	.	.	.	.	.
200	41	80	97	.	.	.	.	.	.	.	.	.	.	.	.	.	.	.	.	.	.
225	46	85	99	.	.	.	.	.	.	.	.	.	.	.	.	.	.	.	.	.	.
250	51	89	99	.	.	.	.	.	.	.	.	.	.	.	.	.	.	.	.	.	.
300	60	94	.	.	.	.	.	.	.	.	.	.	.	.	.	.	.	.	.	.	.
350	67	97	.	.	.	.	.	.	.	.	.	.	.	.	.	.	.	.	.	.	.
400	74	99	.	.	.	.	.	.	.	.	.	.	.	.	.	.	.	.	.	.	.
450	80	99	.	.	.	.	.	.	.	.	.	.	.	.	.	.	.	.	.	.	.
500	84	.	.	.	.	.	.	.	.	.	.	.	.	.	.	.	.	.	.	.	.
600	91	.	.	.	.	.	.	.	.	.	.	.	.	.	.	.	.	.	.	.	.
700	95	.	.	.	.	.	.	.	.	.	.	.	.	.	.	.	.	.	.	.	.
800	97	.	.	.	.	.	.	.	.	.	.	.	.	.	.	.	.	.	.	.	.
900	99	.	.	.	.	.	.	.	.	.	.	.	.	.	.	.	.	.	.	.	.
1000	99	.	.	.	.	.	.	.	.	.	.	.	.	.	.	.	.	.	.	.	.

Table 9.6. Power table for ANCOVA analysis; 2 × 2 design at $r = 0.40$ and alpha = 0.05

n	Hypothesized ES																				
	0.10	0.15	0.20	0.30	0.35	0.40	0.45	0.50	0.55	0.60	0.65	0.70	0.75	0.80	1.00	1.25	1.50	1.75	2.00	2.50	3.00
3	5	5	6	7	8	9	10	11	12	14	15	17	19	21	30	43	57	70	80	94	99
4	5	5	6	8	9	11	12	14	17	19	22	24	27	30	44	62	77	88	95	99	.
5	5	5	6	9	11	13	15	18	21	24	28	32	36	40	57	76	89	96	99	.	.
6	5	6	7	10	13	15	19	22	26	30	34	39	44	48	67	85	95	99	.	.	.
7	5	6	7	12	15	18	22	26	30	35	40	46	51	56	75	91	98	.	.	.	.
8	5	6	8	13	16	20	25	30	35	40	46	52	58	63	82	95	99	.	.	.	.
9	5	6	8	14	18	23	28	33	39	45	52	58	64	69	87	97	.	.	.	.	.
10	5	7	9	16	20	25	31	37	43	50	57	63	69	74	90	98	.	.	.	.	.
11	5	7	10	17	22	28	34	41	48	54	61	68	73	79	93	99	.	.	.	.	.
12	5	7	10	19	24	30	37	44	51	58	65	72	77	82	95	99	.	.	.	.	.
13	6	8	11	20	26	33	40	47	55	62	69	75	81	86	97	.	.	.	.	.	.
14	6	8	11	21	28	35	43	51	58	66	73	79	84	88	98	.	.	.	.	.	.
15	6	8	12	23	30	37	45	54	62	69	76	82	87	90	98	.	.	.	.	.	.
20	7	10	15	30	39	48	58	67	75	82	87	92	95	97	.	.	.	.	.	.	.
25	7	12	18	36	47	58	68	77	84	90	94	96	98	99	.	.	.	.	.	.	.
30	8	14	21	43	55	66	76	84	90	94	97	98	99	.	.	.	.	.	.	.	.
35	9	15	25	49	61	73	82	89	94	97	99	99	.	.	.	.	.	.	.	.	.
40	10	17	28	54	67	79	87	93	96	98	99	.	.	.	.	.	.	.	.	.	.
45	10	19	31	59	73	83	91	95	98	99	.	.	.	.	.	.	.	.	.	.	.
50	11	21	34	64	77	87	93	97	99	.	.	.	.	.	.	.	.	.	.	.	.
55	12	22	37	68	81	90	95	98	99	.	.	.	.	.	.	.	.	.	.	.	.
60	13	24	39	72	84	92	97	99	.	.	.	.	.	.	.	.	.	.	.	.	.
65	13	26	42	75	87	94	98	99	.	.	.	.	.	.	.	.	.	.	.	.	.
70	14	28	45	78	89	95	98	.	.	.	.	.	.	.	.	.	.	.	.	.	.
75	15	29	48	81	91	97	99	.	.	.	.	.	.	.	.	.	.	.	.	.	.
80	16	31	50	84	93	97	99	.	.	.	.	.	.	.	.	.	.	.	.	.	.
90	17	34	55	88	95	99	.	.	.	.	.	.	.	.	.	.	.	.	.	.	.
100	19	38	59	91	97	99	.	.	.	.	.	.	.	.	.	.	.	.	.	.	.
110	20	41	64	93	98	.	.	.	.	.	.	.	.	.	.	.	.	.	.	.	.
120	22	44	67	95	99	.	.	.	.	.	.	.	.	.	.	.	.	.	.	.	.
130	24	47	71	96	99	.	.	.	.	.	.	.	.	.	.	.	.	.	.	.	.
140	25	50	74	97	99	.	.	.	.	.	.	.	.	.	.	.	.	.	.	.	.
150	27	53	77	98	.	.	.	.	.	.	.	.	.	.	.	.	.	.	.	.	.
175	31	59	83	99	.	.	.	.	.	.	.	.	.	.	.	.	.	.	.	.	.
200	34	65	88	.	.	.	.	.	.	.	.	.	.	.	.	.	.	.	.	.	.
225	38	70	91	.	.	.	.	.	.	.	.	.	.	.	.	.	.	.	.	.	.
250	41	74	94	.	.	.	.	.	.	.	.	.	.	.	.	.	.	.	.	.	.
300	48	82	97	.	.	.	.	.	.	.	.	.	.	.	.	.	.	.	.	.	.
350	54	87	98	.	.	.	.	.	.	.	.	.	.	.	.	.	.	.	.	.	.
400	60	91	99	.	.	.	.	.	.	.	.	.	.	.	.	.	.	.	.	.	.
450	65	94	.	.	.	.	.	.	.	.	.	.	.	.	.	.	.	.	.	.	.
500	70	96	.	.	.	.	.	.	.	.	.	.	.	.	.	.	.	.	.	.	.
600	77	98	.	.	.	.	.	.	.	.	.	.	.	.	.	.	.	.	.	.	.
700	83	99	.	.	.	.	.	.	.	.	.	.	.	.	.	.	.	.	.	.	.
800	88	.	.	.	.	.	.	.	.	.	.	.	.	.	.	.	.	.	.	.	.
900	91	.	.	.	.	.	.	.	.	.	.	.	.	.	.	.	.	.	.	.	.
1000	94	.	.	.	.	.	.	.	.	.	.	.	.	.	.	.	.	.	.	.	.

Table 9.7. Power table for ANCOVA analysis; 2 × 2 design at $r = 0.60$ and alpha = 0.05

n	Hypothesized ES																				
	0.10	0.15	0.20	0.30	0.35	0.40	0.45	0.50	0.55	0.60	0.65	0.70	0.75	0.80	1.00	1.25	1.50	1.75	2.00	2.50	3.00
3	5	6	6	8	9	10	11	13	15	16	19	21	23	26	37	53	68	81	90	98	.
4	5	6	6	9	11	13	15	18	20	24	27	31	34	38	55	73	87	95	98	.	.
5	5	6	7	11	13	16	19	23	27	31	35	40	45	50	68	86	95	99	.	.	.
6	5	6	8	12	15	19	23	28	33	38	43	49	54	59	78	93	98	.	.	.	.
7	5	6	8	14	18	22	27	33	38	44	50	56	62	68	86	96	99	.	.	.	.
8	5	7	9	16	20	26	31	37	44	50	57	63	69	75	91	98	.	.	.	.	.
9	5	7	10	18	23	29	35	42	49	56	63	69	75	80	94	99	.	.	.	.	.
10	6	8	11	20	25	32	39	46	54	61	68	74	80	85	96	.	.	.	.	.	.
11	6	8	12	21	28	35	43	51	58	66	73	79	84	88	98	.	.	.	.	.	.
12	6	9	12	23	30	38	46	55	63	70	77	83	87	91	98	.	.	.	.	.	.
13	6	9	13	25	33	41	50	58	67	74	80	86	90	93	99	.	.	.	.	.	.
14	6	9	14	27	35	44	53	62	70	77	84	88	92	95	99	.	.	.	.	.	.
15	6	10	15	29	37	47	56	65	73	80	86	91	94	96	.	.	.	.	.	.	.
20	7	12	19	38	48	59	69	78	85	91	94	97	98	99	.	.	.	.	.	.	.
25	8	14	23	46	58	69	79	87	92	96	98	99	.	.	.	.	.	.	.	.	.
30	9	17	27	53	66	77	86	92	96	98	99	.	.	.	.	.	.	.	.	.	.
35	10	19	31	60	73	84	91	96	98	99	.	.	.	.	.	.	.	.	.	.	.
40	11	21	35	66	79	88	94	98	99	.	.	.	.	.	.	.	.	.	.	.	.
45	12	24	39	71	83	92	96	99	.	.	.	.	.	.	.	.	.	.	.	.	.
50	13	26	42	76	87	94	98	99	.	.	.	.	.	.	.	.	.	.	.	.	.
55	14	28	46	80	90	96	99	.	.	.	.	.	.	.	.	.	.	.	.	.	.
60	15	30	49	83	92	97	99	.	.	.	.	.	.	.	.	.	.	.	.	.	.
65	17	33	53	86	94	98	99	.	.	.	.	.	.	.	.	.	.	.	.	.	.
70	18	35	56	88	96	99	.	.	.	.	.	.	.	.	.	.	.	.	.	.	.
75	19	37	59	90	97	99	.	.	.	.	.	.	.	.	.	.	.	.	.	.	.
80	20	39	61	92	97	99	.	.	.	.	.	.	.	.	.	.	.	.	.	.	.
90	22	43	67	95	99	.	.	.	.	.	.	.	.	.	.	.	.	.	.	.	.
100	24	47	71	96	99	.	.	.	.	.	.	.	.	.	.	.	.	.	.	.	.
110	26	51	75	98	.	.	.	.	.	.	.	.	.	.	.	.	.	.	.	.	.
120	28	54	79	98	.	.	.	.	.	.	.	.	.	.	.	.	.	.	.	.	.
130	30	58	82	99	.	.	.	.	.	.	.	.	.	.	.	.	.	.	.	.	.
140	32	61	85	99	.	.	.	.	.	.	.	.	.	.	.	.	.	.	.	.	.
150	34	64	87	.	.	.	.	.	.	.	.	.	.	.	.	.	.	.	.	.	.
175	39	71	91	.	.	.	.	.	.	.	.	.	.	.	.	.	.	.	.	.	.
200	43	76	95	.	.	.	.	.	.	.	.	.	.	.	.	.	.	.	.	.	.
225	47	81	97	.	.	.	.	.	.	.	.	.	.	.	.	.	.	.	.	.	.
250	52	85	98	.	.	.	.	.	.	.	.	.	.	.	.	.	.	.	.	.	.
300	59	91	99	.	.	.	.	.	.	.	.	.	.	.	.	.	.	.	.	.	.
350	66	94	.	.	.	.	.	.	.	.	.	.	.	.	.	.	.	.	.	.	.
400	72	97	.	.	.	.	.	.	.	.	.	.	.	.	.	.	.	.	.	.	.
450	77	98	.	.	.	.	.	.	.	.	.	.	.	.	.	.	.	.	.	.	.
500	81	99	.	.	.	.	.	.	.	.	.	.	.	.	.	.	.	.	.	.	.
600	87	.	.	.	.	.	.	.	.	.	.	.	.	.	.	.	.	.	.	.	.
700	92	.	.	.	.	.	.	.	.	.	.	.	.	.	.	.	.	.	.	.	.
800	95	.	.	.	.	.	.	.	.	.	.	.	.	.	.	.	.	.	.	.	.
900	97	.	.	.	.	.	.	.	.	.	.	.	.	.	.	.	.	.	.	.	.
1000	98	.	.	.	.	.	.	.	.	.	.	.	.	.	.	.	.	.	.	.	.

Table 9.8. Power table for ANCOVA analysis; 2 × 3 design at $r = 0.40$ and alpha = 0.05

n	Hypothesized ES																				
	0.10	0.15	0.20	0.30	0.35	0.40	0.45	0.50	0.55	0.60	0.65	0.70	0.75	0.80	1.00	1.25	1.50	1.75	2.00	2.50	3.00
3	5	5	6	7	8	9	10	12	13	15	17	19	21	23	34	50	66	80	90	98	.
4	5	6	6	8	10	11	13	16	18	21	24	27	31	34	51	71	86	95	98	.	.
5	5	6	7	10	12	14	17	20	23	27	31	36	40	45	65	84	95	99	.	.	.
6	5	6	7	11	14	17	20	24	29	34	39	44	50	55	76	92	98	.	.	.	.
7	5	6	8	12	16	19	24	29	34	40	46	52	58	64	83	96	99	.	.	.	.
8	5	7	8	14	18	22	27	33	39	46	52	59	65	71	89	98	.	.	.	.	.
9	5	7	9	16	20	25	31	37	44	51	58	65	72	77	93	99	.	.	.	.	.
10	6	7	10	17	22	28	35	42	49	57	64	71	77	82	96	.	.	.	.	.	.
11	6	8	10	19	24	31	38	46	54	62	69	76	82	86	97	.	.	.	.	.	.
12	6	8	11	20	26	34	42	50	58	66	73	80	85	90	98	.	.	.	.	.	.
13	6	8	12	22	29	36	45	54	62	70	77	83	88	92	99	.	.	.	.	.	.
14	6	9	12	23	31	39	48	57	66	74	81	86	91	94	99	.	.	.	.	.	.
15	6	9	13	25	33	42	51	61	69	77	84	89	93	96	.	.	.	.	.	.	.
20	7	11	16	33	43	54	65	75	83	89	93	96	98	99	.	.	.	.	.	.	.
25	8	13	20	41	53	65	76	85	91	95	98	99	.	.	.	.	.	.	.	.	.
30	9	15	23	48	62	74	84	91	95	98	99	.	.	.	.	.	.	.	.	.	.
35	9	16	27	55	69	81	89	95	98	99	.	.	.	.	.	.	.	.	.	.	.
40	10	18	31	61	75	86	93	97	99	.	.	.	.	.	.	.	.	.	.	.	.
45	11	20	34	67	80	90	96	98	99	.	.	.	.	.	.	.	.	.	.	.	.
50	12	22	38	72	85	93	97	99	.	.	.	.	.	.	.	.	.	.	.	.	.
55	13	24	41	76	88	95	98	.	.	.	.	.	.	.	.	.	.	.	.	.	.
60	13	26	44	80	91	97	99	.	.	.	.	.	.	.	.	.	.	.	.	.	.
65	14	28	47	83	93	98	99	.	.	.	.	.	.	.	.	.	.	.	.	.	.
70	15	30	51	86	95	98	.	.	.	.	.	.	.	.	.	.	.	.	.	.	.
75	16	32	54	88	96	99	.	.	.	.	.	.	.	.	.	.	.	.	.	.	.
80	17	34	56	90	97	99	.	.	.	.	.	.	.	.	.	.	.	.	.	.	.
90	19	38	62	94	98	.	.	.	.	.	.	.	.	.	.	.	.	.	.	.	.
100	20	42	67	96	99	.	.	.	.	.	.	.	.	.	.	.	.	.	.	.	.
110	22	46	71	97	.	.	.	.	.	.	.	.	.	.	.	.	.	.	.	.	.
120	24	49	75	98	.	.	.	.	.	.	.	.	.	.	.	.	.	.	.	.	.
130	26	53	79	99	.	.	.	.	.	.	.	.	.	.	.	.	.	.	.	.	.
140	28	56	82	99	.	.	.	.	.	.	.	.	.	.	.	.	.	.	.	.	.
150	29	59	84	.	.	.	.	.	.	.	.	.	.	.	.	.	.	.	.	.	.
175	34	66	90	.	.	.	.	.	.	.	.	.	.	.	.	.	.	.	.	.	.
200	38	72	93	.	.	.	.	.	.	.	.	.	.	.	.	.	.	.	.	.	.
225	42	78	96	.	.	.	.	.	.	.	.	.	.	.	.	.	.	.	.	.	.
250	46	82	97	.	.	.	.	.	.	.	.	.	.	.	.	.	.	.	.	.	.
300	54	89	99	.	.	.	.	.	.	.	.	.	.	.	.	.	.	.	.	.	.
350	61	93	.	.	.	.	.	.	.	.	.	.	.	.	.	.	.	.	.	.	.
400	67	96	.	.	.	.	.	.	.	.	.	.	.	.	.	.	.	.	.	.	.
450	73	98	.	.	.	.	.	.	.	.	.	.	.	.	.	.	.	.	.	.	.
500	77	99	.	.	.	.	.	.	.	.	.	.	.	.	.	.	.	.	.	.	.
600	85	.	.	.	.	.	.	.	.	.	.	.	.	.	.	.	.	.	.	.	.
700	90	.	.	.	.	.	.	.	.	.	.	.	.	.	.	.	.	.	.	.	.
800	94	.	.	.	.	.	.	.	.	.	.	.	.	.	.	.	.	.	.	.	.
900	96	.	.	.	.	.	.	.	.	.	.	.	.	.	.	.	.	.	.	.	.
1000	97	.	.	.	.	.	.	.	.	.	.	.	.	.	.	.	.	.	.	.	.

Table 9.9. Power table for ANCOVA analysis; 2 × 3 design at $r = 0.60$ and alpha = 0.05

n	Hypothesized ES																				
	0.10	0.15	0.20	0.30	0.35	0.40	0.45	0.50	0.55	0.60	0.65	0.70	0.75	0.80	1.00	1.25	1.50	1.75	2.00	2.50	3.00
3	5	6	6	8	9	10	12	14	16	18	21	23	26	29	43	62	79	90	96	.	.
4	5	6	7	10	11	14	16	19	23	26	30	35	39	44	63	83	94	99	.	.	.
5	5	6	7	11	14	17	21	25	30	35	40	45	51	57	77	93	99	.	.	.	.
6	5	6	8	13	17	21	26	31	37	43	49	55	62	68	86	97	.	.	.	.	.
7	5	7	9	15	20	25	30	37	43	50	57	64	70	76	92	99	.	.	.	.	.
8	6	7	10	17	22	28	35	42	50	57	65	71	78	83	96	.	.	.	.	.	.
9	6	8	11	19	25	32	40	48	56	64	71	78	83	88	98	.	.	.	.	.	.
10	6	8	11	21	28	36	44	53	61	69	76	83	88	92	99	.	.	.	.	.	.
11	6	9	12	24	31	39	48	57	66	74	81	87	91	94	99	.	.	.	.	.	.
12	6	9	13	26	34	43	52	62	71	78	85	90	93	96	.	.	.	.	.	.	.
13	7	10	14	28	37	46	56	66	75	82	88	92	95	97	.	.	.	.	.	.	.
14	7	10	15	30	39	50	60	70	78	85	90	94	97	98	.	.	.	.	.	.	.
15	7	11	16	32	42	53	63	73	81	88	92	96	98	99	.	.	.	.	.	.	.
20	8	13	20	42	55	67	77	86	92	96	98	99	.	.	.	.	.	.	.	.	.
25	9	15	25	52	65	77	87	93	97	99	99	.	.	.	.	.	.	.	.	.	.
30	10	18	30	60	74	85	92	97	99	.	.	.	.	.	.	.	.	.	.	.	.
35	11	21	34	67	81	90	96	99	.	.	.	.	.	.	.	.	.	.	.	.	.
40	12	23	39	74	86	94	98	99	.	.	.	.	.	.	.	.	.	.	.	.	.
45	13	26	43	79	90	96	99	.	.	.	.	.	.	.	.	.	.	.	.	.	.
50	14	29	48	83	93	98	99	.	.	.	.	.	.	.	.	.	.	.	.	.	.
55	16	31	52	87	95	99	.	.	.	.	.	.	.	.	.	.	.	.	.	.	.
60	17	34	56	90	97	99	.	.	.	.	.	.	.	.	.	.	.	.	.	.	.
65	18	36	59	92	98	.	.	.	.	.	.	.	.	.	.	.	.	.	.	.	.
70	19	39	63	94	98	.	.	.	.	.	.	.	.	.	.	.	.	.	.	.	.
75	20	41	66	95	99	.	.	.	.	.	.	.	.	.	.	.	.	.	.	.	.
80	21	44	69	96	99	.	.	.	.	.	.	.	.	.	.	.	.	.	.	.	.
90	24	49	74	98	.	.	.	.	.	.	.	.	.	.	.	.	.	.	.	.	.
100	26	53	79	99	.	.	.	.	.	.	.	.	.	.	.	.	.	.	.	.	.
110	28	57	83	99	.	.	.	.	.	.	.	.	.	.	.	.	.	.	.	.	.
120	31	61	86	.	.	.	.	.	.	.	.	.	.	.	.	.	.	.	.	.	.
130	33	65	89	.	.	.	.	.	.	.	.	.	.	.	.	.	.	.	.	.	.
140	35	69	91	.	.	.	.	.	.	.	.	.	.	.	.	.	.	.	.	.	.
150	38	72	93	.	.	.	.	.	.	.	.	.	.	.	.	.	.	.	.	.	.
175	43	79	96	.	.	.	.	.	.	.	.	.	.	.	.	.	.	.	.	.	.
200	48	84	98	.	.	.	.	.	.	.	.	.	.	.	.	.	.	.	.	.	.
225	53	88	99	.	.	.	.	.	.	.	.	.	.	.	.	.	.	.	.	.	.
250	58	91	99	.	.	.	.	.	.	.	.	.	.	.	.	.	.	.	.	.	.
300	66	96	.	.	.	.	.	.	.	.	.	.	.	.	.	.	.	.	.	.	.
350	74	98	.	.	.	.	.	.	.	.	.	.	.	.	.	.	.	.	.	.	.
400	79	99	.	.	.	.	.	.	.	.	.	.	.	.	.	.	.	.	.	.	.
450	84	99	.	.	.	.	.	.	.	.	.	.	.	.	.	.	.	.	.	.	.
500	88	.	.	.	.	.	.	.	.	.	.	.	.	.	.	.	.	.	.	.	.
600	93	.	.	.	.	.	.	.	.	.	.	.	.	.	.	.	.	.	.	.	.
700	96	.	.	.	.	.	.	.	.	.	.	.	.	.	.	.	.	.	.	.	.
800	98	.	.	.	.	.	.	.	.	.	.	.	.	.	.	.	.	.	.	.	.
900	99	.	.	.	.	.	.	.	.	.	.	.	.	.	.	.	.	.	.	.	.
1000	99	.	.	.	.	.	.	.	.	.	.	.	.	.	.	.	.	.	.	.	.

Table 9.10. Power table for ANCOVA analysis; 2 × 4 design at $r = 0.40$ and alpha = 0.05

n	Hypothesized ES																				
	0.10	0.15	0.20	0.30	0.35	0.40	0.45	0.50	0.55	0.60	0.65	0.70	0.75	0.80	1.00	1.25	1.50	1.75	2.00	2.50	3.00
3	5	5	6	7	8	9	11	12	14	16	18	21	23	26	39	58	75	88	95	.	.
4	5	6	6	9	10	12	14	17	20	23	27	31	35	39	58	79	92	98	.	.	.
5	5	6	7	10	12	15	18	22	26	31	35	41	46	52	73	91	98	.	.	.	.
6	5	6	8	12	15	18	22	27	32	38	44	50	56	62	83	96	99	.	.	.	.
7	5	6	8	13	17	21	26	32	38	45	52	59	65	71	90	98	.	.	.	.	.
8	5	7	9	15	19	25	31	37	44	52	59	66	73	79	94	99	.	.	.	.	.
9	6	7	9	17	22	28	35	42	50	58	66	73	79	84	97	.	.	.	.	.	.
10	6	8	10	19	24	31	39	47	55	64	71	78	84	89	98	.	.	.	.	.	.
11	6	8	11	20	27	35	43	52	60	69	76	83	88	92	99	.	.	.	.	.	.
12	6	8	12	22	29	38	47	56	65	73	81	86	91	94	99	.	.	.	.	.	.
13	6	9	12	24	32	41	51	60	69	77	84	90	93	96	.	.	.	.	.	.	.
14	6	9	13	26	34	44	54	64	73	81	87	92	95	97	.	.	.	.	.	.	.
15	6	9	14	28	37	47	58	68	77	84	90	94	97	98	.	.	.	.	.	.	.
20	7	11	18	37	49	61	72	82	89	94	97	99	99	.	.	.	.	.	.	.	.
25	8	13	22	46	60	72	83	90	95	98	99	.	.	.	.	.	.	.	.	.	.
30	9	16	26	54	69	81	90	95	98	99	.	.	.	.	.	.	.	.	.	.	.
35	10	18	30	62	76	87	94	98	99	.	.	.	.	.	.	.	.	.	.	.	.
40	11	20	34	68	82	92	97	99	.	.	.	.	.	.	.	.	.	.	.	.	.
45	12	22	38	74	87	95	98	.	.	.	.	.	.	.	.	.	.	.	.	.	.
50	13	25	42	79	90	97	99	.	.	.	.	.	.	.	.	.	.	.	.	.	.
55	14	27	46	83	93	98	99	.	.	.	.	.	.	.	.	.	.	.	.	.	.
60	14	29	50	86	95	99	.	.	.	.	.	.	.	.	.	.	.	.	.	.	.
65	15	32	53	89	97	99	.	.	.	.	.	.	.	.	.	.	.	.	.	.	.
70	16	34	57	91	98	.	.	.	.	.	.	.	.	.	.	.	.	.	.	.	.
75	17	36	60	93	98	.	.	.	.	.	.	.	.	.	.	.	.	.	.	.	.
80	18	39	63	95	99	.	.	.	.	.	.	.	.	.	.	.	.	.	.	.	.
90	20	43	69	97	99	.	.	.	.	.	.	.	.	.	.	.	.	.	.	.	.
100	22	47	74	98	.	.	.	.	.	.	.	.	.	.	.	.	.	.	.	.	.
110	24	51	78	99	.	.	.	.	.	.	.	.	.	.	.	.	.	.	.	.	.
120	27	55	82	99	.	.	.	.	.	.	.	.	.	.	.	.	.	.	.	.	.
130	29	59	85	.	.	.	.	.	.	.	.	.	.	.	.	.	.	.	.	.	.
140	31	63	88	.	.	.	.	.	.	.	.	.	.	.	.	.	.	.	.	.	.
150	33	66	90	.	.	.	.	.	.	.	.	.	.	.	.	.	.	.	.	.	.
175	38	73	94	.	.	.	.	.	.	.	.	.	.	.	.	.	.	.	.	.	.
200	43	79	97	.	.	.	.	.	.	.	.	.	.	.	.	.	.	.	.	.	.
225	48	84	98	.	.	.	.	.	.	.	.	.	.	.	.	.	.	.	.	.	.
250	52	88	99	.	.	.	.	.	.	.	.	.	.	.	.	.	.	.	.	.	.
300	61	94	.	.	.	.	.	.	.	.	.	.	.	.	.	.	.	.	.	.	.
350	68	97	.	.	.	.	.	.	.	.	.	.	.	.	.	.	.	.	.	.	.
400	74	98	.	.	.	.	.	.	.	.	.	.	.	.	.	.	.	.	.	.	.
450	80	99	.	.	.	.	.	.	.	.	.	.	.	.	.	.	.	.	.	.	.
500	84	.	.	.	.	.	.	.	.	.	.	.	.	.	.	.	.	.	.	.	.
600	90	.	.	.	.	.	.	.	.	.	.	.	.	.	.	.	.	.	.	.	.
700	94	.	.	.	.	.	.	.	.	.	.	.	.	.	.	.	.	.	.	.	.
800	97	.	.	.	.	.	.	.	.	.	.	.	.	.	.	.	.	.	.	.	.
900	98	.	.	.	.	.	.	.	.	.	.	.	.	.	.	.	.	.	.	.	.
1000	99	.	.	.	.	.	.	.	.	.	.	.	.	.	.	.	.	.	.	.	.

Table 9.11. Power table for ANCOVA analysis; 2 × 4 design at $r = 0.60$ and alpha = 0.05

n	Hypothesized ES																				
	0.10	0.15	0.20	0.30	0.35	0.40	0.45	0.50	0.55	0.60	0.65	0.70	0.75	0.80	1.00	1.25	1.50	1.75	2.00	2.50	3.00
3	5	6	6	8	10	11	13	15	17	20	23	26	30	34	50	71	87	95	99	.	.
4	5	6	7	10	12	15	18	21	25	30	34	39	45	50	71	89	98	.	.	.	.
5	5	6	8	12	15	19	23	28	33	39	45	52	58	64	84	97	.	.	.	.	.
6	5	7	9	14	18	23	29	35	41	48	56	63	69	75	92	99	.	.	.	.	.
7	6	7	9	17	21	27	34	41	49	57	64	72	78	83	96	.	.	.	.	.	.
8	6	8	10	19	25	32	39	48	56	64	72	79	85	89	98	.	.	.	.	.	.
9	6	8	11	21	28	36	45	54	63	71	78	85	89	93	99	.	.	.	.	.	.
10	6	9	12	24	31	40	50	59	68	77	83	89	93	96	.	.	.	.	.	.	.
11	6	9	13	26	35	44	54	64	73	81	87	92	95	97	.	.	.	.	.	.	.
12	7	10	14	28	38	48	59	69	78	85	91	94	97	98	.	.	.	.	.	.	.
13	7	10	15	31	41	52	63	73	82	88	93	96	98	99	.	.	.	.	.	.	.
14	7	11	16	33	44	56	67	77	85	91	95	97	99	99	.	.	.	.	.	.	.
15	7	11	17	36	47	59	71	80	88	93	96	98	99	.	.	.	.	.	.	.	.
20	8	14	22	47	61	74	84	91	96	98	99	.	.	.	.	.	.	.	.	.	.
25	9	17	28	58	73	84	92	97	99	.	.	.	.	.	.	.	.	.	.	.	.
30	11	20	33	67	81	91	96	99	.	.	.	.	.	.	.	.	.	.	.	.	.
35	12	23	39	74	87	95	98	.	.	.	.	.	.	.	.	.	.	.	.	.	.
40	13	26	44	81	92	97	99	.	.	.	.	.	.	.	.	.	.	.	.	.	.
45	14	29	49	85	95	99	.	.	.	.	.	.	.	.	.	.	.	.	.	.	.
50	15	32	54	89	97	99	.	.	.	.	.	.	.	.	.	.	.	.	.	.	.
55	17	35	58	92	98	.	.	.	.	.	.	.	.	.	.	.	.	.	.	.	.
60	18	38	62	94	99	.	.	.	.	.	.	.	.	.	.	.	.	.	.	.	.
65	19	41	66	96	99	.	.	.	.	.	.	.	.	.	.	.	.	.	.	.	.
70	21	44	70	97	.	.	.	.	.	.	.	.	.	.	.	.	.	.	.	.	.
75	22	46	73	98	.	.	.	.	.	.	.	.	.	.	.	.	.	.	.	.	.
80	23	49	76	99	.	.	.	.	.	.	.	.	.	.	.	.	.	.	.	.	.
90	26	55	81	99	.	.	.	.	.	.	.	.	.	.	.	.	.	.	.	.	.
100	29	59	86	.	.	.	.	.	.	.	.	.	.	.	.	.	.	.	.	.	.
110	32	64	89	.	.	.	.	.	.	.	.	.	.	.	.	.	.	.	.	.	.
120	34	68	92	.	.	.	.	.	.	.	.	.	.	.	.	.	.	.	.	.	.
130	37	72	94	.	.	.	.	.	.	.	.	.	.	.	.	.	.	.	.	.	.
140	40	76	95	.	.	.	.	.	.	.	.	.	.	.	.	.	.	.	.	.	.
150	42	79	97	.	.	.	.	.	.	.	.	.	.	.	.	.	.	.	.	.	.
175	48	85	98	.	.	.	.	.	.	.	.	.	.	.	.	.	.	.	.	.	.
200	54	90	99	.	.	.	.	.	.	.	.	.	.	.	.	.	.	.	.	.	.
225	60	93	.	.	.	.	.	.	.	.	.	.	.	.	.	.	.	.	.	.	.
250	65	95	.	.	.	.	.	.	.	.	.	.	.	.	.	.	.	.	.	.	.
300	74	98	.	.	.	.	.	.	.	.	.	.	.	.	.	.	.	.	.	.	.
350	80	99	.	.	.	.	.	.	.	.	.	.	.	.	.	.	.	.	.	.	.
400	86	.	.	.	.	.	.	.	.	.	.	.	.	.	.	.	.	.	.	.	.
450	90	.	.	.	.	.	.	.	.	.	.	.	.	.	.	.	.	.	.	.	.
500	93	.	.	.	.	.	.	.	.	.	.	.	.	.	.	.	.	.	.	.	.
600	97	.	.	.	.	.	.	.	.	.	.	.	.	.	.	.	.	.	.	.	.
700	98	.	.	.	.	.	.	.	.	.	.	.	.	.	.	.	.	.	.	.	.
800	99	.	.	.	.	.	.	.	.	.	.	.	.	.	.	.	.	.	.	.	.
900	.	.	.	.	.	.	.	.	.	.	.	.	.	.	.	.	.	.	.	.	.
1000	.	.	.	.	.	.	.	.	.	.	.	.	.	.	.	.	.	.	.	.	.

Table 9.12. Power table for ANCOVA analysis; 3 × 3 design at $r = 0.40$ and alpha = 0.05

n	Hypothesized ES																				
	0.10	0.15	0.20	0.30	0.35	0.40	0.45	0.50	0.55	0.60	0.65	0.70	0.75	0.80	1.00	1.25	1.50	1.75	2.00	2.50	3.00
3	5	6	6	7	8	9	11	12	14	16	18	21	23	26	40	59	77	90	96	.	.
4	5	6	6	9	10	12	14	17	20	23	27	31	35	40	59	81	93	98	.	.	.
5	5	6	7	10	12	15	18	22	26	31	36	41	47	52	74	92	98	.	.	.	.
6	5	6	7	12	14	18	22	27	32	38	44	51	57	63	84	97	.	.	.	.	.
7	5	6	8	13	17	21	26	32	39	45	52	59	66	73	91	99	.	.	.	.	.
8	5	7	9	15	19	24	31	37	45	52	60	67	74	80	95	.	.	.	.	.	.
9	6	7	9	17	22	28	35	43	51	59	67	74	80	86	97	.	.	.	.	.	.
10	6	7	10	18	24	31	39	48	56	65	72	79	85	90	99	.	.	.	.	.	.
11	6	8	11	20	27	35	43	52	61	70	78	84	89	93	99	.	.	.	.	.	.
12	6	8	11	22	29	38	47	57	66	75	82	88	92	95	.	.	.	.	.	.	.
13	6	8	12	24	32	41	51	61	70	79	85	91	94	97	.	.	.	.	.	.	.
14	6	9	13	26	34	44	55	65	74	82	88	93	96	98	.	.	.	.	.	.	.
15	6	9	14	28	37	48	58	69	78	85	91	95	97	99	.	.	.	.	.	.	.
20	7	11	17	37	49	62	73	83	90	95	97	99	.	.	.	.	.	.	.	.	.
25	8	13	21	46	60	73	84	91	96	98	99	.	.	.	.	.	.	.	.	.	.
30	9	15	26	55	70	82	91	96	98	99	.	.	.	.	.	.	.	.	.	.	.
35	10	18	30	62	77	88	95	98	99	.	.	.	.	.	.	.	.	.	.	.	.
40	11	20	34	69	83	92	97	99	.	.	.	.	.	.	.	.	.	.	.	.	.
45	11	22	38	75	88	95	99	.	.	.	.	.	.	.	.	.	.	.	.	.	.
50	12	24	42	80	91	97	99	.	.	.	.	.	.	.	.	.	.	.	.	.	.
55	13	27	46	84	94	98	.	.	.	.	.	.	.	.	.	.	.	.	.	.	.
60	14	29	50	87	96	99	.	.	.	.	.	.	.	.	.	.	.	.	.	.	.
65	15	32	54	90	97	99	.	.	.	.	.	.	.	.	.	.	.	.	.	.	.
70	16	34	57	92	98	.	.	.	.	.	.	.	.	.	.	.	.	.	.	.	.
75	17	36	61	94	99	.	.	.	.	.	.	.	.	.	.	.	.	.	.	.	.
80	18	39	64	96	99	.	.	.	.	.	.	.	.	.	.	.	.	.	.	.	.
90	20	43	70	97	.	.	.	.	.	.	.	.	.	.	.	.	.	.	.	.	.
100	22	48	75	99	.	.	.	.	.	.	.	.	.	.	.	.	.	.	.	.	.
110	24	52	79	99	.	.	.	.	.	.	.	.	.	.	.	.	.	.	.	.	.
120	26	56	83	.	.	.	.	.	.	.	.	.	.	.	.	.	.	.	.	.	.
130	28	60	86	.	.	.	.	.	.	.	.	.	.	.	.	.	.	.	.	.	.
140	31	63	89	.	.	.	.	.	.	.	.	.	.	.	.	.	.	.	.	.	.
150	33	67	91	.	.	.	.	.	.	.	.	.	.	.	.	.	.	.	.	.	.
175	38	74	95	.	.	.	.	.	.	.	.	.	.	.	.	.	.	.	.	.	.
200	43	81	97	.	.	.	.	.	.	.	.	.	.	.	.	.	.	.	.	.	.
225	48	85	99	.	.	.	.	.	.	.	.	.	.	.	.	.	.	.	.	.	.
250	53	89	99	.	.	.	.	.	.	.	.	.	.	.	.	.	.	.	.	.	.
300	61	94	.	.	.	.	.	.	.	.	.	.	.	.	.	.	.	.	.	.	.
350	69	97	.	.	.	.	.	.	.	.	.	.	.	.	.	.	.	.	.	.	.
400	75	99	.	.	.	.	.	.	.	.	.	.	.	.	.	.	.	.	.	.	.
450	81	99	.	.	.	.	.	.	.	.	.	.	.	.	.	.	.	.	.	.	.
500	85	.	.	.	.	.	.	.	.	.	.	.	.	.	.	.	.	.	.	.	.
600	91	.	.	.	.	.	.	.	.	.	.	.	.	.	.	.	.	.	.	.	.
700	95	.	.	.	.	.	.	.	.	.	.	.	.	.	.	.	.	.	.	.	.
800	97	.	.	.	.	.	.	.	.	.	.	.	.	.	.	.	.	.	.	.	.
900	99	.	.	.	.	.	.	.	.	.	.	.	.	.	.	.	.	.	.	.	.
1000	99	.	.	.	.	.	.	.	.	.	.	.	.	.	.	.	.	.	.	.	.

Table 9.13. Power table for ANCOVA analysis; 3 × 3 design at $r = 0.60$ and alpha = 0.05

n	Hypothesized ES																				
	0.10	0.15	0.20	0.30	0.35	0.40	0.45	0.50	0.55	0.60	0.65	0.70	0.75	0.80	1.00	1.25	1.50	1.75	2.00	2.50	3.00
3	5	6	6	8	9	11	13	15	17	20	23	26	30	34	51	73	88	96	99	.	.
4	5	6	7	10	12	15	18	21	25	30	35	40	45	51	72	91	98	.	.	.	.
5	5	6	8	12	15	19	23	28	34	40	46	53	59	65	86	97	.	.	.	.	.
6	5	7	8	14	18	23	29	35	42	49	56	64	70	77	93	99	.	.	.	.	.
7	6	7	9	16	21	27	34	42	50	58	66	73	79	85	97	.	.	.	.	.	.
8	6	8	10	19	25	32	40	48	57	65	73	80	86	90	99	.	.	.	.	.	.
9	6	8	11	21	28	36	45	54	64	72	80	86	91	94	99	.	.	.	.	.	.
10	6	8	12	23	31	40	50	60	69	78	85	90	94	96	.	.	.	.	.	.	.
11	6	9	13	26	35	45	55	65	75	82	89	93	96	98	.	.	.	.	.	.	.
12	7	9	14	28	38	49	60	70	79	86	92	95	98	99	.	.	.	.	.	.	.
13	7	10	15	31	41	53	64	74	83	89	94	97	98	99	.	.	.	.	.	.	.
14	7	10	16	33	45	57	68	78	86	92	96	98	99	.	.	.	.	.	.	.	.
15	7	11	17	36	48	60	72	81	89	94	97	99	99	.	.	.	.	.	.	.	.
20	8	14	22	48	62	75	85	92	97	99	.	.	.	.	.	.	.	.	.	.	.
25	9	16	28	59	74	85	93	97	99	.	.	.	.	.	.	.	.	.	.	.	.
30	10	19	33	68	82	92	97	99	.	.	.	.	.	.	.	.	.	.	.	.	.
35	12	22	39	76	88	96	99	.	.	.	.	.	.	.	.	.	.	.	.	.	.
40	13	25	44	82	93	98	99	.	.	.	.	.	.	.	.	.	.	.	.	.	.
45	14	29	49	87	95	99	.	.	.	.	.	.	.	.	.	.	.	.	.	.	.
50	15	32	54	90	97	99	.	.	.	.	.	.	.	.	.	.	.	.	.	.	.
55	16	35	59	93	98	.	.	.	.	.	.	.	.	.	.	.	.	.	.	.	.
60	18	38	63	95	99	.	.	.	.	.	.	.	.	.	.	.	.	.	.	.	.
65	19	41	67	97	99	.	.	.	.	.	.	.	.	.	.	.	.	.	.	.	.
70	20	44	71	98	.	.	.	.	.	.	.	.	.	.	.	.	.	.	.	.	.
75	22	47	74	98	.	.	.	.	.	.	.	.	.	.	.	.	.	.	.	.	.
80	23	50	77	99	.	.	.	.	.	.	.	.	.	.	.	.	.	.	.	.	.
90	26	55	82	.	.	.	.	.	.	.	.	.	.	.	.	.	.	.	.	.	.
100	29	60	87	.	.	.	.	.	.	.	.	.	.	.	.	.	.	.	.	.	.
110	31	65	90	.	.	.	.	.	.	.	.	.	.	.	.	.	.	.	.	.	.
120	34	69	93	.	.	.	.	.	.	.	.	.	.	.	.	.	.	.	.	.	.
130	37	73	95	.	.	.	.	.	.	.	.	.	.	.	.	.	.	.	.	.	.
140	40	77	96	.	.	.	.	.	.	.	.	.	.	.	.	.	.	.	.	.	.
150	42	80	97	.	.	.	.	.	.	.	.	.	.	.	.	.	.	.	.	.	.
175	49	86	99	.	.	.	.	.	.	.	.	.	.	.	.	.	.	.	.	.	.
200	55	91	99	.	.	.	.	.	.	.	.	.	.	.	.	.	.	.	.	.	.
225	60	94	.	.	.	.	.	.	.	.	.	.	.	.	.	.	.	.	.	.	.
250	66	96	.	.	.	.	.	.	.	.	.	.	.	.	.	.	.	.	.	.	.
300	75	98	.	.	.	.	.	.	.	.	.	.	.	.	.	.	.	.	.	.	.
350	82	99	.	.	.	.	.	.	.	.	.	.	.	.	.	.	.	.	.	.	.
400	87	.	.	.	.	.	.	.	.	.	.	.	.	.	.	.	.	.	.	.	.
450	91	.	.	.	.	.	.	.	.	.	.	.	.	.	.	.	.	.	.	.	.
500	94	.	.	.	.	.	.	.	.	.	.	.	.	.	.	.	.	.	.	.	.
600	97	.	.	.	.	.	.	.	.	.	.	.	.	.	.	.	.	.	.	.	.
700	99	.	.	.	.	.	.	.	.	.	.	.	.	.	.	.	.	.	.	.	.
800	.	.	.	.	.	.	.	.	.	.	.	.	.	.	.	.	.	.	.	.	.
900	.	.	.	.	.	.	.	.	.	.	.	.	.	.	.	.	.	.	.	.	.
1000	.	.	.	.	.	.	.	.	.	.	.	.	.	.	.	.	.	.	.	.	.

Table 9.14. Power table for ANCOVA analysis; 3×4 design at $r = 0.40$ and alpha $= 0.05$

n	Hypothesized ES																				
	0.10	0.15	0.20	0.30	0.35	0.40	0.45	0.50	0.55	0.60	0.65	0.70	0.75	0.80	1.00	1.25	1.50	1.75	2.00	2.50	3.00
3	5	6	6	8	9	10	11	13	15	18	20	23	27	30	47	68	86	95	99	.	.
4	5	6	7	9	11	13	15	19	22	26	30	35	40	46	67	88	97	.	.	.	.
5	5	6	7	11	13	16	20	24	29	35	41	47	53	60	82	96	.	.	.	.	.
6	5	6	8	12	16	20	25	30	37	43	50	58	65	71	91	99	.	.	.	.	.
7	5	7	8	14	18	23	30	36	44	52	59	67	74	80	95	.	.	.	.	.	.
8	6	7	9	16	21	27	34	42	51	59	67	75	81	87	98	.	.	.	.	.	.
9	6	7	10	18	24	31	39	48	57	66	74	81	87	91	99	.	.	.	.	.	.
10	6	8	11	20	27	35	44	54	63	72	80	86	91	95	.	.	.	.	.	.	.
11	6	8	11	22	30	39	49	59	69	77	85	90	94	97	.	.	.	.	.	.	.
12	6	8	12	24	33	43	53	64	74	82	88	93	96	98	.	.	.	.	.	.	.
13	6	9	13	26	36	47	58	68	78	86	91	95	97	99	.	.	.	.	.	.	.
14	6	9	14	29	39	50	62	72	82	89	93	97	98	99	.	.	.	.	.	.	.
15	7	10	15	31	42	54	66	76	85	91	95	98	99	.	.	.	.	.	.	.	.
20	7	12	19	42	56	69	81	89	95	98	99	.	.	.	.	.	.	.	.	.	.
25	8	14	24	52	67	81	90	96	98	99	.	.	.	.	.	.	.	.	.	.	.
30	9	17	28	61	77	88	95	98	.	.	.	.	.	.	.	.	.	.	.	.	.
35	10	19	33	70	84	93	98	99	.	.	.	.	.	.	.	.	.	.	.	.	.
40	11	22	38	76	89	96	99	.	.	.	.	.	.	.	.	.	.	.	.	.	.
45	12	24	43	82	93	98	.	.	.	.	.	.	.	.	.	.	.	.	.	.	.
50	13	27	48	86	96	99	.	.	.	.	.	.	.	.	.	.	.	.	.	.	.
55	14	30	52	90	97	99	.	.	.	.	.	.	.	.	.	.	.	.	.	.	.
60	15	33	56	93	98	.	.	.	.	.	.	.	.	.	.	.	.	.	.	.	.
65	16	35	60	95	99	.	.	.	.	.	.	.	.	.	.	.	.	.	.	.	.
70	17	38	64	96	99	.	.	.	.	.	.	.	.	.	.	.	.	.	.	.	.
75	19	41	68	97	.	.	.	.	.	.	.	.	.	.	.	.	.	.	.	.	.
80	20	43	71	98	.	.	.	.	.	.	.	.	.	.	.	.	.	.	.	.	.
90	22	49	77	99	.	.	.	.	.	.	.	.	.	.	.	.	.	.	.	.	.
100	24	54	82	.	.	.	.	.	.	.	.	.	.	.	.	.	.	.	.	.	.
110	27	58	86	.	.	.	.	.	.	.	.	.	.	.	.	.	.	.	.	.	.
120	29	63	89	.	.	.	.	.	.	.	.	.	.	.	.	.	.	.	.	.	.
130	32	67	92	.	.	.	.	.	.	.	.	.	.	.	.	.	.	.	.	.	.
140	34	71	94	.	.	.	.	.	.	.	.	.	.	.	.	.	.	.	.	.	.
150	37	74	95	.	.	.	.	.	.	.	.	.	.	.	.	.	.	.	.	.	.
175	43	81	98	.	.	.	.	.	.	.	.	.	.	.	.	.	.	.	.	.	.
200	48	87	99	.	.	.	.	.	.	.	.	.	.	.	.	.	.	.	.	.	.
225	54	91	.	.	.	.	.	.	.	.	.	.	.	.	.	.	.	.	.	.	.
250	59	94	.	.	.	.	.	.	.	.	.	.	.	.	.	.	.	.	.	.	.
300	68	97	.	.	.	.	.	.	.	.	.	.	.	.	.	.	.	.	.	.	.
350	76	99	.	.	.	.	.	.	.	.	.	.	.	.	.	.	.	.	.	.	.
400	82	.	.	.	.	.	.	.	.	.	.	.	.	.	.	.	.	.	.	.	.
450	87	.	.	.	.	.	.	.	.	.	.	.	.	.	.	.	.	.	.	.	.
500	91	.	.	.	.	.	.	.	.	.	.	.	.	.	.	.	.	.	.	.	.
600	95	.	.	.	.	.	.	.	.	.	.	.	.	.	.	.	.	.	.	.	.
700	98	.	.	.	.	.	.	.	.	.	.	.	.	.	.	.	.	.	.	.	.
800	99	.	.	.	.	.	.	.	.	.	.	.	.	.	.	.	.	.	.	.	.
900	.	.	.	.	.	.	.	.	.	.	.	.	.	.	.	.	.	.	.	.	.
1000	.	.	.	.	.	.	.	.	.	.	.	.	.	.	.	.	.	.	.	.	.

Table 9.15. Power table for ANCOVA analysis; 3 × 4 design at $r = 0.60$ and alpha = 0.05

n	Hypothesized ES																				
	0.10	0.15	0.20	0.30	0.35	0.40	0.45	0.50	0.55	0.60	0.65	0.70	0.75	0.80	1.00	1.25	1.50	1.75	2.00	2.50	3.00
3	5	6	6	8	10	12	14	16	19	23	26	30	35	39	59	82	94	99	.	.	.
4	5	6	7	10	13	16	20	24	29	34	40	46	52	58	81	96	99	.	.	.	.
5	5	6	8	13	16	21	26	32	38	45	53	60	67	73	92	99	.	.	.	.	.
6	6	7	9	15	20	26	32	40	48	56	64	71	78	84	97	.	.	.	.	.	.
7	6	7	10	18	24	31	39	47	56	65	73	80	86	91	99	.	.	.	.	.	.
8	6	8	11	20	27	36	45	55	64	73	81	87	92	95	.	.	.	.	.	.	.
9	6	8	12	23	31	41	51	61	71	80	86	92	95	97	.	.	.	.	.	.	.
10	6	9	13	26	35	46	57	67	77	85	91	95	97	99	.	.	.	.	.	.	.
11	7	9	14	29	39	51	62	73	82	89	94	97	98	99	.	.	.	.	.	.	.
12	7	10	15	32	43	55	67	78	86	92	96	98	99	.	.	.	.	.	.	.	.
13	7	10	16	35	47	59	71	82	89	94	97	99	.	.	.	.	.	.	.	.	.
14	7	11	17	37	50	64	75	85	92	96	98	99	.	.	.	.	.	.	.	.	.
15	7	12	18	40	54	67	79	88	94	97	99	.	.	.	.	.	.	.	.	.	.
20	8	15	25	54	69	82	91	96	99	.	.	.	.	.	.	.	.	.	.	.	.
25	10	18	31	66	81	91	97	99	.	.	.	.	.	.	.	.	.	.	.	.	.
30	11	21	37	75	88	96	99	.	.	.	.	.	.	.	.	.	.	.	.	.	.
35	12	25	44	83	93	98	.	.	.	.	.	.	.	.	.	.	.	.	.	.	.
40	14	28	50	88	96	99	.	.	.	.	.	.	.	.	.	.	.	.	.	.	.
45	15	32	55	92	98	.	.	.	.	.	.	.	.	.	.	.	.	.	.	.	.
50	16	35	61	95	99	.	.	.	.	.	.	.	.	.	.	.	.	.	.	.	.
55	18	39	66	97	99	.	.	.	.	.	.	.	.	.	.	.	.	.	.	.	.
60	19	43	70	98	.	.	.	.	.	.	.	.	.	.	.	.	.	.	.	.	.
65	21	46	74	99	.	.	.	.	.	.	.	.	.	.	.	.	.	.	.	.	.
70	22	49	78	99	.	.	.	.	.	.	.	.	.	.	.	.	.	.	.	.	.
75	24	53	81	.	.	.	.	.	.	.	.	.	.	.	.	.	.	.	.	.	.
80	26	56	84	.	.	.	.	.	.	.	.	.	.	.	.	.	.	.	.	.	.
90	29	62	89	.	.	.	.	.	.	.	.	.	.	.	.	.	.	.	.	.	.
100	32	67	92	.	.	.	.	.	.	.	.	.	.	.	.	.	.	.	.	.	.
110	35	72	95	.	.	.	.	.	.	.	.	.	.	.	.	.	.	.	.	.	.
120	38	76	96	.	.	.	.	.	.	.	.	.	.	.	.	.	.	.	.	.	.
130	41	80	98	.	.	.	.	.	.	.	.	.	.	.	.	.	.	.	.	.	.
140	45	84	98	.	.	.	.	.	.	.	.	.	.	.	.	.	.	.	.	.	.
150	48	86	99	.	.	.	.	.	.	.	.	.	.	.	.	.	.	.	.	.	.
175	55	92	.	.	.	.	.	.	.	.	.	.	.	.	.	.	.	.	.	.	.
200	62	95	.	.	.	.	.	.	.	.	.	.	.	.	.	.	.	.	.	.	.
225	68	97	.	.	.	.	.	.	.	.	.	.	.	.	.	.	.	.	.	.	.
250	73	98	.	.	.	.	.	.	.	.	.	.	.	.	.	.	.	.	.	.	.
300	82	.	.	.	.	.	.	.	.	.	.	.	.	.	.	.	.	.	.	.	.
350	88	.	.	.	.	.	.	.	.	.	.	.	.	.	.	.	.	.	.	.	.
400	92	.	.	.	.	.	.	.	.	.	.	.	.	.	.	.	.	.	.	.	.
450	95	.	.	.	.	.	.	.	.	.	.	.	.	.	.	.	.	.	.	.	.
500	97	.	.	.	.	.	.	.	.	.	.	.	.	.	.	.	.	.	.	.	.
600	99	.	.	.	.	.	.	.	.	.	.	.	.	.	.	.	.	.	.	.	.
700	.	.	.	.	.	.	.	.	.	.	.	.	.	.	.	.	.	.	.	.	.
800	.	.	.	.	.	.	.	.	.	.	.	.	.	.	.	.	.	.	.	.	.
900	.	.	.	.	.	.	.	.	.	.	.	.	.	.	.	.	.	.	.	.	.
1000	.	.	.	.	.	.	.	.	.	.	.	.	.	.	.	.	.	.	.	.	.

Table 9.16. Power table for repeated measures interaction ANOVA; 2 × 2 mixed design at $r = 0.40$ and alpha = 0.05

n	Hypothesized ES																				
	0.10	0.15	0.20	0.30	0.35	0.40	0.45	0.50	0.55	0.60	0.65	0.70	0.75	0.80	1.00	1.25	1.50	1.75	2.00	2.50	3.00
3	6	7	7	8	9	11	12	13	15	17	18	20	22	25	35	49	63	76	86	96	99
4	6	6	7	9	11	13	15	18	20	23	27	30	34	37	53	72	86	94	98	.	.
5	5	6	7	11	13	16	19	23	27	31	35	40	44	49	68	85	95	99	.	.	.
6	5	6	8	13	16	19	24	28	33	38	43	49	54	60	78	93	98	.	.	.	.
7	5	7	9	15	18	23	28	33	39	45	51	57	63	68	86	97	.	.	.	.	.
8	6	7	10	16	21	26	32	38	45	51	58	64	70	76	91	98	.	.	.	.	.
9	6	8	10	18	24	30	36	43	50	57	64	71	76	81	94	99	.	.	.	.	.
10	6	8	11	20	26	33	40	48	55	63	70	76	81	86	97	.	.	.	.	.	.
11	6	8	12	22	29	36	44	52	60	68	74	80	85	89	98	.	.	.	.	.	.
12	6	9	13	24	31	39	48	56	64	72	78	84	89	92	99	.	.	.	.	.	.
13	6	9	14	26	34	43	51	60	68	76	82	87	91	94	99	.	.	.	.	.	.
14	7	10	14	28	37	46	55	64	72	79	85	90	93	96	.	.	.	.	.	.	.
15	7	10	15	30	39	48	58	67	75	82	88	92	95	97	.	.	.	.	.	.	.
20	8	13	20	39	50	61	72	80	87	92	95	97	99	99	.	.	.	.	.	.	.
25	9	15	24	48	60	72	81	89	94	97	98	99	.	.	.	.	.	.	.	.	.
30	10	18	28	55	69	80	88	94	97	99	99	.	.	.	.	.	.	.	.	.	.
35	11	20	33	62	75	86	92	96	99	99	.	.	.	.	.	.	.	.	.	.	.
40	12	22	37	68	81	90	95	98	99	.	.	.	.	.	.	.	.	.	.	.	.
45	13	25	41	74	85	93	97	99	.	.	.	.	.	.	.	.	.	.	.	.	.
50	14	27	45	78	89	95	98	99	.	.	.	.	.	.	.	.	.	.	.	.	.
55	15	30	48	82	92	97	99	.	.	.	.	.	.	.	.	.	.	.	.	.	.
60	16	32	52	85	94	98	99	.	.	.	.	.	.	.	.	.	.	.	.	.	.
65	17	34	55	88	95	99	.	.	.	.	.	.	.	.	.	.	.	.	.	.	.
70	18	37	58	90	97	99	.	.	.	.	.	.	.	.	.	.	.	.	.	.	.
75	20	39	61	92	97	99	.	.	.	.	.	.	.	.	.	.	.	.	.	.	.
80	21	41	64	93	98	.	.	.	.	.	.	.	.	.	.	.	.	.	.	.	.
90	23	45	69	96	99	.	.	.	.	.	.	.	.	.	.	.	.	.	.	.	.
100	25	50	74	97	99	.	.	.	.	.	.	.	.	.	.	.	.	.	.	.	.
110	27	53	78	98	.	.	.	.	.	.	.	.	.	.	.	.	.	.	.	.	.
120	29	57	81	99	.	.	.	.	.	.	.	.	.	.	.	.	.	.	.	.	.
130	31	60	84	99	.	.	.	.	.	.	.	.	.	.	.	.	.	.	.	.	.
140	34	64	87	.	.	.	.	.	.	.	.	.	.	.	.	.	.	.	.	.	.
150	36	67	89	.	.	.	.	.	.	.	.	.	.	.	.	.	.	.	.	.	.
175	41	73	93	.	.	.	.	.	.	.	.	.	.	.	.	.	.	.	.	.	.
200	45	79	96	.	.	.	.	.	.	.	.	.	.	.	.	.	.	.	.	.	.
225	50	83	97	.	.	.	.	.	.	.	.	.	.	.	.	.	.	.	.	.	.
250	54	87	98	.	.	.	.	.	.	.	.	.	.	.	.	.	.	.	.	.	.
300	62	92	99	.	.	.	.	.	.	.	.	.	.	.	.	.	.	.	.	.	.
350	69	95	.	.	.	.	.	.	.	.	.	.	.	.	.	.	.	.	.	.	.
400	74	97	.	.	.	.	.	.	.	.	.	.	.	.	.	.	.	.	.	.	.
450	79	99	.	.	.	.	.	.	.	.	.	.	.	.	.	.	.	.	.	.	.
500	83	99	.	.	.	.	.	.	.	.	.	.	.	.	.	.	.	.	.	.	.
600	89	.	.	.	.	.	.	.	.	.	.	.	.	.	.	.	.	.	.	.	.
700	93	.	.	.	.	.	.	.	.	.	.	.	.	.	.	.	.	.	.	.	.
800	96	.	.	.	.	.	.	.	.	.	.	.	.	.	.	.	.	.	.	.	.
900	97	.	.	.	.	.	.	.	.	.	.	.	.	.	.	.	.	.	.	.	.
1000	98	.	.	.	.	.	.	.	.	.	.	.	.	.	.	.	.	.	.	.	.

Table 9.17. Power table for repeated measures interaction ANOVA; 2×2 mixed design at $r = 0.60$ and alpha = 0.05

n	Hypothesized ES																				
	0.10	0.15	0.20	0.30	0.35	0.40	0.45	0.50	0.55	0.60	0.65	0.70	0.75	0.80	1.00	1.25	1.50	1.75	2.00	2.50	3.00
3	6	7	8	10	11	13	15	17	19	22	25	27	31	34	48	65	80	90	96	.	.
4	6	7	8	12	14	17	21	24	28	32	37	42	46	51	70	87	96	99	.	.	.
5	6	7	9	14	18	22	27	32	37	43	49	55	60	66	84	96	99	.	.	.	.
6	6	7	10	17	22	27	33	39	46	53	59	66	71	77	92	99	.	.	.	.	.
7	6	8	11	20	26	32	39	46	54	61	68	74	80	85	96	.	.	.	.	.	.
8	6	9	12	23	30	37	45	53	61	68	75	81	86	90	98	.	.	.	.	.	.
9	6	9	14	26	33	42	50	59	67	75	81	86	91	94	99	.	.	.	.	.	.
10	7	10	15	29	37	46	56	64	73	80	86	90	94	96	.	.	.	.	.	.	.
11	7	11	16	31	41	51	60	69	77	84	89	93	96	98	.	.	.	.	.	.	.
12	7	11	17	34	44	55	65	74	81	87	92	95	97	98	.	.	.	.	.	.	.
13	8	12	19	37	48	58	69	77	85	90	94	97	98	99	.	.	.	.	.	.	.
14	8	13	20	40	51	62	72	81	87	92	96	98	99	99	.	.	.	.	.	.	.
15	8	14	21	42	54	65	75	84	90	94	97	98	99	.	.	.	.	.	.	.	.
20	10	17	28	54	67	79	87	93	97	98	99	.	.	.	.	.	.	.	.	.	.
25	11	21	34	64	78	87	94	97	99	.	.	.	.	.	.	.	.	.	.	.	.
30	13	25	40	73	85	93	97	99	.	.	.	.	.	.	.	.	.	.	.	.	.
35	14	28	46	79	90	96	99	.	.	.	.	.	.	.	.	.	.	.	.	.	.
40	16	32	51	85	93	98	99	.	.	.	.	.	.	.	.	.	.	.	.	.	.
45	18	35	56	89	96	99	.	.	.	.	.	.	.	.	.	.	.	.	.	.	.
50	19	39	61	92	97	99	.	.	.	.	.	.	.	.	.	.	.	.	.	.	.
55	21	42	65	94	98	.	.	.	.	.	.	.	.	.	.	.	.	.	.	.	.
60	23	45	69	96	99	.	.	.	.	.	.	.	.	.	.	.	.	.	.	.	.
65	24	48	72	97	99	.	.	.	.	.	.	.	.	.	.	.	.	.	.	.	.
70	26	51	76	98	.	.	.	.	.	.	.	.	.	.	.	.	.	.	.	.	.
75	28	54	78	98	.	.	.	.	.	.	.	.	.	.	.	.	.	.	.	.	.
80	29	57	81	99	.	.	.	.	.	.	.	.	.	.	.	.	.	.	.	.	.
90	32	62	85	99	.	.	.	.	.	.	.	.	.	.	.	.	.	.	.	.	.
100	35	67	89	.	.	.	.	.	.	.	.	.	.	.	.	.	.	.	.	.	.
110	38	71	91	.	.	.	.	.	.	.	.	.	.	.	.	.	.	.	.	.	.
120	41	74	94	.	.	.	.	.	.	.	.	.	.	.	.	.	.	.	.	.	.
130	44	78	95	.	.	.	.	.	.	.	.	.	.	.	.	.	.	.	.	.	.
140	47	81	96	.	.	.	.	.	.	.	.	.	.	.	.	.	.	.	.	.	.
150	50	83	97	.	.	.	.	.	.	.	.	.	.	.	.	.	.	.	.	.	.
175	56	88	99	.	.	.	.	.	.	.	.	.	.	.	.	.	.	.	.	.	.
200	62	92	99	.	.	.	.	.	.	.	.	.	.	.	.	.	.	.	.	.	.
225	67	95	.	.	.	.	.	.	.	.	.	.	.	.	.	.	.	.	.	.	.
250	71	97	.	.	.	.	.	.	.	.	.	.	.	.	.	.	.	.	.	.	.
300	79	99	.	.	.	.	.	.	.	.	.	.	.	.	.	.	.	.	.	.	.
350	85	99	.	.	.	.	.	.	.	.	.	.	.	.	.	.	.	.	.	.	.
400	89	.	.	.	.	.	.	.	.	.	.	.	.	.	.	.	.	.	.	.	.
450	92	.	.	.	.	.	.	.	.	.	.	.	.	.	.	.	.	.	.	.	.
500	95	.	.	.	.	.	.	.	.	.	.	.	.	.	.	.	.	.	.	.	.
600	97	.	.	.	.	.	.	.	.	.	.	.	.	.	.	.	.	.	.	.	.
700	99	.	.	.	.	.	.	.	.	.	.	.	.	.	.	.	.	.	.	.	.
800	99	.	.	.	.	.	.	.	.	.	.	.	.	.	.	.	.	.	.	.	.
900	.	.	.	.	.	.	.	.	.	.	.	.	.	.	.	.	.	.	.	.	.
1000	.	.	.	.	.	.	.	.	.	.	.	.	.	.	.	.	.	.	.	.	.

Table 9.18. Power table for repeated measures interaction ANOVA; 2 × 3 mixed design at $r = 0.40$ and alpha = 0.05

n	Hypothesized ES																				
	0.10	0.15	0.20	0.30	0.35	0.40	0.45	0.50	0.55	0.60	0.65	0.70	0.75	0.80	1.00	1.25	1.50	1.75	2.00	2.50	3.00
3	6	6	7	8	10	11	12	14	16	19	21	24	27	30	44	63	79	90	96	.	.
4	5	6	7	10	12	14	17	20	23	27	31	36	40	45	64	84	95	99	.	.	.
5	5	6	8	12	15	18	22	26	31	36	41	47	53	59	79	94	99	.	.	.	.
6	6	7	8	14	17	22	27	32	38	44	51	57	64	70	88	98	.	.	.	.	.
7	6	7	9	16	20	26	32	38	45	52	59	66	73	78	93	99	.	.	.	.	.
8	6	8	10	18	23	30	37	44	52	60	67	74	80	85	97	.	.	.	.	.	.
9	6	8	11	20	26	34	41	50	58	66	73	80	85	90	98	.	.	.	.	.	.
10	6	8	12	22	30	37	46	55	64	72	79	85	89	93	99	.	.	.	.	.	.
11	6	9	13	25	33	41	51	60	69	76	83	88	92	95	.	.	.	.	.	.	.
12	7	9	14	27	36	45	55	64	73	81	87	91	95	97	.	.	.	.	.	.	.
13	7	10	15	29	38	49	59	68	77	84	90	94	96	98	.	.	.	.	.	.	.
14	7	10	16	31	41	52	63	72	81	87	92	95	97	99	.	.	.	.	.	.	.
15	7	11	17	34	44	55	66	76	84	90	94	97	98	99	.	.	.	.	.	.	.
20	8	14	22	44	57	70	80	88	93	97	98	99	.	.	.	.	.	.	.	.	.
25	9	16	27	54	68	80	89	94	98	99	.	.	.	.	.	.	.	.	.	.	.
30	10	19	32	63	77	87	94	98	99	.	.	.	.	.	.	.	.	.	.	.	.
35	12	22	36	70	83	92	97	99	.	.	.	.	.	.	.	.	.	.	.	.	.
40	13	25	41	76	88	95	98	.	.	.	.	.	.	.	.	.	.	.	.	.	.
45	14	27	46	81	92	97	99	.	.	.	.	.	.	.	.	.	.	.	.	.	.
50	15	30	50	86	95	98	.	.	.	.	.	.	.	.	.	.	.	.	.	.	.
55	16	33	54	89	96	99	.	.	.	.	.	.	.	.	.	.	.	.	.	.	.
60	18	36	58	92	98	99	.	.	.	.	.	.	.	.	.	.	.	.	.	.	.
65	19	38	62	94	98	.	.	.	.	.	.	.	.	.	.	.	.	.	.	.	.
70	20	41	66	95	99	.	.	.	.	.	.	.	.	.	.	.	.	.	.	.	.
75	21	44	69	96	99	.	.	.	.	.	.	.	.	.	.	.	.	.	.	.	.
80	23	46	72	97	.	.	.	.	.	.	.	.	.	.	.	.	.	.	.	.	.
90	25	51	77	99	.	.	.	.	.	.	.	.	.	.	.	.	.	.	.	.	.
100	28	56	82	99	.	.	.	.	.	.	.	.	.	.	.	.	.	.	.	.	.
110	30	60	85	.	.	.	.	.	.	.	.	.	.	.	.	.	.	.	.	.	.
120	32	64	88	.	.	.	.	.	.	.	.	.	.	.	.	.	.	.	.	.	.
130	35	68	91	.	.	.	.	.	.	.	.	.	.	.	.	.	.	.	.	.	.
140	37	71	93	.	.	.	.	.	.	.	.	.	.	.	.	.	.	.	.	.	.
150	40	75	94	.	.	.	.	.	.	.	.	.	.	.	.	.	.	.	.	.	.
175	46	81	97	.	.	.	.	.	.	.	.	.	.	.	.	.	.	.	.	.	.
200	51	86	99	.	.	.	.	.	.	.	.	.	.	.	.	.	.	.	.	.	.
225	56	90	99	.	.	.	.	.	.	.	.	.	.	.	.	.	.	.	.	.	.
250	61	93	.	.	.	.	.	.	.	.	.	.	.	.	.	.	.	.	.	.	.
300	69	97	.	.	.	.	.	.	.	.	.	.	.	.	.	.	.	.	.	.	.
350	76	98	.	.	.	.	.	.	.	.	.	.	.	.	.	.	.	.	.	.	.
400	82	99	.	.	.	.	.	.	.	.	.	.	.	.	.	.	.	.	.	.	.
450	86	.	.	.	.	.	.	.	.	.	.	.	.	.	.	.	.	.	.	.	.
500	90	.	.	.	.	.	.	.	.	.	.	.	.	.	.	.	.	.	.	.	.
600	95	.	.	.	.	.	.	.	.	.	.	.	.	.	.	.	.	.	.	.	.
700	97	.	.	.	.	.	.	.	.	.	.	.	.	.	.	.	.	.	.	.	.
800	99	.	.	.	.	.	.	.	.	.	.	.	.	.	.	.	.	.	.	.	.
900	99	.	.	.	.	.	.	.	.	.	.	.	.	.	.	.	.	.	.	.	.
1000	.	.	.	.	.	.	.	.	.	.	.	.	.	.	.	.	.	.	.	.	.

Table 9.19. Power table for repeated measures interaction ANOVA; 2 × 3 mixed design at $r = 0.60$ and alpha = 0.05

n	Hypothesized ES																				
	0.10	0.15	0.20	0.30	0.35	0.40	0.45	0.50	0.55	0.60	0.65	0.70	0.75	0.80	1.00	1.25	1.50	1.75	2.00	2.50	3.00
3	6	6	7	10	12	14	16	19	22	26	30	34	38	43	61	81	93	98	.	.	.
4	6	7	8	13	16	19	23	28	33	39	45	51	57	62	82	95	99	.	.	.	.
5	6	7	9	16	20	25	31	37	44	51	58	65	71	77	93	99	.	.	.	.	.
6	6	8	11	19	25	31	38	46	54	62	69	76	82	87	97	.	.	.	.	.	.
7	6	8	12	22	29	37	45	54	63	71	78	84	89	93	99	.	.	.	.	.	.
8	6	9	13	26	34	43	52	61	70	78	84	90	93	96	.	.	.	.	.	.	.
9	7	10	15	29	38	48	58	68	76	84	89	93	96	98	.	.	.	.	.	.	.
10	7	11	16	32	42	53	64	73	82	88	93	96	98	99	.	.	.	.	.	.	.
11	7	11	17	35	47	58	69	78	86	91	95	97	99	99	.	.	.	.	.	.	.
12	8	12	19	39	51	62	73	82	89	94	97	98	99	.	.	.	.	.	.	.	.
13	8	13	20	42	54	67	77	86	92	96	98	99	.	.	.	.	.	.	.	.	.
14	8	14	22	45	58	70	81	89	94	97	99	99	.	.	.	.	.	.	.	.	.
15	9	15	23	48	62	74	84	91	95	98	99	.	.	.	.	.	.	.	.	.	.
20	10	19	31	62	76	87	93	97	99	.	.	.	.	.	.	.	.	.	.	.	.
25	12	23	38	73	85	93	98	99	.	.	.	.	.	.	.	.	.	.	.	.	.
30	14	27	45	81	92	97	99	.	.	.	.	.	.	.	.	.	.	.	.	.	.
35	16	31	52	87	95	99	.	.	.	.	.	.	.	.	.	.	.	.	.	.	.
40	17	35	58	91	97	99	.	.	.	.	.	.	.	.	.	.	.	.	.	.	.
45	19	39	63	94	99	.	.	.	.	.	.	.	.	.	.	.	.	.	.	.	.
50	21	43	68	96	99	.	.	.	.	.	.	.	.	.	.	.	.	.	.	.	.
55	23	47	73	98	.	.	.	.	.	.	.	.	.	.	.	.	.	.	.	.	.
60	25	51	77	99	.	.	.	.	.	.	.	.	.	.	.	.	.	.	.	.	.
65	27	54	80	99	.	.	.	.	.	.	.	.	.	.	.	.	.	.	.	.	.
70	29	58	83	99	.	.	.	.	.	.	.	.	.	.	.	.	.	.	.	.	.
75	30	61	86	.	.	.	.	.	.	.	.	.	.	.	.	.	.	.	.	.	.
80	32	64	88	.	.	.	.	.	.	.	.	.	.	.	.	.	.	.	.	.	.
90	36	70	92	.	.	.	.	.	.	.	.	.	.	.	.	.	.	.	.	.	.
100	40	74	94	.	.	.	.	.	.	.	.	.	.	.	.	.	.	.	.	.	.
110	43	79	96	.	.	.	.	.	.	.	.	.	.	.	.	.	.	.	.	.	.
120	47	82	97	.	.	.	.	.	.	.	.	.	.	.	.	.	.	.	.	.	.
130	50	85	98	.	.	.	.	.	.	.	.	.	.	.	.	.	.	.	.	.	.
140	53	88	99	.	.	.	.	.	.	.	.	.	.	.	.	.	.	.	.	.	.
150	56	90	99	.	.	.	.	.	.	.	.	.	.	.	.	.	.	.	.	.	.
175	63	94	.	.	.	.	.	.	.	.	.	.	.	.	.	.	.	.	.	.	.
200	69	97	.	.	.	.	.	.	.	.	.	.	.	.	.	.	.	.	.	.	.
225	75	98	.	.	.	.	.	.	.	.	.	.	.	.	.	.	.	.	.	.	.
250	79	99	.	.	.	.	.	.	.	.	.	.	.	.	.	.	.	.	.	.	.
300	86	.	.	.	.	.	.	.	.	.	.	.	.	.	.	.	.	.	.	.	.
350	91	.	.	.	.	.	.	.	.	.	.	.	.	.	.	.	.	.	.	.	.
400	95	.	.	.	.	.	.	.	.	.	.	.	.	.	.	.	.	.	.	.	.
450	97	.	.	.	.	.	.	.	.	.	.	.	.	.	.	.	.	.	.	.	.
500	98	.	.	.	.	.	.	.	.	.	.	.	.	.	.	.	.	.	.	.	.
600	99	.	.	.	.	.	.	.	.	.	.	.	.	.	.	.	.	.	.	.	.
700	.	.	.	.	.	.	.	.	.	.	.	.	.	.	.	.	.	.	.	.	.
800	.	.	.	.	.	.	.	.	.	.	.	.	.	.	.	.	.	.	.	.	.
900	.	.	.	.	.	.	.	.	.	.	.	.	.	.	.	.	.	.	.	.	.
1000	.	.	.	.	.	.	.	.	.	.	.	.	.	.	.	.	.	.	.	.	.

Table 9.20. Power table for repeated measures interaction ANOVA; 2 × 4 mixed design at $r = 0.40$ and alpha = 0.05

n	Hypothesized ES																				
	0.10	0.15	0.20	0.30	0.35	0.40	0.45	0.50	0.55	0.60	0.65	0.70	0.75	0.80	1.00	1.25	1.50	1.75	2.00	2.50	3.00
3	5	6	7	9	10	12	13	16	18	21	24	28	31	35	52	73	88	96	99	.	.
4	5	6	7	10	13	15	19	22	27	31	36	41	47	52	73	91	98	.	.	.	.
5	5	6	8	13	16	20	24	30	35	41	48	54	61	67	87	97	.	.	.	.	.
6	6	7	9	15	19	24	30	37	44	51	58	65	72	78	94	99	.	.	.	.	.
7	6	7	10	17	23	29	36	44	52	60	67	74	80	86	97	.	.	.	.	.	.
8	6	8	11	20	26	33	42	50	59	67	75	81	87	91	99	.	.	.	.	.	.
9	6	8	12	22	30	38	47	56	65	74	81	87	91	95	.	.	.	.	.	.	.
10	6	9	13	25	33	42	52	62	71	79	86	91	94	97	.	.	.	.	.	.	.
11	7	9	14	28	37	47	57	67	76	84	89	94	96	98	.	.	.	.	.	.	.
12	7	10	15	30	40	51	62	72	81	87	92	96	98	99	.	.	.	.	.	.	.
13	7	10	16	33	43	55	66	76	84	90	94	97	99	99	.	.	.	.	.	.	.
14	7	11	17	35	47	59	70	80	87	93	96	98	99	.	.	.	.	.	.	.	.
15	7	12	18	38	50	62	73	83	90	94	97	99	99	.	.	.	.	.	.	.	.
20	9	15	24	50	64	77	87	93	97	99	.	.	.	.	.	.	.	.	.	.	.
25	10	18	30	61	75	87	94	97	99	.	.	.	.	.	.	.	.	.	.	.	.
30	11	21	35	70	84	93	97	99	.	.	.	.	.	.	.	.	.	.	.	.	.
35	12	24	41	77	89	96	99	.	.	.	.	.	.	.	.	.	.	.	.	.	.
40	14	27	46	83	93	98	.	.	.	.	.	.	.	.	.	.	.	.	.	.	.
45	15	31	52	88	96	99	.	.	.	.	.	.	.	.	.	.	.	.	.	.	.
50	16	34	56	91	98	99	.	.	.	.	.	.	.	.	.	.	.	.	.	.	.
55	18	37	61	94	99	.	.	.	.	.	.	.	.	.	.	.	.	.	.	.	.
60	19	40	65	96	99	.	.	.	.	.	.	.	.	.	.	.	.	.	.	.	.
65	20	43	69	97	99	.	.	.	.	.	.	.	.	.	.	.	.	.	.	.	.
70	22	46	73	98	.	.	.	.	.	.	.	.	.	.	.	.	.	.	.	.	.
75	23	49	76	99	.	.	.	.	.	.	.	.	.	.	.	.	.	.	.	.	.
80	25	52	79	99	.	.	.	.	.	.	.	.	.	.	.	.	.	.	.	.	.
90	28	57	84	.	.	.	.	.	.	.	.	.	.	.	.	.	.	.	.	.	.
100	31	63	88	.	.	.	.	.	.	.	.	.	.	.	.	.	.	.	.	.	.
110	33	67	91	.	.	.	.	.	.	.	.	.	.	.	.	.	.	.	.	.	.
120	36	71	93	.	.	.	.	.	.	.	.	.	.	.	.	.	.	.	.	.	.
130	39	75	95	.	.	.	.	.	.	.	.	.	.	.	.	.	.	.	.	.	.
140	42	78	96	.	.	.	.	.	.	.	.	.	.	.	.	.	.	.	.	.	.
150	45	81	97	.	.	.	.	.	.	.	.	.	.	.	.	.	.	.	.	.	.
175	51	87	99	.	.	.	.	.	.	.	.	.	.	.	.	.	.	.	.	.	.
200	57	92	.	.	.	.	.	.	.	.	.	.	.	.	.	.	.	.	.	.	.
225	63	95	.	.	.	.	.	.	.	.	.	.	.	.	.	.	.	.	.	.	.
250	68	97	.	.	.	.	.	.	.	.	.	.	.	.	.	.	.	.	.	.	.
300	77	99	.	.	.	.	.	.	.	.	.	.	.	.	.	.	.	.	.	.	.
350	83	.	.	.	.	.	.	.	.	.	.	.	.	.	.	.	.	.	.	.	.
400	88	.	.	.	.	.	.	.	.	.	.	.	.	.	.	.	.	.	.	.	.
450	92	.	.	.	.	.	.	.	.	.	.	.	.	.	.	.	.	.	.	.	.
500	94	.	.	.	.	.	.	.	.	.	.	.	.	.	.	.	.	.	.	.	.
600	98	.	.	.	.	.	.	.	.	.	.	.	.	.	.	.	.	.	.	.	.
700	99	.	.	.	.	.	.	.	.	.	.	.	.	.	.	.	.	.	.	.	.
800	.	.	.	.	.	.	.	.	.	.	.	.	.	.	.	.	.	.	.	.	.
900	.	.	.	.	.	.	.	.	.	.	.	.	.	.	.	.	.	.	.	.	.
1000	.	.	.	.	.	.	.	.	.	.	.	.	.	.	.	.	.	.	.	.	.

Table 9.21. Power table for repeated measures interaction ANOVA; 2 × 4 mixed design at $r = 0.60$ and alpha = 0.05

n	Hypothesized ES																				
	0.10	0.15	0.20	0.30	0.35	0.40	0.45	0.50	0.55	0.60	0.65	0.70	0.75	0.80	1.00	1.25	1.50	1.75	2.00	2.50	3.00
3	6	6	7	10	13	15	18	22	26	30	35	40	45	51	71	90	98	.	.	.	.
4	6	7	8	14	17	22	27	32	39	45	52	59	65	72	90	98	.	.	.	.	.
5	6	7	10	17	22	28	35	43	51	59	66	73	80	85	97	.	.	.	.	.	.
6	6	8	11	21	28	35	44	53	62	70	77	84	89	93	99	.	.	.	.	.	.
7	6	9	13	25	33	42	52	62	71	79	85	90	94	97	.	.	.	.	.	.	.
8	7	10	14	29	38	48	59	69	78	85	91	95	97	98	.	.	.	.	.	.	.
9	7	10	16	32	43	55	66	76	84	90	94	97	99	99	.	.	.	.	.	.	.
10	7	11	17	36	48	60	71	81	88	93	97	98	99	.	.	.	.	.	.	.	.
11	8	12	19	40	53	65	76	85	92	96	98	99	.	.	.	.	.	.	.	.	.
12	8	13	21	44	57	70	81	89	94	97	99	.	.	.	.	.	.	.	.	.	.
13	8	14	23	47	61	74	84	91	96	98	99	.	.	.	.	.	.	.	.	.	.
14	9	15	24	51	65	78	87	94	97	99	.	.	.	.	.	.	.	.	.	.	.
15	9	16	26	54	69	81	90	95	98	99	.	.	.	.	.	.	.	.	.	.	.
20	11	20	35	69	83	92	97	99	.	.	.	.	.	.	.	.	.	.	.	.	.
25	13	25	43	80	91	97	99	.	.	.	.	.	.	.	.	.	.	.	.	.	.
30	15	30	51	87	96	99	.	.	.	.	.	.	.	.	.	.	.	.	.	.	.
35	17	35	58	92	98	.	.	.	.	.	.	.	.	.	.	.	.	.	.	.	.
40	19	40	65	95	99	.	.	.	.	.	.	.	.	.	.	.	.	.	.	.	.
45	21	44	71	97	.	.	.	.	.	.	.	.	.	.	.	.	.	.	.	.	.
50	23	49	76	99	.	.	.	.	.	.	.	.	.	.	.	.	.	.	.	.	.
55	25	53	80	99	.	.	.	.	.	.	.	.	.	.	.	.	.	.	.	.	.
60	28	57	84	.	.	.	.	.	.	.	.	.	.	.	.	.	.	.	.	.	.
65	30	61	87	.	.	.	.	.	.	.	.	.	.	.	.	.	.	.	.	.	.
70	32	65	89	.	.	.	.	.	.	.	.	.	.	.	.	.	.	.	.	.	.
75	34	68	91	.	.	.	.	.	.	.	.	.	.	.	.	.	.	.	.	.	.
80	36	71	93	.	.	.	.	.	.	.	.	.	.	.	.	.	.	.	.	.	.
90	40	77	96	.	.	.	.	.	.	.	.	.	.	.	.	.	.	.	.	.	.
100	44	81	97	.	.	.	.	.	.	.	.	.	.	.	.	.	.	.	.	.	.
110	48	85	98	.	.	.	.	.	.	.	.	.	.	.	.	.	.	.	.	.	.
120	52	88	99	.	.	.	.	.	.	.	.	.	.	.	.	.	.	.	.	.	.
130	56	91	99	.	.	.	.	.	.	.	.	.	.	.	.	.	.	.	.	.	.
140	59	93	.	.	.	.	.	.	.	.	.	.	.	.	.	.	.	.	.	.	.
150	63	95	.	.	.	.	.	.	.	.	.	.	.	.	.	.	.	.	.	.	.
175	70	97	.	.	.	.	.	.	.	.	.	.	.	.	.	.	.	.	.	.	.
200	76	99	.	.	.	.	.	.	.	.	.	.	.	.	.	.	.	.	.	.	.
225	82	99	.	.	.	.	.	.	.	.	.	.	.	.	.	.	.	.	.	.	.
250	86	.	.	.	.	.	.	.	.	.	.	.	.	.	.	.	.	.	.	.	.
300	92	.	.	.	.	.	.	.	.	.	.	.	.	.	.	.	.	.	.	.	.
350	95	.	.	.	.	.	.	.	.	.	.	.	.	.	.	.	.	.	.	.	.
400	98	.	.	.	.	.	.	.	.	.	.	.	.	.	.	.	.	.	.	.	.
450	99	.	.	.	.	.	.	.	.	.	.	.	.	.	.	.	.	.	.	.	.
500	99	.	.	.	.	.	.	.	.	.	.	.	.	.	.	.	.	.	.	.	.
600	.	.	.	.	.	.	.	.	.	.	.	.	.	.	.	.	.	.	.	.	.
700	.	.	.	.	.	.	.	.	.	.	.	.	.	.	.	.	.	.	.	.	.
800	.	.	.	.	.	.	.	.	.	.	.	.	.	.	.	.	.	.	.	.	.
900	.	.	.	.	.	.	.	.	.	.	.	.	.	.	.	.	.	.	.	.	.
1000	.	.	.	.	.	.	.	.	.	.	.	.	.	.	.	.	.	.	.	.	.

Table 9.22. Power table for repeated measures interaction ANOVA; 3 × 3 mixed design at $r = 0.40$ and alpha = 0.05

n	Hypothesized ES																				
	0.10	0.15	0.20	0.30	0.35	0.40	0.45	0.50	0.55	0.60	0.65	0.70	0.75	0.80	1.00	1.25	1.50	1.75	2.00	2.50	3.00
3	6	6	7	8	10	11	13	15	17	20	23	27	30	34	51	73	88	96	99	.	.
4	5	6	7	10	12	15	18	22	26	30	35	41	46	52	73	91	98	.	.	.	.
5	5	6	8	12	15	19	24	29	34	41	47	54	60	67	87	98	.	.	.	.	.
6	6	7	9	15	19	24	30	36	43	51	58	65	72	78	94	99	.	.	.	.	.
7	6	7	10	17	22	28	35	43	51	60	67	75	81	86	98	.	.	.	.	.	.
8	6	8	10	19	26	33	41	50	59	68	75	82	87	92	99	.	.	.	.	.	.
9	6	8	11	22	29	38	47	56	66	74	82	87	92	95	.	.	.	.	.	.	.
10	6	9	12	24	33	42	52	62	72	80	86	91	95	97	.	.	.	.	.	.	.
11	6	9	13	27	36	47	57	68	77	84	90	94	97	98	.	.	.	.	.	.	.
12	7	10	14	30	40	51	62	73	81	88	93	96	98	99	.	.	.	.	.	.	.
13	7	10	16	32	43	55	67	77	85	91	95	98	99	.	.	.	.	.	.	.	.
14	7	11	17	35	47	59	71	80	88	93	97	98	99	.	.	.	.	.	.	.	.
15	7	11	18	38	50	63	74	84	91	95	98	99	.	.	.	.	.	.	.	.	.
20	8	14	23	50	65	78	87	94	97	99	.	.	.	.	.	.	.	.	.	.	.
25	10	17	29	61	76	88	94	98	99	.	.	.	.	.	.	.	.	.	.	.	.
30	11	20	35	71	85	93	98	99	.	.	.	.	.	.	.	.	.	.	.	.	.
35	12	24	41	78	90	97	99	.	.	.	.	.	.	.	.	.	.	.	.	.	.
40	13	27	47	84	94	98	.	.	.	.	.	.	.	.	.	.	.	.	.	.	.
45	15	30	52	89	97	99	.	.	.	.	.	.	.	.	.	.	.	.	.	.	.
50	16	34	57	92	98	.	.	.	.	.	.	.	.	.	.	.	.	.	.	.	.
55	17	37	62	95	99	.	.	.	.	.	.	.	.	.	.	.	.	.	.	.	.
60	19	40	66	96	99	.	.	.	.	.	.	.	.	.	.	.	.	.	.	.	.
65	20	43	70	97	.	.	.	.	.	.	.	.	.	.	.	.	.	.	.	.	.
70	22	46	74	98	.	.	.	.	.	.	.	.	.	.	.	.	.	.	.	.	.
75	23	49	77	99	.	.	.	.	.	.	.	.	.	.	.	.	.	.	.	.	.
80	25	52	80	99	.	.	.	.	.	.	.	.	.	.	.	.	.	.	.	.	.
90	27	58	85	.	.	.	.	.	.	.	.	.	.	.	.	.	.	.	.	.	.
100	30	63	89	.	.	.	.	.	.	.	.	.	.	.	.	.	.	.	.	.	.
110	33	68	92	.	.	.	.	.	.	.	.	.	.	.	.	.	.	.	.	.	.
120	36	72	94	.	.	.	.	.	.	.	.	.	.	.	.	.	.	.	.	.	.
130	39	76	96	.	.	.	.	.	.	.	.	.	.	.	.	.	.	.	.	.	.
140	42	80	97	.	.	.	.	.	.	.	.	.	.	.	.	.	.	.	.	.	.
150	45	83	98	.	.	.	.	.	.	.	.	.	.	.	.	.	.	.	.	.	.
175	52	89	99	.	.	.	.	.	.	.	.	.	.	.	.	.	.	.	.	.	.
200	58	93	.	.	.	.	.	.	.	.	.	.	.	.	.	.	.	.	.	.	.
225	64	95	.	.	.	.	.	.	.	.	.	.	.	.	.	.	.	.	.	.	.
250	69	97	.	.	.	.	.	.	.	.	.	.	.	.	.	.	.	.	.	.	.
300	78	99	.	.	.	.	.	.	.	.	.	.	.	.	.	.	.	.	.	.	.
350	84	.	.	.	.	.	.	.	.	.	.	.	.	.	.	.	.	.	.	.	.
400	89	.	.	.	.	.	.	.	.	.	.	.	.	.	.	.	.	.	.	.	.
450	93	.	.	.	.	.	.	.	.	.	.	.	.	.	.	.	.	.	.	.	.
500	95	.	.	.	.	.	.	.	.	.	.	.	.	.	.	.	.	.	.	.	.
600	98	.	.	.	.	.	.	.	.	.	.	.	.	.	.	.	.	.	.	.	.
700	99	.	.	.	.	.	.	.	.	.	.	.	.	.	.	.	.	.	.	.	.
800	.	.	.	.	.	.	.	.	.	.	.	.	.	.	.	.	.	.	.	.	.
900	.	.	.	.	.	.	.	.	.	.	.	.	.	.	.	.	.	.	.	.	.
1000	.	.	.	.	.	.	.	.	.	.	.	.	.	.	.	.	.	.	.	.	.

Table 9.23. Power table for repeated measures interaction ANOVA; 3 × 3 mixed design at $r = 0.60$ and alpha = 0.05

n	Hypothesized ES																				
	0.10	0.15	0.20	0.30	0.35	0.40	0.45	0.50	0.55	0.60	0.65	0.70	0.75	0.80	1.00	1.25	1.50	1.75	2.00	2.50	3.00
3	6	6	7	10	12	15	18	21	25	29	34	39	44	49	71	90	98	.	.	.	.
4	6	7	8	13	17	21	26	31	38	44	51	58	65	71	90	99	.	.	.	.	.
5	6	7	10	17	22	28	35	42	50	58	66	74	80	85	97	.	.	.	.	.	.
6	6	8	11	20	27	35	43	52	62	70	78	84	89	93	99	.	.	.	.	.	.
7	6	9	12	24	32	42	52	62	71	79	86	91	95	97	.	.	.	.	.	.	.
8	7	9	14	28	38	48	59	70	79	86	91	95	97	99	.	.	.	.	.	.	.
9	7	10	15	32	43	55	66	76	85	91	95	97	99	99	.	.	.	.	.	.	.
10	7	11	17	36	48	60	72	82	89	94	97	99	99	.	.	.	.	.	.	.	.
11	8	12	19	40	53	66	77	86	92	96	98	99	.	.	.	.	.	.	.	.	.
12	8	13	20	44	57	71	81	90	95	98	99	.	.	.	.	.	.	.	.	.	.
13	8	14	22	47	62	75	85	92	96	99	99	.	.	.	.	.	.	.	.	.	.
14	9	14	24	51	66	79	88	94	98	99	.	.	.	.	.	.	.	.	.	.	.
15	9	15	26	55	70	82	91	96	98	99	.	.	.	.	.	.	.	.	.	.	.
20	11	20	34	70	84	93	97	99	.	.	.	.	.	.	.	.	.	.	.	.	.
25	13	25	43	81	92	97	99	.	.	.	.	.	.	.	.	.	.	.	.	.	.
30	14	30	51	88	96	99	.	.	.	.	.	.	.	.	.	.	.	.	.	.	.
35	17	35	59	93	98	.	.	.	.	.	.	.	.	.	.	.	.	.	.	.	.
40	19	40	65	96	99	.	.	.	.	.	.	.	.	.	.	.	.	.	.	.	.
45	21	44	71	98	.	.	.	.	.	.	.	.	.	.	.	.	.	.	.	.	.
50	23	49	77	99	.	.	.	.	.	.	.	.	.	.	.	.	.	.	.	.	.
55	25	54	81	99	.	.	.	.	.	.	.	.	.	.	.	.	.	.	.	.	.
60	27	58	85	.	.	.	.	.	.	.	.	.	.	.	.	.	.	.	.	.	.
65	29	62	88	.	.	.	.	.	.	.	.	.	.	.	.	.	.	.	.	.	.
70	32	65	90	.	.	.	.	.	.	.	.	.	.	.	.	.	.	.	.	.	.
75	34	69	92	.	.	.	.	.	.	.	.	.	.	.	.	.	.	.	.	.	.
80	36	72	94	.	.	.	.	.	.	.	.	.	.	.	.	.	.	.	.	.	.
90	40	78	96	.	.	.	.	.	.	.	.	.	.	.	.	.	.	.	.	.	.
100	45	82	98	.	.	.	.	.	.	.	.	.	.	.	.	.	.	.	.	.	.
110	49	86	99	.	.	.	.	.	.	.	.	.	.	.	.	.	.	.	.	.	.
120	53	89	99	.	.	.	.	.	.	.	.	.	.	.	.	.	.	.	.	.	.
130	56	92	.	.	.	.	.	.	.	.	.	.	.	.	.	.	.	.	.	.	.
140	60	94	.	.	.	.	.	.	.	.	.	.	.	.	.	.	.	.	.	.	.
150	63	95	.	.	.	.	.	.	.	.	.	.	.	.	.	.	.	.	.	.	.
175	71	98	.	.	.	.	.	.	.	.	.	.	.	.	.	.	.	.	.	.	.
200	77	99	.	.	.	.	.	.	.	.	.	.	.	.	.	.	.	.	.	.	.
225	83	.	.	.	.	.	.	.	.	.	.	.	.	.	.	.	.	.	.	.	.
250	87	.	.	.	.	.	.	.	.	.	.	.	.	.	.	.	.	.	.	.	.
300	93	.	.	.	.	.	.	.	.	.	.	.	.	.	.	.	.	.	.	.	.
350	96	.	.	.	.	.	.	.	.	.	.	.	.	.	.	.	.	.	.	.	.
400	98	.	.	.	.	.	.	.	.	.	.	.	.	.	.	.	.	.	.	.	.
450	99	.	.	.	.	.	.	.	.	.	.	.	.	.	.	.	.	.	.	.	.
500	.	.	.	.	.	.	.	.	.	.	.	.	.	.	.	.	.	.	.	.	.
600	.	.	.	.	.	.	.	.	.	.	.	.	.	.	.	.	.	.	.	.	.
700	.	.	.	.	.	.	.	.	.	.	.	.	.	.	.	.	.	.	.	.	.
800	.	.	.	.	.	.	.	.	.	.	.	.	.	.	.	.	.	.	.	.	.
900	.	.	.	.	.	.	.	.	.	.	.	.	.	.	.	.	.	.	.	.	.
1000	.	.	.	.	.	.	.	.	.	.	.	.	.	.	.	.	.	.	.	.	.

Table 9.24. Power table for repeated measures interaction ANOVA; 3 × 4 mixed design at $r = 0.40$ and alpha = 0.05

n	Hypothesized ES																				
	0.10	0.15	0.20	0.30	0.35	0.40	0.45	0.50	0.55	0.60	0.65	0.70	0.75	0.80	1.00	1.25	1.50	1.75	2.00	2.50	3.00
3	5	6	7	9	10	12	14	17	20	23	27	31	36	40	61	83	95	99	.	.	.
4	5	6	7	11	13	17	20	25	30	35	41	47	54	60	82	96	.	.	.	.	.
5	6	7	8	13	17	21	27	33	40	47	55	62	69	76	93	99	.	.	.	.	.
6	6	7	9	16	21	27	34	41	50	58	66	74	80	86	98	.	.	.	.	.	.
7	6	8	10	19	25	32	41	50	59	68	76	83	88	92	99	.	.	.	.	.	.
8	6	8	11	21	29	38	47	57	67	76	83	89	93	96	.	.	.	.	.	.	.
9	6	9	12	24	33	43	54	64	74	82	88	93	96	98	.	.	.	.	.	.	.
10	6	9	13	27	37	48	59	70	80	87	92	96	98	99	.	.	.	.	.	.	.
11	7	10	14	30	41	53	65	76	84	91	95	98	99	.	.	.	.	.	.	.	.
12	7	10	16	34	45	58	70	80	88	93	97	99	99	.	.	.	.	.	.	.	.
13	7	11	17	37	49	62	74	84	91	95	98	99	.	.	.	.	.	.	.	.	.
14	7	11	18	40	53	66	78	87	93	97	99	.	.	.	.	.	.	.	.	.	.
15	8	12	19	43	57	70	82	90	95	98	99	.	.	.	.	.	.	.	.	.	.
20	9	15	26	57	72	85	93	97	99	.	.	.	.	.	.	.	.	.	.	.	.
25	10	19	33	69	83	93	98	99	.	.	.	.	.	.	.	.	.	.	.	.	.
30	11	23	40	78	91	97	99	.	.	.	.	.	.	.	.	.	.	.	.	.	.
35	13	26	46	85	95	99	.	.	.	.	.	.	.	.	.	.	.	.	.	.	.
40	14	30	52	90	97	99	.	.	.	.	.	.	.	.	.	.	.	.	.	.	.
45	16	34	58	94	99	.	.	.	.	.	.	.	.	.	.	.	.	.	.	.	.
50	17	38	64	96	99	.	.	.	.	.	.	.	.	.	.	.	.	.	.	.	.
55	19	42	69	98	.	.	.	.	.	.	.	.	.	.	.	.	.	.	.	.	.
60	21	45	73	99	.	.	.	.	.	.	.	.	.	.	.	.	.	.	.	.	.
65	22	49	77	99	.	.	.	.	.	.	.	.	.	.	.	.	.	.	.	.	.
70	24	52	81	99	.	.	.	.	.	.	.	.	.	.	.	.	.	.	.	.	.
75	25	56	84	.	.	.	.	.	.	.	.	.	.	.	.	.	.	.	.	.	.
80	27	59	87	.	.	.	.	.	.	.	.	.	.	.	.	.	.	.	.	.	.
90	31	65	91	.	.	.	.	.	.	.	.	.	.	.	.	.	.	.	.	.	.
100	34	70	94	.	.	.	.	.	.	.	.	.	.	.	.	.	.	.	.	.	.
110	37	75	96	.	.	.	.	.	.	.	.	.	.	.	.	.	.	.	.	.	.
120	41	79	97	.	.	.	.	.	.	.	.	.	.	.	.	.	.	.	.	.	.
130	44	83	98	.	.	.	.	.	.	.	.	.	.	.	.	.	.	.	.	.	.
140	47	86	99	.	.	.	.	.	.	.	.	.	.	.	.	.	.	.	.	.	.
150	50	89	99	.	.	.	.	.	.	.	.	.	.	.	.	.	.	.	.	.	.
175	58	93	.	.	.	.	.	.	.	.	.	.	.	.	.	.	.	.	.	.	.
200	65	96	.	.	.	.	.	.	.	.	.	.	.	.	.	.	.	.	.	.	.
225	71	98	.	.	.	.	.	.	.	.	.	.	.	.	.	.	.	.	.	.	.
250	76	99	.	.	.	.	.	.	.	.	.	.	.	.	.	.	.	.	.	.	.
300	84	.	.	.	.	.	.	.	.	.	.	.	.	.	.	.	.	.	.	.	.
350	90	.	.	.	.	.	.	.	.	.	.	.	.	.	.	.	.	.	.	.	.
400	94	.	.	.	.	.	.	.	.	.	.	.	.	.	.	.	.	.	.	.	.
450	96	.	.	.	.	.	.	.	.	.	.	.	.	.	.	.	.	.	.	.	.
500	98	.	.	.	.	.	.	.	.	.	.	.	.	.	.	.	.	.	.	.	.
600	99	.	.	.	.	.	.	.	.	.	.	.	.	.	.	.	.	.	.	.	.
700	.	.	.	.	.	.	.	.	.	.	.	.	.	.	.	.	.	.	.	.	.
800	.	.	.	.	.	.	.	.	.	.	.	.	.	.	.	.	.	.	.	.	.
900	.	.	.	.	.	.	.	.	.	.	.	.	.	.	.	.	.	.	.	.	.
1000	.	.	.	.	.	.	.	.	.	.	.	.	.	.	.	.	.	.	.	.	.

Table 9.25. Power table for repeated measures interaction ANOVA; 3 × 4 mixed design at $r = 0.60$ and alpha = 0.05

n	Hypothesized ES																				
	0.10	0.15	0.20	0.30	0.35	0.40	0.45	0.50	0.55	0.60	0.65	0.70	0.75	0.80	1.00	1.25	1.50	1.75	2.00	2.50	3.00
3	6	6	7	11	13	16	20	24	29	34	40	46	52	59	81	96	.	.	.	.	.
4	6	7	9	14	19	24	30	37	44	52	60	67	74	81	96	.	.	.	.	.	.
5	6	8	10	18	24	32	40	49	58	67	75	82	88	92	99	.	.	.	.	.	.
6	6	8	12	23	31	40	50	60	70	79	86	91	95	97	.	.	.	.	.	.	.
7	6	9	13	27	37	48	59	70	79	87	92	96	98	99	.	.	.	.	.	.	.
8	7	10	15	32	43	55	67	78	86	92	96	98	99	.	.	.	.	.	.	.	.
9	7	11	17	36	49	62	74	84	91	95	98	99	.	.	.	.	.	.	.	.	.
10	7	12	19	41	55	68	80	88	94	97	99	.	.	.	.	.	.	.	.	.	.
11	8	13	21	45	60	74	84	92	96	99	.	.	.	.	.	.	.	.	.	.	.
12	8	14	23	50	65	78	88	94	98	99	.	.	.	.	.	.	.	.	.	.	.
13	9	15	25	54	69	82	91	96	99	.	.	.	.	.	.	.	.	.	.	.	.
14	9	16	27	58	74	86	93	98	99	.	.	.	.	.	.	.	.	.	.	.	.
15	9	17	29	62	77	89	95	98	.	.	.	.	.	.	.	.	.	.	.	.	.
20	11	22	39	77	90	97	99	.	.	.	.	.	.	.	.	.	.	.	.	.	.
25	13	28	49	87	96	99	.	.	.	.	.	.	.	.	.	.	.	.	.	.	.
30	16	33	58	93	99	.	.	.	.	.	.	.	.	.	.	.	.	.	.	.	.
35	18	39	66	97	.	.	.	.	.	.	.	.	.	.	.	.	.	.	.	.	.
40	20	45	73	98	.	.	.	.	.	.	.	.	.	.	.	.	.	.	.	.	.
45	23	50	79	99	.	.	.	.	.	.	.	.	.	.	.	.	.	.	.	.	.
50	25	55	84	.	.	.	.	.	.	.	.	.	.	.	.	.	.	.	.	.	.
55	28	60	87	.	.	.	.	.	.	.	.	.	.	.	.	.	.	.	.	.	.
60	30	65	91	.	.	.	.	.	.	.	.	.	.	.	.	.	.	.	.	.	.
65	33	69	93	.	.	.	.	.	.	.	.	.	.	.	.	.	.	.	.	.	.
70	35	73	95	.	.	.	.	.	.	.	.	.	.	.	.	.	.	.	.	.	.
75	38	76	96	.	.	.	.	.	.	.	.	.	.	.	.	.	.	.	.	.	.
80	41	79	97	.	.	.	.	.	.	.	.	.	.	.	.	.	.	.	.	.	.
90	45	84	99	.	.	.	.	.	.	.	.	.	.	.	.	.	.	.	.	.	.
100	50	89	99	.	.	.	.	.	.	.	.	.	.	.	.	.	.	.	.	.	.
110	55	92	.	.	.	.	.	.	.	.	.	.	.	.	.	.	.	.	.	.	.
120	59	94	.	.	.	.	.	.	.	.	.	.	.	.	.	.	.	.	.	.	.
130	63	96	.	.	.	.	.	.	.	.	.	.	.	.	.	.	.	.	.	.	.
140	67	97	.	.	.	.	.	.	.	.	.	.	.	.	.	.	.	.	.	.	.
150	71	98	.	.	.	.	.	.	.	.	.	.	.	.	.	.	.	.	.	.	.
175	78	99	.	.	.	.	.	.	.	.	.	.	.	.	.	.	.	.	.	.	.
200	84	.	.	.	.	.	.	.	.	.	.	.	.	.	.	.	.	.	.	.	.
225	89	.	.	.	.	.	.	.	.	.	.	.	.	.	.	.	.	.	.	.	.
250	92	.	.	.	.	.	.	.	.	.	.	.	.	.	.	.	.	.	.	.	.
300	96	.	.	.	.	.	.	.	.	.	.	.	.	.	.	.	.	.	.	.	.
350	98	.	.	.	.	.	.	.	.	.	.	.	.	.	.	.	.	.	.	.	.
400	99	.	.	.	.	.	.	.	.	.	.	.	.	.	.	.	.	.	.	.	.
450	.	.	.	.	.	.	.	.	.	.	.	.	.	.	.	.	.	.	.	.	.
500	.	.	.	.	.	.	.	.	.	.	.	.	.	.	.	.	.	.	.	.	.
600	.	.	.	.	.	.	.	.	.	.	.	.	.	.	.	.	.	.	.	.	.
700	.	.	.	.	.	.	.	.	.	.	.	.	.	.	.	.	.	.	.	.	.
800	.	.	.	.	.	.	.	.	.	.	.	.	.	.	.	.	.	.	.	.	.
900	.	.	.	.	.	.	.	.	.	.	.	.	.	.	.	.	.	.	.	.	.
1000	.	.	.	.	.	.	.	.	.	.	.	.	.	.	.	.	.	.	.	.	.

Table 9.26. Sample size table for between subjects analysis; $p = 0.05$

Chart A. 2 × 2 design at alpha = 0.05																			
Power	Hypothesized ES																		
	0.20	0.30	0.35	0.40	0.45	0.50	0.55	0.60	0.65	0.70	0.75	0.80	1.00	1.25	1.50	1.75	2.00	2.50	3.00
0.80	193	87	64	49	39	32	27	23	20	17	15	13	9	6	5	4	4	3	3
0.90	258	116	85	66	52	43	35	30	26	22	20	18	12	8	6	5	4	3	3
Chart B. 2 × 3 design at alpha = 0.05																			
Power	Hypothesized ES																		
	0.20	0.30	0.35	0.40	0.45	0.50	0.55	0.60	0.65	0.70	0.75	0.80	1.00	1.25	1.50	1.75	2.00	2.50	3.00
0.80	160	72	53	41	33	27	22	19	16	14	13	11	8	6	4	4	3	3	2
0.90	210	94	70	54	43	35	29	25	21	19	16	15	10	7	5	4	4	3	3
Chart C. 2 × 4 design at alpha = 0.05																			
Power	Hypothesized ES																		
	0.20	0.30	0.35	0.40	0.45	0.50	0.55	0.60	0.65	0.70	0.75	0.80	1.00	1.25	1.50	1.75	2.00	2.50	3.00
0.80	136	61	45	35	28	23	19	16	14	13	11	10	7	5	4	3	3	2	2
0.90	177	80	59	45	36	30	25	21	18	16	14	12	9	6	5	4	3	3	2
Chart D. 3 × 3 design at alpha = 0.05																			
Power	Hypothesized ES																		
	0.20	0.30	0.35	0.40	0.45	0.50	0.55	0.60	0.65	0.70	0.75	0.80	1.00	1.25	1.50	1.75	2.00	2.50	3.00
0.80	133	60	44	34	27	23	19	16	14	12	11	10	7	5	4	3	3	2	2
0.90	172	77	57	44	35	29	24	20	18	15	14	12	8	6	5	4	3	3	2
Chart E. 3 × 4 design at alpha = 0.05																			
Power	Hypothesized ES																		
	0.20	0.30	0.35	0.40	0.45	0.50	0.55	0.60	0.65	0.70	0.75	0.80	1.00	1.25	1.50	1.75	2.00	2.50	3.00
0.80	114	52	38	30	24	19	16	14	12	11	9	9	6	4	4	3	3	2	2
0.90	146	66	49	38	30	25	21	18	15	13	12	11	7	5	4	3	3	2	2

Table 9.27. Sample size table for between subjects analysis; $p = 0.01$

Chart A. 2 × 2 design at alpha = 0.01

Power	Hypothesized ES																		
	0.20	0.30	0.35	0.40	0.45	0.50	0.55	0.60	0.65	0.70	0.75	0.80	1.00	1.25	1.50	1.75	2.00	2.50	3.00
0.80	290	130	96	74	59	48	40	34	29	25	22	20	13	9	7	6	5	4	3
0.90	369	165	122	94	74	61	50	43	37	32	28	25	17	11	8	7	6	4	4

Chart B. 2 × 3 design at alpha = 0.01

Power	Hypothesized ES																		
	0.20	0.30	0.35	0.40	0.45	0.50	0.55	0.60	0.65	0.70	0.75	0.80	1.00	1.25	1.50	1.75	2.00	2.50	3.00
0.80	231	104	77	59	47	38	32	27	23	21	18	16	11	8	6	5	4	3	3
0.90	290	130	96	74	59	48	40	34	29	25	22	20	13	9	7	6	5	4	3

Chart C. 2 × 4 design at alpha = 0.01

Power	Hypothesized ES																		
	0.20	0.30	0.35	0.40	0.45	0.50	0.55	0.60	0.65	0.70	0.75	0.80	1.00	1.25	1.50	1.75	2.00	2.50	3.00
0.80	194	87	64	50	40	32	27	23	20	17	15	14	9	7	5	4	4	3	3
0.90	241	108	80	62	49	40	33	28	24	21	19	17	11	8	6	5	4	3	3

Chart D. 3 × 3 design at alpha = 0.01

Power	Hypothesized ES																		
	0.20	0.30	0.35	0.40	0.45	0.50	0.55	0.60	0.65	0.70	0.75	0.80	1.00	1.25	1.50	1.75	2.00	2.50	3.00
0.80	187	84	62	48	38	31	26	22	19	17	15	13	9	6	5	4	4	3	3
0.90	231	104	77	59	47	38	32	27	23	20	18	16	11	8	6	5	4	3	3

Chart E. 3 × 4 design at alpha = 0.01

Power	Hypothesized ES																		
	0.20	0.30	0.35	0.40	0.45	0.50	0.55	0.60	0.65	0.70	0.75	0.80	1.00	1.25	1.50	1.75	2.00	2.50	3.00
0.80	158	71	53	41	33	27	22	19	16	14	13	11	8	6	4	4	3	3	2
0.90	194	87	65	50	40	32	27	23	20	17	15	14	9	7	5	4	4	3	3

Table 9.28. Sample size table for between subjects analysis; $p = 0.10$

Chart A. 2 × 2 design at alpha = 0.10

Power	Hypothesized ES																		
	0.20	0.30	0.35	0.40	0.45	0.50	0.55	0.60	0.65	0.70	0.75	0.80	1.00	1.25	1.50	1.75	2.00	2.50	3.00
0.80	150	68	50	39	31	25	21	18	15	14	12	11	7	5	4	3	3	2	2
0.90	209	94	69	53	42	35	29	24	21	18	16	14	10	7	5	4	4	3	2

Chart B. 2 × 3 design at alpha = 0.10

Power	Hypothesized ES																		
	0.20	0.30	0.35	0.40	0.45	0.50	0.55	0.60	0.65	0.70	0.75	0.80	1.00	1.25	1.50	1.75	2.00	2.50	3.00
0.80	127	57	43	33	26	22	18	15	13	12	10	9	6	5	4	3	3	2	2
0.90	173	78	57	44	35	27	24	20	18	15	14	12	8	6	4	4	3	3	2

Chart C. 2 × 4 design at alpha = 0.10

Power	Hypothesized ES																		
	0.20	0.30	0.35	0.40	0.45	0.50	0.55	0.60	0.65	0.70	0.75	0.80	1.00	1.25	1.50	1.75	2.00	2.50	3.00
0.80	110	50	37	28	23	19	16	13	12	10	9	8	6	4	3	3	3	2	2
0.90	148	66	49	38	30	25	21	18	15	13	12	11	7	5	4	3	3	2	2

Chart D. 3 × 3 design at alpha = 0.10

Power	Hypothesized ES																		
	0.20	0.30	0.35	0.40	0.45	0.50	0.55	0.60	0.65	0.70	0.75	0.80	1.00	1.25	1.50	1.75	2.00	2.50	3.00
0.80	108	49	36	28	22	18	15	13	11	10	9	8	6	4	3	3	3	2	2
0.90	144	65	48	37	30	24	20	17	15	13	12	10	7	5	4	3	3	2	2

Chart E. 3 × 4 design at alpha = 0.10

Power	Hypothesized ES																		
	0.20	0.30	0.35	0.40	0.45	0.50	0.55	0.60	0.65	0.70	0.75	0.80	1.00	1.25	1.50	1.75	2.00	2.50	3.00
0.80	93	42	31	24	20	16	14	12	10	9	8	7	5	4	3	3	2	2	2
0.90	123	56	41	32	25	21	17	15	13	11	10	9	6	5	4	3	3	2	2

Table 9.29. Sample size table for ANCOVA analysis; $r = 0.40$

Chart A. 2 × 2 design at alpha = 0.05																			
Power	Hypothesized ES																		
	0.20	0.30	0.35	0.40	0.45	0.50	0.55	0.60	0.65	0.70	0.75	0.80	1.00	1.25	1.50	1.75	2.00	2.50	3.00
0.80	162	73	54	42	34	27	23	20	17	15	13	12	8	6	5	4	3	3	3
0.90	218	98	72	56	44	36	30	26	22	19	17	15	10	7	6	5	4	3	3
Chart B. 2 × 3 design at alpha = 0.05																			
Power	Hypothesized ES																		
	0.20	0.30	0.35	0.40	0.45	0.50	0.55	0.60	0.65	0.70	0.75	0.80	1.00	1.25	1.50	1.75	2.00	2.50	3.00
0.80	135	61	45	35	28	23	19	16	14	13	11	10	7	5	4	3	3	3	2
0.90	177	79	59	45	36	30	25	21	18	16	14	13	9	6	5	4	4	3	3
Chart C. 2 × 4 design at alpha = 0.05																			
Power	Hypothesized ES																		
	0.20	0.30	0.35	0.40	0.45	0.50	0.55	0.60	0.65	0.70	0.75	0.80	1.00	1.25	1.50	1.75	2.00	2.50	3.00
0.80	115	52	39	30	24	20	17	14	12	11	10	9	6	5	4	3	3	2	2
0.90	149	67	50	39	31	25	21	18	16	14	12	11	8	5	4	4	3	3	2
Chart D. 3 × 3 design at alpha = 0.05																			
Power	Hypothesized ES																		
	0.20	0.30	0.35	0.40	0.45	0.50	0.55	0.60	0.65	0.70	0.75	0.80	1.00	1.25	1.50	1.75	2.00	2.50	3.00
0.80	112	51	38	29	23	19	16	14	12	11	9	9	6	4	4	3	3	2	2
0.90	145	65	48	37	30	24	21	17	15	13	12	11	7	5	4	4	3	3	2
Chart E. 3 × 4 design at alpha = 0.05																			
Power	Hypothesized ES																		
	0.20	0.30	0.35	0.40	0.45	0.50	0.55	0.60	0.65	0.70	0.75	0.80	1.00	1.25	1.50	1.75	2.00	2.50	3.00
0.80	96	44	32	25	20	17	14	12	11	9	8	7	5	4	3	3	3	2	2
0.90	123	56	41	32	26	21	18	15	13	11	10	9	6	5	4	3	3	2	2

Table 9.30. Sample size table for ANCOVA analysis; $r = 0.60$

Chart A. 2 × 2 design at alpha = 0.05																			
Power	Hypothesized ES																		
	0.20	0.30	0.35	0.40	0.45	0.50	0.55	0.60	0.65	0.70	0.75	0.80	1.00	1.25	1.50	1.75	2.00	2.50	3.00
0.80	124	56	42	32	26	21	18	15	13	12	10	9	7	5	4	3	3	3	3
0.90	166	75	55	43	34	28	24	20	17	15	13	12	8	6	5	4	4	3	3

Chart B. 2 × 3 design at alpha = 0.05																			
Power	Hypothesized ES																		
	0.20	0.30	0.35	0.40	0.45	0.50	0.55	0.60	0.65	0.70	0.75	0.80	1.00	1.25	1.50	1.75	2.00	2.50	3.00
0.80	103	47	35	27	22	18	15	13	11	10	9	8	6	4	4	3	3	2	2
0.90	135	61	45	35	28	23	19	17	14	13	11	10	7	5	4	4	3	3	2

Chart C. 2 × 4 design at alpha = 0.05																			
Power	Hypothesized ES																		
	0.20	0.30	0.35	0.40	0.45	0.50	0.55	0.60	0.65	0.70	0.75	0.80	1.00	1.25	1.50	1.75	2.00	2.50	3.00
0.80	88	40	30	23	19	15	13	11	10	9	8	7	5	4	3	3	3	2	2
0.90	114	52	38	30	24	20	17	14	12	11	10	9	6	5	4	3	3	2	2

Chart D. 3 × 3 design at alpha = 0.05																			
Power	Hypothesized ES																		
	0.20	0.30	0.35	0.40	0.45	0.50	0.55	0.60	0.65	0.70	0.75	0.80	1.00	1.25	1.50	1.75	2.00	2.50	3.00
0.80	86	39	29	23	18	15	13	11	10	8	8	7	5	4	3	3	3	2	2
0.90	111	50	37	29	23	19	16	14	12	11	9	8	6	4	4	3	3	2	2

Chart E. 3 × 4 design at alpha = 0.05																			
Power	Hypothesized ES																		
	0.20	0.30	0.35	0.40	0.45	0.50	0.55	0.60	0.65	0.70	0.75	0.80	1.00	1.25	1.50	1.75	2.00	2.50	3.00
0.80	74	34	25	20	16	13	11	10	8	7	7	6	4	3	3	3	2	2	2
0.90	94	43	32	25	20	16	14	12	10	9	8	7	5	4	3	3	3	2	2

Table 9.31. Sample size table for ANCOVA; $p = 0.01$, $r = 0.40$

Chart A. 2 × 2 ANCOVA at alpha = 0.01, $r = 0.40$																			
Power	Hypothesized ES																		
	0.20	0.30	0.35	0.40	0.45	0.50	0.55	0.60	0.65	0.70	0.75	0.80	1.00	1.25	1.50	1.75	2.00	2.50	3.00
0.80	244	110	81	63	50	41	34	29	25	22	19	17	12	8	6	5	5	4	3
0.90	311	139	103	79	63	51	43	36	31	27	24	21	14	10	8	6	5	4	4
Chart B. 2 × 3 ANCOVA at alpha = 0.01, $r = 0.40$																			
Power	Hypothesized ES																		
	0.20	0.30	0.35	0.40	0.45	0.50	0.55	0.60	0.65	0.70	0.75	0.80	1.00	1.25	1.50	1.75	2.00	2.50	3.00
0.80	195	88	65	50	40	33	27	23	20	18	16	14	10	7	5	5	4	3	3
0.90	244	110	81	62	50	41	34	29	25	22	19	17	12	8	6	5	4	4	3
Chart C. 2 × 4 ANCOVA at alpha = 0.01, $r = 0.40$																			
Power	Hypothesized ES																		
	0.20	0.30	0.35	0.40	0.45	0.50	0.55	0.60	0.65	0.70	0.75	0.80	1.00	1.25	1.50	1.75	2.00	2.50	3.00
0.80	163	73	54	42	34	28	23	20	17	15	13	12	8	6	5	4	4	3	3
0.90	203	91	67	52	42	34	28	24	21	18	16	14	10	7	5	5	4	3	3
Chart D. 3 × 3 ANCOVA at alpha = 0.01, $r = 0.40$																			
Power	Hypothesized ES																		
	0.20	0.30	0.35	0.40	0.45	0.50	0.55	0.60	0.65	0.70	0.75	0.80	1.00	1.25	1.50	1.75	2.00	2.50	3.00
0.80	157	71	53	41	33	27	22	19	17	14	13	12	8	6	5	4	3	3	3
0.90	195	87	65	50	40	33	27	23	20	18	16	14	10	7	5	4	4	3	3
Chart E. 3 × 4 ANCOVA at alpha = 0.01, $r = 0.40$																			
Power	Hypothesized ES																		
	0.20	0.30	0.35	0.40	0.45	0.50	0.55	0.60	0.65	0.70	0.75	0.80	1.00	1.25	1.50	1.75	2.00	2.50	3.00
0.80	133	60	45	35	28	23	19	16	14	12	11	10	7	5	4	3	3	3	2
0.90	164	74	55	42	34	28	23	20	17	15	13	12	8	6	5	4	3	3	3

Table 9.32. Sample size table for ANCOVA; $p = 0.01$, $r = 0.60$

Chart A. 2 × 2 ANCOVA at alpha = 0.01, $r = 0.60$

Power	Hypothesized ES																		
	0.20	0.30	0.35	0.40	0.45	0.50	0.55	0.60	0.65	0.70	0.75	0.80	1.00	1.25	1.50	1.75	2.00	2.50	3.00
0.80	186	84	62	48	39	32	26	23	20	17	15	14	10	7	5	5	4	3	3
0.90	237	107	79	61	49	40	33	28	24	21	19	17	12	8	6	5	5	4	3

Chart B. 2 × 3 ANCOVA at alpha = 0.01, $r = 0.60$

Power	Hypothesized ES																		
	0.20	0.30	0.35	0.40	0.45	0.50	0.55	0.60	0.65	0.70	0.75	0.80	1.00	1.25	1.50	1.75	2.00	2.50	3.00
0.80	149	67	50	39	31	25	21	18	16	14	12	11	8	6	5	4	3	3	3
0.90	186	84	62	48	38	31	26	22	19	17	15	13	9	7	5	4	4	3	3

Chart C. 2 × 4 ANCOVA at alpha = 0.01, $r = 0.60$

Power	Hypothesized ES																		
	0.20	0.30	0.35	0.40	0.45	0.50	0.55	0.60	0.65	0.70	0.75	0.80	1.00	1.25	1.50	1.75	2.00	2.50	3.00
0.80	125	56	42	33	26	21	18	15	13	12	11	10	7	5	4	4	3	3	2
0.90	155	70	52	40	32	26	22	19	16	14	13	11	8	6	5	4	3	3	3

Chart D. 3 × 3 ANCOVA at alpha = 0.01, $r = 0.60$

Power	Hypothesized ES																		
	0.20	0.30	0.35	0.40	0.45	0.50	0.55	0.60	0.65	0.70	0.75	0.80	1.00	1.25	1.50	1.75	2.00	2.50	3.00
0.80	120	54	40	31	25	21	17	15	13	11	10	9	7	5	4	3	3	3	2
0.90	149	67	50	39	31	25	21	18	16	14	12	11	8	6	4	4	3	3	3

Chart E. 3 × 4 ANCOVA at alpha = 0.01, $r = 0.60$

Power	Hypothesized ES																		
	0.20	0.30	0.35	0.40	0.45	0.50	0.55	0.60	0.65	0.70	0.75	0.80	1.00	1.25	1.50	1.75	2.00	2.50	3.00
0.80	102	46	34	27	22	18	15	13	11	10	9	8	6	4	4	3	3	2	2
0.90	125	57	42	33	26	21	18	15	13	12	10	9	7	5	4	3	3	3	2

Table 9.33. Sample size table for ANCOVA; $p = 0.10$, $r = 0.40$

Chart A. 2 × 2 ANCOVA at alpha = 0.10, $r = 0.40$

Power	Hypothesized ES																		
	0.20	0.30	0.35	0.40	0.45	0.50	0.55	0.60	0.65	0.70	0.75	0.80	1.00	1.25	1.50	1.75	2.00	2.50	3.00
0.80	127	57	42	33	26	22	18	16	13	12	11	9	7	5	4	3	3	3	2
0.90	176	79	59	45	36	30	25	21	18	16	14	13	9	6	5	4	3	3	3

Chart B. 2 × 3 ANCOVA at alpha = 0.10, $r = 0.40$

Power	Hypothesized ES																		
	0.20	0.30	0.35	0.40	0.45	0.50	0.55	0.60	0.65	0.70	0.75	0.80	1.00	1.25	1.50	1.75	2.00	2.50	3.00
0.80	107	49	36	28	22	18	16	13	12	10	9	8	6	4	4	3	3	2	2
0.90	146	66	49	38	30	25	21	18	15	13	12	11	7	5	4	4	3	3	2

Chart C. 2 × 4 ANCOVA at alpha = 0.10, $r = 0.40$

Power	Hypothesized ES																		
	0.20	0.30	0.35	0.40	0.45	0.50	0.55	0.60	0.65	0.70	0.75	0.80	1.00	1.25	1.50	1.75	2.00	2.50	3.00
0.80	93	42	31	24	20	16	14	12	10	9	8	7	5	4	3	3	3	2	2
0.90	124	56	42	32	26	21	18	15	13	12	10	9	6	5	4	3	3	2	2

Chart D. 3 × 3 ANCOVA at alpha = 0.10, $r = 0.40$

Power	Hypothesized ES																		
	0.20	0.30	0.35	0.40	0.45	0.50	0.55	0.60	0.65	0.70	0.75	0.80	1.00	1.25	1.50	1.75	2.00	2.50	3.00
0.80	91	41	31	24	19	16	13	11	10	9	8	7	5	4	3	3	2	2	2
0.90	121	55	41	31	25	21	17	15	13	11	10	9	6	5	4	3	3	2	2

Chart E. 3 × 4 ANCOVA at alpha = 0.10, $r = 0.40$

Power	Hypothesized ES																		
	0.20	0.30	0.35	0.40	0.45	0.50	0.55	0.60	0.65	0.70	0.75	0.80	1.00	1.25	1.50	1.75	2.00	2.50	3.00
0.80	79	36	27	21	17	14	12	10	9	8	7	6	5	3	3	3	2	2	2
0.90	104	47	35	27	22	18	15	13	11	10	9	8	6	4	3	3	3	2	2

Table 9.34. Sample size table for ANCOVA; $p = 0.10$, $r = 0.60$

Chart A. 2 × 2 ANCOVA at alpha = 0.10, $r = 0.60$

Power	Hypothesized ES																		
	0.20	0.30	0.35	0.40	0.45	0.50	0.55	0.60	0.65	0.70	0.75	0.80	1.00	1.25	1.50	1.75	2.00	2.50	3.00
0.80	97	44	33	25	20	17	14	12	11	9	8	8	5	4	3	3	3	2	2
0.90	135	61	45	35	28	23	23	19	16	14	12	11	10	7	5	4	3	3	2

Chart B. 2 × 3 ANCOVA at alpha = 0.10, $r = 0.60$

Power	Hypothesized ES																		
	0.20	0.30	0.35	0.40	0.45	0.50	0.55	0.60	0.65	0.70	0.75	0.80	1.00	1.25	1.50	1.75	2.00	2.50	3.00
0.80	82	37	28	22	17	14	12	11	9	8	7	7	5	4	3	3	3	2	2
0.90	111	50	37	29	23	19	16	14	12	11	9	8	6	4	4	3	3	2	2

Chart C. 2 × 4 ANCOVA at alpha = 0.10, $r = 0.60$

Power	Hypothesized ES																		
	0.20	0.30	0.35	0.40	0.45	0.50	0.55	0.60	0.65	0.70	0.75	0.80	1.00	1.25	1.50	1.75	2.00	2.50	3.00
0.80	71	32	24	19	15	13	11	9	8	7	6	6	4	3	3	3	2	2	2
0.90	95	43	32	25	20	17	14	12	10	9	8	7	5	4	3	3	3	2	2

Chart D. 3 × 3 ANCOVA at alpha = 0.10, $r = 0.60$

Power	Hypothesized ES																		
	0.20	0.30	0.35	0.40	0.45	0.50	0.55	0.60	0.65	0.70	0.75	0.80	1.00	1.25	1.50	1.75	2.00	2.50	3.00
0.80	70	32	24	19	15	12	11	9	8	7	6	6	4	3	3	2	2	2	2
0.90	93	42	31	24	20	16	14	12	10	9	8	7	5	4	3	3	3	2	2

Chart E. 3 × 4 ANCOVA at alpha = 0.10, $r = 0.60$

Power	Hypothesized ES																		
	0.20	0.30	0.35	0.40	0.45	0.50	0.55	0.60	0.65	0.70	0.75	0.80	1.00	1.25	1.50	1.75	2.00	2.50	3.00
0.80	60	28	21	16	13	11	9	8	7	6	6	5	4	3	3	2	2	2	2
0.90	79	36	27	21	17	14	12	10	9	8	7	6	5	3	3	3	2	2	2

Table 9.35. Sample size table for repeated measures interaction ANOVA; mixed design, $r = 0.40$

Chart A. 2 × 2 design at alpha = 0.05, r = 0.40																			
Power	Hypothesized ES																		
	0.20	0.30	0.35	0.40	0.45	0.50	0.55	0.60	0.65	0.70	0.75	0.80	1.00	1.25	1.50	1.75	2.00	2.50	3.00
0.80	117	53	40	31	25	20	17	15	13	11	10	9	7	5	4	4	3	3	3
0.90	156	71	52	41	32	27	22	19	17	15	13	12	8	6	5	4	4	3	3

Chart B. 2 × 3 design at alpha = 0.05, r = 0.40																			
Power	Hypothesized ES																		
	0.20	0.30	0.35	0.40	0.45	0.50	0.55	0.60	0.65	0.70	0.75	0.80	1.00	1.25	1.50	1.75	2.00	2.50	3.00
0.80	97	44	33	25	21	17	14	12	11	10	9	8	6	4	4	3	3	2	2
0.90	127	57	43	33	26	22	18	16	14	12	11	10	7	5	4	3	3	3	2

Chart C. 2 × 4 design at alpha = 0.05, r = 0.40																			
Power	Hypothesized ES																		
	0.20	0.30	0.35	0.40	0.45	0.50	0.55	0.60	0.65	0.70	0.75	0.80	1.00	1.25	1.50	1.75	2.00	2.50	3.00
0.80	83	38	28	22	18	15	12	11	9	8	7	7	5	4	3	3	3	2	2
0.90	107	49	36	28	23	19	16	13	12	10	9	8	6	4	4	3	3	2	2

Chart D. 3 × 3 design at alpha = 0.05, r = 0.40																			
Power	Hypothesized ES																		
	0.20	0.30	0.35	0.40	0.45	0.50	0.55	0.60	0.65	0.70	0.75	0.80	1.00	1.25	1.50	1.75	2.00	2.50	3.00
0.80	81	37	27	21	17	14	12	11	9	8	7	7	5	4	3	3	3	2	2
0.90	104	47	35	27	22	18	15	13	11	10	9	8	6	4	4	3	3	2	2

Chart E. 3 × 4 design at alpha = 0.05, r = 0.40																			
Power	Hypothesized ES																		
	0.20	0.30	0.35	0.40	0.45	0.50	0.55	0.60	0.65	0.70	0.75	0.80	1.00	1.25	1.50	1.75	2.00	2.50	3.00
0.80	69	32	24	19	15	12	11	9	8	7	6	6	4	3	3	3	2	2	2
0.90	89	40	30	23	19	16	13	11	10	9	8	7	5	4	3	3	3	2	2

Table 9.36. Sample size table for repeated measures interaction ANOVA; mixed design, $r = 0.60$

Chart A. 2×2 design at alpha = 0.05, $r = 0.60$

Power	Hypothesized ES																		
	0.20	0.30	0.35	0.40	0.45	0.50	0.55	0.60	0.65	0.70	0.75	0.80	1.00	1.25	1.50	1.75	2.00	2.50	3.00
0.80	79	36	27	21	17	14	12	11	9	8	8	7	5	4	4	3	3	3	2
0.90	105	48	36	28	22	18	16	13	12	10	9	8	6	5	4	4	3	3	3

Chart B. 2×3 design at alpha = 0.05, $r = 0.60$

Power	Hypothesized ES																		
	0.20	0.30	0.35	0.40	0.45	0.50	0.55	0.60	0.65	0.70	0.75	0.80	1.00	1.25	1.50	1.75	2.00	2.50	3.00
0.80	65	30	22	18	14	12	10	9	8	7	6	6	4	3	3	3	3	2	2
0.90	85	39	29	23	18	15	13	11	10	9	8	7	5	4	3	3	3	2	2

Chart C. 2×4 design at alpha = 0.05, $r = 0.60$

Power	Hypothesized ES																		
	0.20	0.30	0.35	0.40	0.45	0.50	0.55	0.60	0.65	0.70	0.75	0.80	1.00	1.25	1.50	1.75	2.00	2.50	3.00
0.80	56	26	19	15	12	10	9	8	7	6	6	5	4	3	3	2	2	2	2
0.90	72	33	25	19	16	13	11	9	8	7	7	6	5	4	3	3	2	2	2

Chart D. 3×3 design at alpha = 0.05, $r = 0.60$

Power	Hypothesized ES																		
	0.20	0.30	0.35	0.40	0.45	0.50	0.55	0.60	0.65	0.70	0.75	0.80	1.00	1.25	1.50	1.75	2.00	2.50	3.00
0.80	54	25	19	15	12	10	9	8	7	6	6	5	4	3	3	2	2	2	2
0.90	70	32	24	19	15	13	11	9	8	7	7	6	4	4	3	3	2	2	2

Chart E. 3×4 design at alpha = 0.05, $r = 0.60$

Power	Hypothesized ES																		
	0.20	0.30	0.35	0.40	0.45	0.50	0.55	0.60	0.65	0.70	0.75	0.80	1.00	1.25	1.50	1.75	2.00	2.50	3.00
0.80	47	22	16	13	11	9	8	7	6	5	5	4	3	3	2	2	2	2	2
0.90	60	27	21	16	13	11	9	8	7	6	6	5	4	3	3	2	2	2	2

Table 9.37. Sample size table for repeated measures interaction ANOVA; mixed design, $r=0.40$

Chart A. 2×2 design at alpha = 0.01, $r=0.40$

Power	Hypothesized ES																		
	0.20	0.30	0.35	0.40	0.45	0.50	0.55	0.60	0.65	0.70	0.75	0.80	1.00	1.25	1.50	1.75	2.00	2.50	3.00
0.80	176	80	59	46	37	30	26	22	19	17	15	14	10	7	6	5	4	4	3
0.90	223	101	75	58	46	38	32	27	24	21	18	16	12	8	7	6	5	4	4

Chart B. 2×3 design at alpha = 0.01, $r=0.40$

Power	Hypothesized ES																		
	0.20	0.30	0.35	0.40	0.45	0.50	0.55	0.60	0.65	0.70	0.75	0.80	1.00	1.25	1.50	1.75	2.00	2.50	3.00
0.80	140	63	47	37	29	24	20	17	15	13	12	11	8	6	5	4	4	3	3
0.90	175	79	59	45	36	30	25	21	19	16	14	13	9	7	5	4	4	3	3

Chart C. 2×4 design at alpha = 0.01, $r=0.40$

Power	Hypothesized ES																		
	0.20	0.30	0.35	0.40	0.45	0.50	0.55	0.60	0.65	0.70	0.75	0.80	1.00	1.25	1.50	1.75	2.00	2.50	3.00
0.80	117	53	40	31	25	20	17	15	13	11	10	9	7	5	4	3	3	3	2
0.90	146	66	49	38	30	25	21	18	16	14	12	11	8	6	5	4	3	3	3

Chart D. 3×3 design at alpha = 0.01, $r=0.40$

Power	Hypothesized ES																		
	0.20	0.30	0.35	0.40	0.45	0.50	0.55	0.60	0.65	0.70	0.75	0.80	1.00	1.25	1.50	1.75	2.00	2.50	3.00
0.80	113	51	38	30	24	20	17	14	13	11	10	9	7	5	4	4	3	3	3
0.90	140	63	47	36	29	24	20	17	15	13	12	11	8	6	5	4	3	3	3

Chart E. 3×4 design at alpha = 0.01, $r=0.40$

Power	Hypothesized ES																		
	0.20	0.30	0.35	0.40	0.45	0.50	0.55	0.60	0.65	0.70	0.75	0.80	1.00	1.25	1.50	1.75	2.00	2.50	3.00
0.80	96	44	33	25	20	17	14	12	11	10	9	8	6	4	4	3	3	2	2
0.90	118	53	40	31	25	20	17	15	13	11	10	9	6	5	4	3	3	3	2

Table 9.38. Sample size table for repeated measures interaction ANOVA; mixed design, $r = 0.60$

Chart A. 2×2 design at alpha = 0.01, $r = 0.60$

Power	Hypothesized ES																		
	0.20	0.30	0.35	0.40	0.45	0.50	0.55	0.60	0.65	0.70	0.75	0.80	1.00	1.25	1.50	1.75	2.00	2.50	3.00
0.80	118	54	40	32	25	21	18	16	14	12	11	10	7	6	5	4	4	3	3
0.90	150	68	51	39	32	26	22	19	17	15	13	12	9	7	5	5	4	4	3

Chart B. 2×3 design at alpha = 0.01, $r = 0.60$

Power	Hypothesized ES																		
	0.20	0.30	0.35	0.40	0.45	0.50	0.55	0.60	0.65	0.70	0.75	0.80	1.00	1.25	1.50	1.75	2.00	2.50	3.00
0.80	94	43	32	25	20	17	14	12	11	10	9	8	6	5	4	3	3	3	3
0.90	117	53	40	31	25	21	17	15	13	12	10	9	7	5	4	4	3	3	3

Chart C. 2×4 design at alpha = 0.01, $r = 0.60$

Power	Hypothesized ES																		
	0.20	0.30	0.35	0.40	0.45	0.50	0.55	0.60	0.65	0.70	0.75	0.80	1.00	1.25	1.50	1.75	2.00	2.50	3.00
0.80	79	36	27	21	17	14	12	10	9	8	7	7	5	4	3	3	3	2	2
0.90	98	44	33	26	21	17	15	13	11	10	9	8	6	4	4	3	3	3	2

Chart D. 3×3 design at alpha = 0.01, $r = 0.60$

Power	Hypothesized ES																		
	0.20	0.30	0.35	0.40	0.45	0.50	0.55	0.60	0.65	0.70	0.75	0.80	1.00	1.25	1.50	1.75	2.00	2.50	3.00
0.80	76	35	26	21	17	14	12	10	9	8	7	7	5	4	3	3	3	2	2
0.90	94	43	32	25	20	17	14	12	11	10	9	8	6	4	4	3	3	3	2

Chart E. 3×4 design at alpha = 0.01, $r = 0.60$

Power	Hypothesized ES																		
	0.20	0.30	0.35	0.40	0.45	0.50	0.55	0.60	0.65	0.70	0.75	0.80	1.00	1.25	1.50	1.75	2.00	2.50	3.00
0.80	64	30	22	18	14	12	10	9	8	7	6	6	4	3	3	3	2	2	2
0.90	79	36	27	21	17	14	12	10	9	8	7	7	5	4	3	3	3	2	2

Table 9.39. Sample size table for repeated measures interaction ANOVA; mixed design, $r = 0.40$

Chart A. 2×2 design at alpha = 0.10, $r = 0.40$

Power	Hypothesized ES																		
	0.20	0.30	0.35	0.40	0.45	0.50	0.55	0.60	0.65	0.70	0.75	0.80	1.00	1.25	1.50	1.75	2.00	2.50	3.00
0.80	91	41	31	24	19	16	14	12	10	9	8	7	5	4	3	3	3	3	2
0.90	126	57	42	33	26	22	18	16	14	12	11	10	7	5	4	4	3	3	2

Chart B. 2×3 design at alpha = 0.10, $r = 0.40$

Power	Hypothesized ES																		
	0.20	0.30	0.35	0.40	0.45	0.50	0.55	0.60	0.65	0.70	0.75	0.80	1.00	1.25	1.50	1.75	2.00	2.50	3.00
0.80	77	35	26	20	17	14	12	10	9	8	7	6	5	4	3	3	2	2	2
0.90	105	47	35	27	22	18	15	13	11	10	9	8	6	4	4	3	3	2	2

Chart C. 2×4 design at alpha = 0.10, $r = 0.40$

Power	Hypothesized ES																		
	0.20	0.30	0.35	0.40	0.45	0.50	0.55	0.60	0.65	0.70	0.75	0.80	1.00	1.25	1.50	1.75	2.00	2.50	3.00
0.80	67	30	23	18	14	12	10	9	8	7	6	6	4	3	3	2	2	2	2
0.90	89	41	30	23	19	16	13	11	10	9	8	7	5	4	3	3	3	2	2

Chart D. 3×3 design at alpha = 0.10, $r = 0.40$

Power	Hypothesized ES																		
	0.20	0.30	0.35	0.40	0.45	0.50	0.55	0.60	0.65	0.70	0.75	0.80	1.00	1.25	1.50	1.75	2.00	2.50	3.00
0.80	65	30	22	18	14	12	10	9	8	7	6	6	4	3	3	3	2	2	2
0.90	87	40	30	23	18	15	13	11	10	9	8	7	5	4	3	3	3	2	2

Chart E. 3×4 design at alpha = 0.10, $r = 0.40$

Power	Hypothesized ES																		
	0.20	0.30	0.35	0.40	0.45	0.50	0.55	0.60	0.65	0.70	0.75	0.80	1.00	1.25	1.50	1.75	2.00	2.50	3.00
0.80	57	26	20	15	12	10	9	8	7	6	5	5	4	3	3	2	2	2	2
0.90	75	34	25	20	16	13	11	10	8	7	7	6	4	3	3	3	2	2	2

Table 9.40. Sample size table for repeated measures interaction ANOVA; mixed design, $r = 0.60$

Chart A. 2 × 2 design at alpha = 0.10, r = 0.60																			
Power	Hypothesized ES																		
	0.20	0.30	0.35	0.40	0.45	0.50	0.55	0.60	0.65	0.70	0.75	0.80	1.00	1.25	1.50	1.75	2.00	2.50	3.00
0.80	61	28	21	17	14	11	10	8	7	7	6	6	4	3	3	3	3	2	2
0.90	85	39	29	23	18	15	13	11	10	9	8	7	5	4	3	3	3	2	2

Chart B. 2 × 3 design at alpha = 0.10, r = 0.60																			
Power	Hypothesized ES																		
	0.20	0.30	0.35	0.40	0.45	0.50	0.55	0.60	0.65	0.70	0.75	0.80	1.00	1.25	1.50	1.75	2.00	2.50	3.00
0.80	52	24	18	14	12	10	8	7	6	6	5	5	4	3	3	2	2	2	2
0.90	70	32	24	19	15	13	11	9	8	7	7	6	4	3	3	3	2	2	2

Chart C. 2 × 4 design at alpha = 0.10, r = 0.60																			
Power	Hypothesized ES																		
	0.20	0.30	0.35	0.40	0.45	0.50	0.55	0.60	0.65	0.70	0.75	0.80	1.00	1.25	1.50	1.75	2.00	2.50	3.00
0.80	45	21	16	12	10	8	7	6	6	5	5	4	3	3	2	2	2	2	2
0.90	60	28	21	16	13	11	9	8	7	6	6	5	4	3	3	2	2	2	2

Chart D. 3 × 3 design at alpha = 0.10, r = 0.60																			
Power	Hypothesized ES																		
	0.20	0.30	0.35	0.40	0.45	0.50	0.55	0.60	0.65	0.70	0.75	0.80	1.00	1.25	1.50	1.75	2.00	2.50	3.00
0.80	44	21	16	12	10	8	7	6	6	5	5	4	3	3	2	2	2	2	2
0.90	59	27	20	16	13	11	9	8	7	6	6	5	4	3	3	2	2	2	2

Chart E. 3 × 4 design at alpha = 0.10, r = 0.60																			
Power	Hypothesized ES																		
	0.20	0.30	0.35	0.40	0.45	0.50	0.55	0.60	0.65	0.70	0.75	0.80	1.00	1.25	1.50	1.75	2.00	2.50	3.00
0.80	38	18	14	11	9	7	6	6	5	5	4	4	3	3	2	2	2	2	2
0.90	50	23	17	14	11	9	8	7	6	6	5	5	4	3	2	2	2	2	2

10 Power analysis for more complex designs

In Chapters 4 through 9 we have presented power and sample size tables that can be addressed directly with no preliminary adjustments to the effect size or *N*. Unfortunately the complexity of some research designs introduces scenarios in which this convenience is not always feasible, hence in this chapter we present a number of tables and algorithms by which both the ES and *N* can be adjusted to permit the reader to adapt the tables presented in previous chapters. Unfortunately, our guidelines in this chapter will not be quite as explicit as those provided to this point – because the techniques for modeling the power of complex designs are not well developed and there is little consensus regarding the appropriateness for those that do exist. Our advice for employing the modeling procedures presented in this chapter, therefore, is to approach the task from the perspective that any accruing results will truly be estimates (as all power/sample size estimates are). It is also a good practice, when communicating these results, to present both a conservative estimate (based upon the procedures employing relatively fewer assumptions presented in previous chapters) along with these modeled estimates.

To this point, then, we have provided tables applicable to designs involving a continuous dependent variable coupled with (a) a single grouping variable involving from two to five groups (with and without repeated measures), (b) the interaction between two grouping variables when no more than one of these variables involves a repeated measure, and (c) the addition of one or more covariates to all of these designs which do not involve repeated measures. Common effects that are not applicable to these tables include those experiments that employ (a) more than two grouping variables, (b) two or more repeated measures, and (c) the addition of one or more covariates to a mixed design.

Main effects in two-factor designs

In Chapter 9 we considered the interaction term produced by a two-way ANOVA/ANCOVA. At first glance it might appear that the power of the two main effects could be computed using either the one-way ANOVA/

ANCOVA or *t*-test tables already presented, but unfortunately the *N*/group must be adjusted prior to doing this based upon the differing degrees of freedom associated with a factorial as opposed to a one-way ANOVA.

As discussed previously, the most common scenario for this type of design (at least from a between subject perspective) involves a treatment factor combined with a subject characteristic (or attribute) for which the treatment is hypothesized to be differentially effective. (Alternatively the second independent variable can be employed simply because it is known to be related to the dependent variable, and hence is included as a post hoc blocking variable for its potential to increase statistical power.) It is also quite possible, however, for both grouping variables to represent different genres of experimental interventions – an example of which will be provided below.

One way to calculate the *N*/group for, say factor A in a two-factor experiment, is simply to multiply the *N*/cell used to calculate the power of the interaction in Chapter 9 by the number of levels contained in the second factor (B). Thus, in an A(2) × B(4) design employing 10 subjects per cell, the *N*/group for the main effect for factor A would be 10 × 4 = 40 and the *N*/group for factor B would be 10 × 2 = 20. This strategy does not take into consideration that the degree of freedom for the error term is slightly different for a factorial design as opposed to a one-way ANOVA, since the degrees of freedom for both the second factor and the interaction must be subtracted therefrom. While this will not prove to be a major factor in designs employing moderate to large numbers of subjects, it can prove substantive for smaller values of *N*, especially those with more than two groups per factor.

A better approximation, therefore, is provided by the following formula:

Formula 10.1. Calculating the *N*/group for a factorial design

$$N/\text{group} = \left(\frac{df_{error}}{df_{numerator} + 1}\right) + 1$$

This in turn results in the following calculations for a 2 × 4 design employing 10 subjects per cell:

$df_{error} = (\#\ \text{cells} \times N/\text{cell}) - \#\ \text{cells} = (8 \times 10) - 8 = 72$,
$df_{numerator} = (\#\ \text{groups} - 1)$ or 1 df for factor A and 3 df for factor B,
$N/\text{group A} = [72/(1 + 1)] + 1 = 37$ and $N/\text{group B} = [72/(3 + 1)] + 1 = 19$.

Note that these *N* values are quite close to the values that would be generated by simply multiplying the *N*/cell by 2 and 4 respectively as described above. This procedure is simply one mechanism by which the additional degrees of freedom associated with the other main effect (one in the case of B and three in the case of A) and interaction (three degrees of freedom) may be taken into account.

For convenience we have provided Table 10.1 involving a number of common *N* values and factorial designs. To use this table, the *N*/group is located at the intersection of the *N*/cell row corresponding to the design (e.g., 2×2, 2×3) and the factor (i.e., A vs. B). It should be noted that the results produced by this strategy (as well as most of the strategies proposed in this chapter) will not be exact, since the one-way ANOVA/ANCOVA tables presented in the previous chapters were set up using the *N*/group and degree of freedom for error specifically based upon a single factor, which in the case of factor A would be 78 and for factor B would be 76. The lack of precision resulting from this fact is trivial, however (except for the very smallest of trials), and largely avoids the potentially larger errors involved noted by Bradley *et al.* (1996).

Example. Suppose an animal study is proposed to provide a preliminary test of an experimental weight loss drug hypothesized to be especially effective when coupled with high intensity exercise. Although there would be several reasonable design options, let us assume that the investigator decided to employ two grouping variables: (a) animals that received the drug vs. those that did not and (b) animals that received high intensity exercise (i.e., were required to run for extended periods of time on a treadmill to avoid a noxious stimulus) vs. those required to run only a few minutes per day vs. those who were not required to exercise at all. These conditions thus produce a perfectly crossed 2 (drug vs. no drug)×3 (high intensity exercise vs. low intensity exercise vs. no exercise) design that would test two main effect and one interaction hypotheses:

(1) *Main effect A*: Animals that receive drug A will lose more weight than animals that do not.
(2) *Main effect B*: Animals that receive high intensity exercise will lose more weight than animals that receive either low intensity or no exercise.
(3) *Interaction AB*: Animals that receive drug A and who receive high intensity exercise will lose more weight than animals who receive the drug but do not receive high intensity exercise or who do not receive the drug at all.

Converting grams of weight lost to standardized means with the weakest condition set at 0.0 allows us to hypothesize the following results:

Chart 10.1. *Hypothetical results for a drug trial conducted under different exercise conditions*

	High intensity exercise	Low intensity exercise	No exercise
Drug A	1.2	0.6	0.6
No drug	0.2	0.0	0.0

Before we proceed, it should be noted that it is a relatively rare study employing a factorial design in which all of the main effects and interactions are of scientific interest. Of the three above hypotheses, for example, the drug main effect would probably be of some interest, although its interpretation would depend upon the characteristics of any drug-by-exercise interaction that occurred. (If no such interaction existed, then the drug main effect would be of primary interest.) The exercise main effect, on the other hand, would probably be of little scientific interest since it does not test the effects of exercise upon weight loss in rats given that half of the entire sample also received a drug designed to cause weight loss. (For a completely uncontaminated test of the hypothesis that these three levels of exercise would result in weight loss in rats, the investigator would be better served to examine only the mean differences represented by the second (no drug) row of Chart 10.1 via a one-way ANOVA employing half of the overall sample size.)

Thus, in most cases, as discussed in Chapter 9, it would be the interaction hypothesis (hence the interaction ES) that would be of primary interest in a study such as this and we have shown how such an interaction can be computed by either (a) the use of Formula 9.3 in the Technical appendix or (b) collapsing the 2×3 interaction table represented above into a 2×2 table that best reflects the hypothesis or that results in the largest ES.

This is not to say, however, that main effects such as these could not be of substantive interest as long as they are interpreted properly – which would be to consider them in the perspective of the interaction. As we have just mentioned, if there were no interaction between the drug and exercise factors (i.e., if the interaction ES were truly zero), then the drug main effect would be of interest. Alternatively, if there were no interaction *and* no drug effect, then the exercise main effect would not be contaminated.

Estimating the power of between subject main effects. To estimate the power of the main effects represented in Chart 10.1, the investigator would access Template 10.1 and follow the indicated steps. Step 1 asks the user (a) to select the shell that reflects the design of the study (which in this case would be the 2×3 table), (b) to fill in the interior rows and columns (i.e. A1–A2×B1–B3) with standardized means that best reflect the hypothesized patterns (based upon preliminary data or previous relevant research literature), and (c) to calculate each row and column mean based upon averaging the appropriate standardized cell means. (Note, that the process is identical to the steps in Template 9.1 with the exception of calculating the column/row means.)

While probably obvious, the grand or total means for each row and column are computed, which would occur as follows (see Chart 10.2):

(1) The column means, collapsing across the drug factor, would be 0.7 for high intensity exercise (i.e., 1.2+0.2=1.4/2=0.7) and 0.3 (0.6+0.0=0.6/2=0.3) for both low intensity and no exercise.
(2) The row means, collapsing across exercise level, for drug A vs. no drug would be 0.8 (1.2+0.6+0.6=2.4/3=0.8) and 0.067 (0.2+0+0=0.2/3=0.067), which can safely be rounded to 0.1.

Chart 10.2. *Row and column totals computed for Chart 10.1*

	B1 High intensity exercise	B2 Low intensity exercise	B3 No exercise	Total
A1 Drug A	1.2	0.6	0.6	0.8
A2 No drug	0.2	0.0	0.0	0.067 or 0.1
Total	0.7	0.3	0.3	

Step 2 requests that the ES values for two main effects be computed by simply subtracting the smallest row total from the largest for main effect A and the smallest column total from its largest counterpart for main effect B. This would produce a hypothesized ES for the drug main effect of 0.7 and an ES for the exercise main effect of 0.4:

main effect A ES = largest row mean (0.8) − smallest row mean (0.1) = 0.7

main effect B ES = largest row mean (0.7) − smallest row mean (0.3) = 0.4

Step 3 dictates that the ES for the interaction be computed by following the instructions in Template 9.1 in Chapter 9. This involves collapsing the above 2×3 table into a 2×2 shell (which as discussed above would mean combining cells B2 and B3) and performing the following computation:

Interaction ES = {[(B1–B2/B3) for A1] – [(B1–B2/B3) for A2]}/2 = [(1.2–0.6) – (0.2–0.0)]/2 = (0.6–0.2)/2 = 0.2

At this point, all that is necessary to compute the power for the interaction is to continue to follow the instructions in Template 9.1. Thus, assuming that 20 animals per cell were to be employed, the investigator would access Table 9.2 and find the intersection of the 0.2 ES column and the *N*/cell = 20 row, producing a power level of 0.14, obviously not a very encouraging result.

For the main effects in our example, the tables in Chapter 4 (for the two-group drug comparison) and Chapter 6 (for the overall three-group exercise comparison) can be used for power and sample size analyses once the *N*/group is adjusted for the fact that this is truly a factorial ANOVA rather than a one-way analysis. We have provided Table 10.1 for convenience which converts the *N*/cell available in a two-factor study such as depicted in Chart 10.1 to the *N*/group parameter required for a number of the more common factorial designs (i.e., 2×2, 2×3, 3×3, 2×4, and 3×4) for a relatively wide range of *N* values.

To illustrate, let us begin with factor A, the two-level drug variable (drug A vs. no drug). To ascertain the *N*/group to be employed in Table 4.1, step 4 instructs us that all we need do is locate the intersection of the *N*/cell row of 20 and the "2×3:A" column in Table 10.1, which yields an *N*/group of 58. (The "A" always denotes the first factor in the design, the "B" the second factor.) The *N*/group for the exercise factor is correspondingly found at the intersection of the same row and the "2×3:B" column, or 39.

To obtain the power for main effect A, step 6 of Template 10.1 sends us to the appropriate template in the previous chapter that deals with the design employed, which is Template 4.1 for the two-group factor that in turn indicates the power for this main effect is located at the intersection of the *N*/group = 58 row and the ES = 0.70 column of Table 4.1, which produces an interpolated value of 0.96. For the required sample size to achieve an 80% chance of obtaining statistical significance, Template 4.1 sends us to Table 4.2, where the *N*/group of 34 is located at the intersection of the 0.80 power row and the 0.70 ES column. *It is important to remember that this N/group must be translated back to an N/cell for present purposes, which can be done by employing Table 10.1 backwards (i.e., by locating the closest value to 34 in the 2×3:A column and reading the N/cell in the left-most column, which is 12).*

Since we already have the *N*/group for the exercise main effect (39), it is necessary to return to Chapter 6 to obtain its power, where, once the pattern of means is specified (high dispersion), the power (0.43) power of the main effect can be obtained directly from Table 6.2. Should the

N/group be desired, the template directs us to Table 6.9, yielding a value of 90 for a desired power level of 0.80. Run backwards through Table 10.1, this in turn yields an approximate *N*/cell of 45. Should the power available or sample size requirements for a multiple comparison procedure be desired, the appropriate steps in the Chapter 6 templates would be followed.

Estimating the power of complex designs employing three or more factors

From a power analytic perspective, the more complex the design becomes, the more parameters must be hypothesized and hence the more tenuous the results of the power analytic process become. One way around this problem is to simplify the design artificially for power/sample size estimation purposes. This will normally result in a more conservative estimate, but only slightly so since the adjustments made based upon additional factors, interactions, and variations in covariate–dependent variable estimates do not exert a great deal of influence thereon. Let us therefore illustrate how this process can be accomplished for a three-factor design.

Example. A meta-analysis was performed on the relative efficacy of *N*-acetylcysteine with respect to reducing the number of severe bronchial episodes experienced over a fixed time interval. Two additional variables (severity of the disease and length of treatment) were identified that could potentially affect this outcome. Testing all three variables simultaneously, therefore, would produce a 2 (drug: *N*-acetylcyteine vs. placebo) × 2 (severity: high vs. low airway obstruction) × 2 (length of treatment: one month vs. two months) between subject design. This model would consequently result in the ability to answer seven questions, the first four being of potential scientific interest:

(1) Is *N*-acetylcysteine more effective than placebo irrespective of length of treatment or severity of the disease (main effect A)?
(2) Is *N*-acetylcysteine more effective after three months of treatment than at the end of one month as compared to a placebo (AC interaction)?
(3) Is *N*-acetylcysteine relatively more effective for individuals with lower severity of airway obstruction than individuals with a high degree of obstruction (as compared to placebo) (AB interaction)?
(4) Is *N*-acetylcysteine administered for longer periods of time to patients with relatively less airway obstruction more effective than *N*-acetylcysteine administered for shorter periods of time to patients with severe airway obstruction (as compared to placebo) (ABC interaction)?

Three additional terms are tested, but these would probably not be particularly interesting:

(5) Do patients with severe airway obstruction experience more bronchial episodes than individuals with relatively less severe obstruction (main effect B)?
(6) Do patients undergoing an extended treatment regimen experience fewer bronchial episodes than individuals administered a more brief treatment regimen (main effect C – note that effects such as this are confounded, since half of the patients receiving each length of treatment actually receive nothing but a placebo)?
(7) Do patients with relatively greater airway obstruction benefit more from an extended regimen than individuals with less severe obstruction (BC interaction – note the same confound exists for this effect as described in (6) above)?

The first step in estimating the power for these seven effects is, as always, to hypothesize their ES values. The most direct method of doing this would be for the investigator to fill in the mean numbers of bronchial episodes he/she hypothesizes would occur within each of the eight cells represented by this 2×2×2 design. This can be a complicated process which definitely merits a considerable amount of thought, not to mention knowledge of the literature. For illustrative purposes, however, let us assume that this process produced the following results:

Chart 10.3. *Number of bronchial episodes for hypothesized three-factor design*

		C1 1 Month	C2 2 Months
A1 Drug	B1 High obstruction	6	4
	B2 Low obstruction	4	2
A2 Placebo	B1 High obstruction	8	8
	B2 Low obstruction	6	6

What is being hypothesized here, then, is a scenario in which patients with more severe disease will experience more adverse bronchial episodes than individuals with less severe disease states. The investigator is also hypothesizing a rather dramatic effect for *N*-acetylcysteine in comparison to a placebo, especially given a longer treatment interval.

As was discussed in Chapter 9, the next step in the process is to compute standardized means for the eight cells and then to convert the lowest of these means to zero. In this case, since lower dependent variable means are the "desired" outcomes, it is helpful to subtract the "weakest" condition (or the highest standardized cell mean which would be the raw mean of 8.0 converted to a standardized mean of 6.4 (8.0/1.25)) from all of the other cells, producing the following results (note that the sign is ignored):

Chart 10.4. *Standardized means for the hypothetical 2 × 2 × 2 design*

		C1 1 Month	C2 2 Months
A1 Drug	B1 High obstruction	4.8 − 6.4 = 1.6	3.2 − 6.4 = 3.2
	B2 Low obstruction	3.2 − 6.4 = 3.2	1.6 − 6.4 = 4.8
A2 Placebo	B1 High obstruction	6.4 − 6.4 = 0.0	6.4 − 6.4 = 0.0
	B2 Low obstruction	4.8 − 6.4 = 1.6	4.8 − 6.4 = 1.6

For ease of interpretation from a power analytic viewpoint, all the effects except the three-way interaction can now be interpreted by collapsing these results into two-way tables and the ES values for the main effects and interactions can be computed as previously illustrated. There are three possible two-factor interactions in a three-way ANOVA/ANCOVA, although in this case the investigator would probably be primarily interested only in those involving factor A (drug vs. placebo). It is important, however, that all of these combinations be considered carefully at the planning phase of a trial, since one or more may be interesting enough for the investigator to design his/her study to ensure sufficient power specifically for these effects:

Chart 10.5. *The three-factor design (Chart 10.4) collapsed as two-factor designs*
Scenario A. *The A (drug) × B (severity) interaction collapsed across C (duration of treatment)*

	B1 High obstruction	B2 Low obstruction	Difference
A1 Drug	(1.6 + 3.2)/2 = 2.4	(3.2 + 4.8)/2 = 4.0	(4.0 − 2.4) = 1.6
A2 Placebo	(0.0 + 0.0)/2 = 0.0	(1.6 + 1.6)/2 = 1.6	(1.6 − 0.0) = 1.6
Difference	2.4 − 0.0 = 2.4	4.0 − 1.6 = 2.4	ES = (1.6 − 1.6)/2 = 0.0

Scenario B. *The A (drug) × C (duration of treatment) interaction collapsed across B (severity)*

	C1 1 Month	C2 2 Months	Difference
A1 Drug	(1.6 + 3.2)/2 = 2.4	(3.2 + 4.8)/2 = 4.0	(4.0 − 2.4) = 1.6
A2 Placebo	(0.0 + 1.6)/2 = 0.8	(0.0 + 1.6)/2 = 0.8	(0.8 − 0.8) = 0.0
Difference	2.4 − 0.8 = 1.6	4.0 − 0.8 = 3.2	ES = (1.6 − 0.0)/2 = 0.8

Scenario C. *The B (severity) × C (duration of treatment) interaction collapsed across A (drug)*

	C1 1 Month	C2 2 Months	Difference
B1 High obstruction	(1.6 + 0.0)/2 = 0.8	(3.2 + 0.0)/2 = 1.6	(1.6 − 0.8) = 0.8
B2 Low obstruction	(3.2 + 1.6)/2 = 2.4	(4.8 + 1.6)/2 = 3.2	(3.2 − 2.4) = 0.8
Difference	2.4 − 0.8 = 1.6	3.2 − 1.6 = 1.6	ES = (0.8 − 0.8)/2 = 0.0

A close examination of Chart 10.5 indicates that a relatively dramatic drug × treatment duration interaction is being hypothesized here. By employing Templates 9.1 and 9.2 in the previous chapter, the investigator will be directed to either Table 9.1 to obtain the power available for testing this interaction or Table 9.26 (if the desired sample size is required for power levels of 0.80 or 0.90). *It is important to remember, however, that the values of N/cell in these collapsed tables actually represent twice as many subjects as did the two-factor power and sample size tables presented in Chapter 9, although of course the investigator always has the option of deleting a factor should it prove uninteresting upon reflection.* It is also important to remember that just as it was necessary to adjust the *N*/group for main effects in factorial designs based upon the additional terms employed, strictly speaking it is also important to adjust the *N*/cell for a two-factor design based upon the additional degrees of freedom associated with the extra main effect and three interactions (two two-factor and one three-factor) emanating from the use of three independent variables, hence the results from the tables in Chapter 9 will not be extremely precise. It is our opinion, however, that we have achieved a point at which there is an acceptable trade-off with respect to precision and the complexities involved in attempting to achieve it. For those individuals who do desire a greater degree of precision, however, Formula 10.1 can be used to adjust either the *N*/cell or the *N*/group based upon any number of independent variables and Formula 6.1 in the Technical appendix can be employed to calculate the exact power emanating therefrom.

A note on main effects in a three-factor ANOVA. The same logic holds for assessing the power of main effects in more complex designs. In our example above, the main effects for duration of treatment and the experimental drug could be assessed using standardized means in Chart 10.6 (since each factor contained only two groups):

Chart 10.6. *Treatment (drug) and duration of treatment main effects*

	B1 1 Month	B2 2 Months	Treatment main effect
A1 Drug	(1.6 + 3.2)/2 = 2.4	(3.2 + 4.8)/2 = 4.0	(4.0 + 2.4)/2 = 3.2
A2 Placebo	(0.0 + 1.6)/2 = 0.8	(0.0 + 1.6)/2 = 0.8	(0.8 + 0.8)/2 = 0.8
Duration main effect	(2.4 + 0.8)/2 = 1.6	(4.0 + 0.8)/2 = 2.4	ES A = 3.2 − 0.8 = 2.4 ES B = 2.4 − 1.6 = 0.8

Should the power of either of these effects be desired, an estimate may be obtained by accessing Table 4.1 (since there are only two levels per factor). (The *N*/group may be adjusted based upon Formula 10.1.)

Thus, if 20 subjects per cell were projected for the above design, the df_{error} would be equal to $(20 \times 8) - 8$ (because there are three factors each possessing two groups) or 152 and the $df_{numerator}$ for each factor would be $2 - 1 = 1$. Hence, the *N*/group that could be inputted into Table 4.1 would be $152/2 + 1 = 77$, which would produce approximately the same results as would be the case if the *N*/cell were simply multiplied by the relevant number of factors (4) over which the main effect means in Chart 10.5 (Scenario B) were collapsed. (It should be noted, however, that when more than two groups are represented in the various factors, it becomes more important to adjust the *N*/group.)

Three-way interactions. It is rare that it is essential for the power (or required *N*/cell) to be required for a three-way interaction. Should this be necessary, a very gross method of estimating the hypothesized ES for such an interaction can be obtained by assessing the difference between the ES values of the two relevant two-way interactions upon which this term is based. Thus for the ABC interaction implied by question (4) above ("Is *N*-acetylcysteine administered for longer periods of time to patients with relatively less airway obstruction more effective than *N*-acetylcysteine administered for shorter periods of time to patients with severe airway obstruction"), the investigator is basically positing that the ES for the two-way, uncollapsed BC interactions for the drug will be different *and in a different direction* from the two-way, uncollapsed BC interaction for the placebo. (Said another way, the pattern of means in the top four cells in Chart 10.4 will be different from the pattern of means in the bottom four cells.) Upon examination, this is obviously not the case, since the ES values for both the placebo BC interaction and its BC drug counterpart are 0.0. If there were a substantive difference between the two two-way interaction ES values, then (a) the actual ES could be computed via Formula 9.3 in the Technical appendix, (b) the *N*/group could be computed via Formula 10.1, and (c) the power available for the contrast could be obtained from Formula 6.1 in the Technical appendix.

In general it is worth repeating, however, that the more complex the design becomes, the more tenuous the power estimates become due to the number of assumptions required *and* the approximations we have introduced for convenience sake. With this in mind it is better to conduct a tenuous power analysis, even if this involves nothing more than computing an ES and entering it into the table that most closely approximates the model involved, than to design an experiment without doing so. Our practical advice here is for a researcher to be most hesitant to design a study primarily designed to test a three-way interaction, especially from a between subject perspective because of both (a) the low degree of power normally available for such effects and (b) the difficulties involved in hypothesizing the exact pattern of means that is likely to accrue from the myriad cells contained therein. If such an experiment is planned, the investigator should be cognizant of the fact he/she has probably reached a point at which a power analysis becomes more of a heuristic than a statistical tool.[2]

One-way between subjects ANCOVA designs with *r* values other than 0.40 and 0.60

The tables in Chapter 7 are set up for only two values of the covariate–dependent variable correlation, 0.40 and 0.60. If an investigator wishes to model a different correlation coefficient, Table 10.2 is provided for this purpose. (Table 10.3 represents a different type of conversion table for within subject designs to facilitate modeling for repeated measures.) All that needs to be done to model a different covariate–dependent *r* is to locate the adjusted ES at the intersection of the hypothesized covariate–dependent variable correlation row and the *unadjusted* ES column. This value can then be inputted into the appropriate Chapter 6 template, which will then produce either the power or required sample size for an ANCOVA design involving the desired estimate for the covariate–dependent variable correlation. Although we will illustrate this process for power, the same procedures are followed to determine the required *N*/group with the exception that Template 6.3 instead of 6.2 is employed.

As an example, let us recast the three-group example provided in Chapter 6 to one in which an available covariate was hypothesized to correlate 0.70 with the dependent variable (weight loss). Assuming that the remainder of the parameters did not change, Table 10.2 can be employed to adjust the original ES value(s) and then the process proceeds identically to the procedures outlined in Templates 6.1 and 6.2.

To review the original parameters, it will be remembered that three groups with 50 subjects each were employed ((1) treatment-as-usual, (2) an attention placebo, and (3) an exercise plus education intervention) in order to compare their effectiveness with respect to the amount of weight lost with the ES differences between the three groups being:

ES (2) − (1) = 0.2
ES (3) − (1) = 0.5
ES (3) − (2) = 0.3

Following the steps outlined in Templates 6.1 and 6.2 indicated that the ES for the *F*-ratio is 0.5, the ES differences represented a low/medium dispersion pattern and that the intersection between the ES column of 0.5 and the *N*/group row of 50 in Table 6.1 indicated that the power for this study would have been 0.60.

The simplest way to perform this analysis based upon the addition of a hypothesized covariate–dependent variable relationship of 0.70, would thus be to adjust the original ES of 0.50 via the use of Table 10.2 and then follow the templates presented in Chapter 6 as though a one-way between subject ANOVA was to be employed. Locating the intersection between the 0.50 column and the $r = 0.70$ row, then, provides an adjusted ES of 0.70. Finding the intersection of this ES column and the *N*/row of 50 in Table 6.1 indicates that the power for the overall *F*-ratio would be 0.89 instead of the original 0.60. It is also interesting to compare this figure to the power that would have been obtained if Table 7.4 had been employed which assumes an *r* of 0.60. This value, it will be noted would have been 0.80, which while not dramatically different from the value obtained by using Table 10.2, does reflect a substantive change in power.

Continuing this analysis for the power of the pairwise contrasts would require that the other two ES differences (i.e., 0.2 (group 2 vs. group 1) and 0.3 (group 3 vs. group 2)) also be adjusted via Table 10.2, which would yield adjusted ES values of 0.28 and 0.42 in addition to the 0.70 adjustment for group 3 vs. group 1. Assuming the use of the Tukey HSD multiple comparison procedure, Table 6.12 would result in interpolated power estimates of 0.19, 0.41, and 0.87 respectively, which again are higher than the values that would have accrued via the use of Table 7.26 and the use of a hypothesized *r* of 0.60.

Caveat. Two caveats are in order. In the first place, the ANOVA tables in Chapter 6 are not adjusted for the one degree of freedom per covariate loss to the error term associated with ANCOVA as are the tables in Chapter 7. This is usually trivial, however, for all but the smallest studies.

The second caveat involves the fact that minor variations from the 0.40 and 0.60 options provided in Chapter 7 truly have very little practical implications for the resulting power/sample size estimates as an examination of Table 10.2 will illustrate. However, for differences of 0.10 or greater, especially for higher correlations, it is probably worthwhile to make the indicated adjustment.

One-way within subject designs with *r* values other than 0.40 and 0.60

As with the tables for between subject ANCOVA designs, the tables in Chapter 8 are only set up for hypothesized correlations among the dependent observations of 0.40 and 0.60. Given the fact that within subject correlations can exhibit a more dramatic effect upon power than is true for the covariate–dependent variable relationship, any hypothesized divergence from these two values of r is relatively more important to model. For this reason, Table 10.3 is provided to adjust the ES values based upon small increments in correlations ranging between 0.20 and 0.80. These values can then be modeled to determine their effects upon both the power of the overall F-ratio and the pairwise contrasts for studies involving from three to five groups. To facilitate this process, we have linked the use of Table 10.3 to the tables in Chapters 8 and 9 relevant to a repeated measures correlation coefficient of 0.40.

Specifically, Table 10.3 is provided to indicate how the ES values in both the MCP and overall F tables can be adjusted to permit the use of the appropriate $r = 0.40$ tables. As an example, if the within subject r were hypothesized to be 0.20 rather than 0.40, the intersection of the 0.20 row and the unadjusted ES column would indicate what value would need to be *subtracted* from the ES that would normally be employed in the appropriate use of the Chapter 8 or 9, $r = 0.40$ tables. In other words, if the actual hypothesized ES were 0.60, Table 10.3 indicates that an adjustment of 0.10 would be *subtracted* (because of the negative sign) from this value prior (i.e., $0.60 - 0.10 = 0.50$) and that this interpolated value (0.50) would be entered into the appropriate power or sample size table.

Example. In way of illustration, let us use the actual four-group, medium dispersion acupuncture example provided in Chapter 8 in which the basic parameters were hypothesized as follows:

(1) four groups,
(2) ANOVA ES of 1.0,
(3) individual pairwise ES values ranging from 0.3 to 1.0,

(4) medium mean dispersion pattern, and
(5) N/group = 20.

Let us assume, however, that the correlation across the repeated observations was hypothesized to be 0.50 instead of the modeled 0.40 and 0.60. Table 10.3 could therefore be accessed to determine the adjusted increase in the overall F-ratio ES of 1.0 in comparison to the $r = 0.40$ table (Table 8.9 as indicated in Template 8.2). This value (located at the intersection of the 1.0 ES column and the $r = 0.50$ row) is found to be +0.12, which when *added* to the 1.0 ES in Table 8.9 yields an interpolated power level of 0.97 (obviously not a substantive difference). The same process applied to the pairwise comparisons (using the Tukey HSD procedure) provides similar alterations:

Contrast	Unadjusted ES	Adjusted ES	Unadjusted power	Adjusted power
2 − 1	0.3	0.33	0.08	0.10
3 − 2	0.3	0.33	0.08	0.10
4 − 3	0.4	0.45	0.16	0.22
3 − 1	0.6	0.68	0.31	0.55
4 − 2	0.7	0.79	0.58	0.72
4 − 1	1.0	1.12	0.92	0.95

While these particular examples do not reflect a major difference in power, they are also not completely trivial in nature. It should also be noted that the higher the r, the more dramatic the power/sample size adjustments, hence they should be considered if there is good reason to hypothesize an r among repeated observations different than 0.40 or 0.60.

Other techniques for adjusting the ES values of main effects and interactions

There are occasions when it is desirable to adjust the ES values emanating from an experiment based upon the hypothetical addition of one or more other design components capable of reducing the original model's error term. One example includes the use of a post hoc blocking variable (i.e., one which was not employed in subject selection for the study in the first place[1]) that was not originally included as part of trial.

Interestingly, other between subject terms such as this (including interactions with the original grouping variable) basically affect power identically to the addition of a covariate (except that the N/group and the N/cell must be adjusted slightly given the fact that only one degree of freedom is lost from the error term with the addition of a covariate but more are typically lost with the addition of a blocking variable).

Since categorical variables such as the three levels of exercise employed in Chart 10.1, or a dichotomous characteristic such as gender, are somewhat difficult to conceptualize in terms of r or r^2, the easiest way to estimate the hypothetical reduction in the error term is to hypothesize an ES for these additional variables and then convert them to either r or r^2 using the following formula:

Formula 10.2. Translation of an ES (*d*) to r^2

$$r^2 = \frac{\text{ES}^2}{\text{ES}^2 + 4}$$

Once accomplished, a categorical independent variable (and its hypothesized interaction) can then be used in the same way that a covariate's dependent variable correlation is used to adjust the study's primary ES of interest (although the changes in the error term's degrees of freedom should be taken into account as described later). To facilitate this process, we have provided Table 10.4 which allows ES values to be expressed in terms of either r^2 or r. In fact, since squared correlation coefficients are additive, and since there are occasions when multiple ES values are theoretically additive, Table 10.4 can be employed to facilitate this process via the substitution of the sum of two or more squared correlation coefficients (i.e., by replacing r^2 with R^2).

To illustrate one simple application of Table 10.4, let us suppose that a two-group trial was planned with a hypothesized ES of 0.50. If 50 subjects per group were available, Table 4.1 would indicate that the power for this experiment would be 0.70. What if, however, our researcher wished to know what effect a post hoc blocking variable such as gender would have upon such a trial? Assuming that he/she estimated that the ES for males vs. females would also be 0.50, Table 10.4 could be used to convert this additional ES (0.50) to a correlation coefficient, which would be 0.24 and could be treated identically to the addition of a covariate possessing that correlation assuming no interaction between gender and the intervention was hypothesized. Locating the intersection between the closest row value ($r =$ 0.25) in Table 10.2 and the ES column of 0.50, the investigator would find that the adjusted ES for the intervention would be 0.52 which would produce a power level of 0.73, which in this case would probably have little practical import upon the planning of this particular study. If, however, gender were expected to interact with the grouping variable and the interaction ES was hypothesized to also be 0.50, these two ES values could be added together once they were converted to r^2 and the resulting sum could be used to estimate the adjusted ES by employing Table 10.4 backwards.

The steps employed in this process are as follows:

(1) The ES values for any terms that are to be used to adjust the primary ES are converted to r^2 via the use of Table 10.4.
(2) All of the relevant r^2 values are added together.
(3) The sum (which is now technically an R^2) is then located under the r^2 column in Table 10.4 and the Pearson r is read in the same row under the r column variable.
(4) This value is then used in Table 10.2 to find the adjusted ES.

Thus, in our original example in which the treatment ES was hypothesized to be 0.50 as was the gender and the treatment × gender interaction, the computation of the adjusted ES would proceed as follows:

(1) gender r^2 + gender × treatment interaction r^2 = adjusted R^2,
(2) converting the adjusted R^2 to r via Table 10.4 produced the value that will be used via Table 10.2 to produce the study's adjusted ES, or $0.06+0.06=0.12$,
(3) converted to $r=0.35$,
(4) used to adjust the original ES of 0.50 via Table 10.2 producing an adjusted ES of 0.53,
(5) which would yield an interpolated power value of 0.75 using Table 4.1. (If the original design had employed a covariate, this ES would have been used in the appropriate table in Chapter 7.)

Modeling more complex designs employing simpler analogs

By now it is probably obvious that we are treading in gray areas at the border between accepted procedures and techniques that have not yet been fully developed. The only viable option at this point, therefore, is to model these more complex designs employing one or more simple analogs. This can be a time consuming process, but hopefully will be facilitated by the aids provided in this book. It is also a worthwhile process because it often points the way for refining one's hypotheses and for making design changes that improve the quality of the planned trial.

As an example, let us hypothesize a mixed, two-factor design employing a covariate. The first consideration is what type of covariate is being employed. For mixed designs there are two genres: (a) those measured once prior to the implementation of the study and (b) those measured each time the treatment (when it is represented as a repeated measure) is implemented. In the first scenario, the covariate serves to increase the power of the between subject treatment variable only, having no substantive impact upon the power of the repeated measure nor of the interaction.

To illustrate, let us consider a preliminary experiment designed to evaluate the effects of a quick acting (and transitory) hypertension drug upon systolic blood pressure under different, experimentally manipulated levels of stressful activities. Let us assume that three groups were used to accomplish the latter. These were (a) two computerized tasks, one of which constituted a relatively simple problem solving exercise (low induced stress) and one with high stress potential possessing a built in algorithm that precluded its successful completion within the allotted time and (b) a condition in which subjects simply sat and listened to pleasant music.

Instead of randomly assigning different subjects to each of the six resulting cells (the two drug conditions × the three tasks), however, it was proposed that healthy college students be randomly assigned to either receive the blood pressure medication or its placebo, although all subjects would perform all three tasks (i.e., the high stress task, its low stress counterpart, and simply sitting in a resting position for the allotted time period). Obviously the order in which subjects performed the tasks would be randomly assigned in some counterbalanced way and the experimenter would take pains to assure that his/her subjects' blood pressure was given a chance to return to baseline values following each assessment (and prior to the beginning of each new task). This would thus produce the following mixed 2 (drug vs. placebo) × 3 (stress level) design in which the second factor was represented as a repeated measure or within subject factor:

Chart 10.7. *Mixed subject design in which both factors are manipulated but the second factor is a repeated measure*

	High stress task	Low stress task	No activity
BP drug			
Placebo			

Now obviously a covariate, such as blood pressure, could be measured once immediately before administration of the drug as would be done if this were a between subjects design. Such a covariate, however, would serve only to adjust the overall drug vs. placebo main effect (i.e., the average drug pressure collapsed across all three stress levels) but would have no effect upon the stress factor or its crucial interaction with the BP drug (because its values could not vary across all three stress levels for each subject and hence it would have no way of impacting the power of either the stress main effect or the drug by task interaction). Should this particular design be employed, then, the drug main effect could be adjusted for the effects of the covariate via the use of Table 10.2 and this adjusted ES could be employed in subsequent power/sample size analyses.

What, however, if each of the three tasks could be preceded by a unique covariate, such as baseline blood pressure (BP) levels producing the mixed design depicted in Chart 10.8?

Chart 10.8. *Mixed design in which both factors are manipulated with the task represented as a within subject factor and baseline blood pressure measures serving as unique covariates for each condition*

	Covariate 1 (baseline 1)	High stress task	Covariate 2 (baseline 2)	Low stress task	Covariate 3 (baseline 3)	No activity
Drug						
Placebo						

Here, the covariate affects the power of the within subject (type of task) main effect and its interaction with the drug factor, as well as the between subject drug factor. The data emanating from this design could be appropriately analyzed in one of three ways:

(1) Classically, the recommended approach would be a 2 (drug) × 3 (baseline BP for the three tasks) × 3 (post task BP) mixed ANOVA in which the final two factors represented repeated measures. This design, however, employs four error terms (unless the three within subject terms can be pooled) and multiple correlations (e.g., across baseline/post-task assessment and across performances on the three tasks). It also has the inherent disadvantage of testing the primary hypothesis via a three-way interaction.

(2) Perhaps because of the complexity of this analysis (and the general distaste of experienced data analysts for three-way interactions), the analytic option most likely to be employed would probably be a 2 (drug) × 3 (type of task) mixed ANOVA with the dependent variable being *changes* or *difference scores* in BP calculated by subtracting baseline values measured prior to the beginning of each task from BP levels following same. Here, the primary hypothesis would be tested by the two-way interaction, which on the surface would require fewer assumptions from a power analytic perspective. The problem, however, resides in estimating the correlation among

these *differences* scores across tasks, since measures such as this tend to be relatively unreliable.

(3) Finally, the data could be analyzed via a 2 (drug) × 3 (type of task) ANCOVA in which each baseline measure would serve as the covariate for the subjects as they participated in the indicated task. The problem with this analysis, however, also resides in the difficulty of estimating two separate correlations (i.e., for the covariate and for the repeated measures) and hence would be relatively tenuous.

The question becomes, then, how should the investigator go about estimating the power likely to be available for a design such as this? Our advice is to employ multiple modeling strategies (none of which will probably be optimal in and of themselves) irrespective of the final analytic option that will be ultimately employed.

Difference scores, for example, could be modeled relatively easily if the investigator had a reasonable idea of what the baseline-to-post intervention BP correlations were likely to be. Standardizing each score (and thus setting the standard deviation of each measure to 1.0) would permit the standard deviation of these difference scores to be assessed quite easily via Formula 10.3, in which case the appropriate ES values for the two main effects and the interaction could be calculated as previously discussed. The problem with this analysis, as indicated above, resides in the difficulty of estimating the correlation across tasks based upon these difference scores. The investigator could therefore conduct a pilot study to estimate this relationship or he/she could assume a relatively low correlation (certainly no more than $r = 0.40$) and then analyze the within subject main effect and interaction as already discussed.

Formula 10.3. Standard deviation of difference scores

$$\text{SD differences} = \sqrt{2 - 2r}$$

Another option would be to break the design into its most relevant pieces and estimate the power of these fractions. Thus, the power of the separate 2 (drug vs. placebo) × 2 (baseline BP vs. post task BP) interactions emanating from each task could be assessed. (Alternatively, the power of separate two-group ANCOVAs could be estimated for each task employing the relevant baseline measure as the covariate.) Obviously this would be a very gross approach, but the investigator would be heartened if there were sufficient power for one of these analyses, which would indicate that the drug would at least be effective for one level of stress. The process of hypothesizing the ES values for each of these contrasts might also be enlightening, since if there were little difference in the ES values between the three tasks (and the pattern of means were in the same direction), then the investigator would definitely want to collapse the hypothesized means across tasks and assess the power of the resulting ES for the 2 (drug vs. placebo) × 2 (baseline vs. post task BP) interaction.

While power calculations such as these are admittedly imperfect, they can be invaluable in the design phase of an experiment. As one example, if the differences between ES values for the separate 2×2 task interactions did not seem to be large enough to result in statistical significance, then the investigator would most certainly want to ensure that there was sufficient power for the 2×2 interaction collapsed across tasks. If there were not, it would be *essential* to address this issue, either by increasing the sample size or perhaps by deleting (or altering) a task that might be dissipating the 2×2 collapsed (across tasks) interaction ES.

The results of a power analysis such as this should be reported exactly the way in which it was conducted. The investigator should, in other words, report the modeling procedures employed and the relevant power levels obtained along with a frank admission that the results were primarily heuristic in nature.

Obviously there are many design permutations whose power analysis has not been considered in this or the previous chapters. We recommend this same basic approach for dealing with them, however, which is (a) simply to select the model from the options that we have discussed that most closely matches the one that will actually be used and then estimate the power based upon this less complicated version, (b) to break the complex design into its constituent parts and assess the power of the more interesting of these, or (c) to opt for a less complex design in the first place. In most cases the first two options will produce an underestimate of the actual power available (or an overestimate of the sample size required), but usually the resulting imprecision will not be severe and *if the investigator explicitly states the basis of his/her calculations (which we always advocate)*, a funding agency or an IRB is exceedingly unlikely to object to the resulting approximation.

Summary

Techniques for extending the tables presented in previous chapters to more complex designs are presented. These include the calculation of power and sample size for main effects in factorial designs, extensions to higher level (i.e., three or more factor designs), modeling procedures for covariate and within subject correlations other than 0.40 and 0.60, and the addition of a covariate to a mixed design.

All of the processes discussed provide estimates rather than exact values, but as we have emphasized throughout this book, a power/sample size analysis is by definition an estimate anyway given the number of hypotheses and assumptions necessitated. We therefore advise the investigator to describe quite explicitly the assumptions and techniques employed in any such analysis, especially those involving the more complex scenarios discussed in this chapter, which will then give the research consumer a basis upon which to evaluate the results. We have found that sophisticated reviewers and colleagues appreciate the vantage point this additional information provides, even when they do not completely agree with some of the assumptions/hypotheses themselves.

Endnotes

1 When a blocking variable is employed in the original sampling design or in the random assignment process, it does not necessarily increase the power of the other terms since the error term is also increased along with the overall variance. When such a variable is identified after the data have been collected (or the inclusion of this variable does not increase the overall variance if identified a priori), then the error term will not be increased and power will be favorably affected. From a post hoc perspective, however, the investigator should take steps to ensure that any effect due to this blocking variable is not a chance occurrence.

2 When it is advantageous to calculate the power of a three-way interaction, a gross indicator of its ES can be obtained by subtracting the ES values of the two collapsed 2×2 interactions hypothesized to best reflect this higher level interaction.

Template 10.1. Two-factor power and sample size template

Step 1. Choose one of the following table shells and fill in the standardized means that are expected to make up the interior cells. (These means should be hypothesized prior to filling in the row and column totals.) This is done by subtracting the smallest cell mean from itself and all of the other cells, thereby providing a zero value for the lowest mean. All cell means are then divided by the estimated standard deviation to produce standardized means. Compute the grand means for each row and column and fill them in the appropriate "Mean" cells. (As an example, the grand mean for level A1 in the 2×2 shell would be computed by adding the B1 and B2 ES values and dividing by 2.)

2×2	B1	B2	Mean
A1			
A2			
Mean			

2×3	B1	B2	B3	Mean
A1				
A2				
Mean				

2×4	B1	B2	B3	B4	Mean
A1					
A2					
Mean					

3×3	B1	B2	B3	Mean
A1				
A2				
A3				
Mean				

3×4	B1	B2	B3	B4	Mean
A1					
A2					
A3					
Mean					

Step 2. Compute main effect ES values by simply subtracting the smallest row mean from the largest row mean and the smallest column mean from the largest column mean.

Main effect A ES = largest "A" row mean − smallest "A" row mean =
Main effect B ES = largest "B" row mean − smallest "B" row mean =

Step 3. Compute the interaction ES by following the instructions in Template 9.1.

Interaction ES =

Step 4. If no further adjustments to the ES values emanating from steps 2 and 3 are indicated, proceed to step 5. Otherwise, employ one of the procedures discussed in this chapter to adjust one or more of the ES values emanating from steps 2 and 3 prior to proceeding to step 5 if the design is to include:

(a) use of a covariate other than 0.40 or 0.60 for a between subject effect,
(b) use of a within subject *r* other than 0.40 or 0.60 for a within subject effect,
(c) use of a post hoc grouping variable or other between design element, or
(d) addition of a covariate to a within subject effect.

Step 5. Calculate the *N*/group for each main effect as follows:

(a) For main effects, locate the intersection between the *N*/cell row for each factor and the appropriate column associated with the design/factor column in Table 10.1.

(b) For the interaction, simply enter the *N*/cell or the desired power level.
N/group (main effect A) = or desired power =
N/group (main effect B) = or desired power =
N/cell (interaction A × B) = or desired power =

Step 6. Proceed to calculate the power/required sample size as dictated in previous chapters using the appropriate adjusted (step 4) or unadjusted (steps 2 and 3) ES values as appropriate.

Table 10.1. *N*/group for between subject main effects A and B

N/cell	Design									
	2×2		2×3		2×4		3×3		3×4	
	A	B	A	B	A	B	A	B	A	B
3	5	5	7	5	9	5	7	7	9	7
4	7	7	10	7	13	7	10	10	13	10
5	9	9	13	9	17	9	13	13	17	13
6	11	11	16	11	21	11	16	16	21	16
7	13	13	19	13	25	13	19	19	25	19
8	15	15	22	15	29	15	22	22	29	22
9	17	17	25	17	33	17	25	25	33	25
10	19	19	28	19	37	19	28	28	37	28
11	21	21	31	21	41	21	31	31	41	31
12	23	23	34	23	45	23	34	34	45	34
13	25	25	37	25	49	25	37	37	49	37
14	27	27	40	27	53	27	40	40	53	40
15	29	29	43	29	57	29	43	43	57	43
16	31	31	46	31	61	31	46	46	61	46
17	33	33	49	33	65	33	49	49	65	49
18	35	35	52	35	69	35	52	52	69	52
19	37	37	55	37	73	37	55	55	73	55
20	39	39	58	39	77	39	58	58	77	58
25	49	49	73	49	97	49	73	73	97	73
30	59	59	88	59	117	59	88	88	117	88
35	69	69	103	69	137	69	103	103	137	103
40	79	79	118	79	157	79	118	118	157	118
45	89	89	133	89	177	89	133	133	177	133
50	99	99	148	99	197	99	148	148	197	148
55	109	109	163	109	217	109	163	163	217	163
60	119	119	178	119	237	119	178	178	237	178
65	129	129	193	129	257	129	193	193	257	193
70	139	139	208	139	277	139	208	208	277	208
75	149	149	223	149	297	149	223	223	297	223
80	159	159	238	159	317	159	238	238	317	238
85	169	169	253	169	337	169	253	253	337	253
90	179	179	268	179	357	179	268	268	357	268
95	189	189	283	189	377	189	283	283	377	283
100	199	199	298	199	397	199	298	298	397	298

Table 10.2. ANCOVA conversion for effect sizes

r	Hypothesized ES																		
	0.20	0.30	0.35	0.40	0.45	0.50	0.55	0.60	0.65	0.70	0.75	0.80	1.00	1.25	1.50	1.75	2.00	2.50	3.00
0.200	0.20	0.31	0.36	0.41	0.46	0.51	0.56	0.61	0.66	0.71	0.77	0.82	1.02	1.28	1.53	1.79	2.04	2.55	3.06
0.225	0.21	0.31	0.36	0.41	0.46	0.51	0.56	0.62	0.67	0.72	0.77	0.82	1.03	1.28	1.54	1.80	2.05	2.57	3.08
0.250	0.21	0.31	0.36	0.41	0.46	0.52	0.57	0.62	0.67	0.72	0.77	0.83	1.03	1.29	1.55	1.81	2.07	2.58	3.10
0.275	0.21	0.31	0.36	0.42	0.47	0.52	0.57	0.62	0.68	0.73	0.78	0.83	1.04	1.30	1.56	1.82	2.08	2.60	3.12
0.300	0.21	0.31	0.37	0.42	0.47	0.52	0.58	0.63	0.68	0.73	0.79	0.84	1.05	1.31	1.57	1.83	2.10	2.62	3.14
0.325	0.21	0.32	0.37	0.42	0.48	0.53	0.58	0.63	0.69	0.74	0.79	0.85	1.06	1.32	1.59	1.85	2.11	2.64	3.17
0.350	0.21	0.32	0.37	0.43	0.48	0.53	0.59	0.64	0.69	0.75	0.80	0.85	1.07	1.33	1.60	1.87	2.14	2.67	3.20
0.375	0.22	0.32	0.38	0.43	0.49	0.54	0.59	0.65	0.70	0.76	0.81	0.86	1.08	1.35	1.62	1.89	2.16	2.70	3.24
0.400	0.22	0.33	0.38	0.44	0.49	0.55	0.60	0.65	0.71	0.76	0.82	0.87	1.09	1.36	1.64	1.91	2.18	2.73	3.27
0.425	0.22	0.33	0.39	0.44	0.50	0.55	0.61	0.66	0.72	0.77	0.83	0.88	1.10	1.38	1.66	1.93	2.21	2.76	3.31
0.450	0.22	0.34	0.39	0.45	0.50	0.56	0.62	0.67	0.73	0.78	0.84	0.90	1.12	1.40	1.68	1.96	2.24	2.80	3.36
0.475	0.23	0.34	0.40	0.45	0.51	0.57	0.63	0.68	0.74	0.80	0.85	0.91	1.14	1.42	1.70	1.99	2.27	2.84	3.41
0.500	0.23	0.35	0.40	0.46	0.52	0.58	0.64	0.69	0.75	0.81	0.87	0.92	1.15	1.44	1.73	2.02	2.31	2.89	3.46
0.525	0.23	0.35	0.41	0.47	0.53	0.59	0.65	0.70	0.76	0.82	0.88	0.94	1.17	1.47	1.76	2.06	2.35	2.94	3.52
0.550	0.24	0.36	0.42	0.48	0.54	0.60	0.66	0.72	0.78	0.84	0.90	0.96	1.20	1.50	1.80	2.10	2.39	2.99	3.59
0.575	0.24	0.37	0.43	0.49	0.55	0.61	0.67	0.73	0.79	0.86	0.92	0.98	1.22	1.53	1.83	2.14	2.44	3.06	3.67
0.600	0.25	0.38	0.44	0.50	0.56	0.63	0.69	0.75	0.81	0.88	0.94	1.00	1.25	1.56	1.88	2.19	2.50	3.13	3.75
0.625	0.26	0.38	0.45	0.51	0.58	0.64	0.70	0.77	0.83	0.90	0.96	1.02	1.28	1.60	1.92	2.24	2.56	3.20	3.84
0.650	0.26	0.39	0.46	0.53	0.59	0.66	0.72	0.79	0.86	0.92	0.99	1.05	1.32	1.64	1.97	2.30	2.63	3.29	3.95
0.675	0.27	0.41	0.47	0.54	0.61	0.68	0.75	0.81	0.88	0.95	1.02	1.08	1.36	1.69	2.03	2.37	2.71	3.39	4.07
0.700	0.28	0.42	0.49	0.56	0.63	0.70	0.77	0.84	0.91	0.98	1.05	1.12	1.40	1.75	2.10	2.45	2.80	3.50	4.20
0.725	0.29	0.44	0.51	0.58	0.65	0.73	0.80	0.87	0.94	1.02	1.09	1.16	1.45	1.81	2.18	2.54	2.90	3.63	4.36
0.750	0.30	0.45	0.53	0.60	0.68	0.76	0.83	0.91	0.98	1.06	1.13	1.21	1.51	1.89	2.27	2.65	3.02	3.78	4.54
0.775	0.32	0.47	0.55	0.63	0.71	0.79	0.87	0.95	1.03	1.11	1.19	1.27	1.58	1.98	2.37	2.77	3.16	3.96	4.75
0.800	0.33	0.50	0.58	0.67	0.75	0.83	0.92	1.00	1.08	1.17	1.25	1.33	1.67	2.08	2.50	2.92	3.33	4.17	5.00

Table 10.3. Repeated measures adjustments for values of r other than 0.40

r	Hypothesized ES																		
	0.20	0.30	0.35	0.40	0.45	0.50	0.55	0.60	0.65	0.70	0.75	0.80	1.00	1.25	1.50	1.75	2.00	2.50	3.00
0.200	−0.04	−0.05	−0.06	−0.07	−0.08	−0.09	−0.10	−0.10	−0.11	−0.12	−0.13	−0.14	−0.17	−0.21	−0.26	−0.30	−0.34	−0.43	−0.52
0.225	−0.03	−0.05	−0.05	−0.07	−0.07	−0.08	−0.09	−0.09	−0.10	−0.10	−0.12	−0.12	−0.15	−0.19	−0.24	−0.27	−0.31	−0.39	−0.46
0.250	−0.03	−0.04	−0.05	−0.06	−0.06	−0.07	−0.07	−0.08	−0.09	−0.09	−0.10	−0.11	−0.14	−0.17	−0.21	−0.24	−0.27	−0.34	−0.41
0.275	−0.03	−0.04	−0.04	−0.05	−0.05	−0.06	−0.06	−0.07	−0.08	−0.08	−0.09	−0.09	−0.12	−0.14	−0.18	−0.20	−0.23	−0.29	−0.35
0.300	−0.02	−0.03	−0.03	−0.04	−0.04	−0.05	−0.05	−0.05	−0.06	−0.06	−0.07	−0.07	−0.09	−0.12	−0.15	−0.17	−0.19	−0.24	−0.28
0.325	−0.02	−0.02	−0.02	−0.03	−0.03	−0.04	−0.04	−0.04	−0.05	−0.05	−0.06	−0.06	−0.07	−0.09	−0.11	−0.13	−0.15	−0.19	−0.22
0.350	−0.01	−0.02	−0.02	−0.02	−0.02	−0.03	−0.03	−0.03	−0.03	−0.03	−0.04	−0.04	−0.05	−0.06	−0.08	−0.09	−0.10	−0.13	−0.15
0.375	−0.01	−0.01	−0.01	−0.01	−0.01	−0.02	−0.01	−0.01	−0.02	−0.01	−0.02	−0.02	−0.03	−0.03	−0.04	−0.05	−0.05	−0.07	−0.08
0.400	0.00	0.00	0.00	0.00	0.00	0.00	0.00	0.00	0.00	0.00	0.00	0.00	0.00	0.00	0.00	0.00	0.00	0.00	0.00
0.425	+0.00	+0.01	+0.01	+0.01	+0.01	+0.01	+0.02	+0.02	+0.02	+0.02	+0.02	+0.03	+0.03	+0.04	+0.04	+0.05	+0.06	+0.07	+0.09
0.450	+0.01	+0.01	+0.02	+0.02	+0.03	+0.02	+0.03	+0.04	+0.04	+0.04	+0.04	+0.05	+0.06	+0.08	+0.08	+0.10	+0.12	+0.14	+0.18
0.475	+0.02	+0.02	+0.03	+0.03	+0.04	+0.04	+0.05	+0.06	+0.06	+0.07	+0.07	+0.07	+0.09	+0.12	+0.13	+0.16	+0.18	+0.22	+0.27
0.500	+0.02	+0.03	+0.04	+0.05	+0.06	+0.06	+0.07	+0.08	+0.08	+0.09	+0.09	+0.10	+0.12	+0.16	+0.18	+0.21	+0.25	+0.31	+0.37
0.525	+0.03	+0.05	+0.06	+0.06	+0.07	+0.08	+0.09	+0.10	+0.10	+0.12	+0.12	+0.13	+0.16	+0.20	+0.24	+0.28	+0.32	+0.40	+0.48
0.550	+0.04	+0.06	+0.07	+0.08	+0.09	+0.10	+0.11	+0.12	+0.13	+0.14	+0.15	+0.16	+0.20	+0.25	+0.30	+0.35	+0.40	+0.50	+0.60
0.575	+0.05	+0.07	+0.09	+0.09	+0.11	+0.12	+0.13	+0.15	+0.16	+0.17	+0.18	+0.20	+0.24	+0.31	+0.36	+0.42	+0.49	+0.60	+0.73
0.600	+0.06	+0.08	+0.10	+0.11	+0.13	+0.14	+0.16	+0.18	+0.19	+0.21	+0.22	+0.23	+0.29	+0.37	+0.43	+0.51	+0.58	+0.72	+0.87
0.625	+0.07	+0.10	+0.12	+0.13	+0.15	+0.17	+0.19	+0.21	+0.22	+0.24	+0.25	+0.28	+0.34	+0.43	+0.51	+0.60	+0.69	+0.85	+1.03
0.650	+0.08	+0.12	+0.14	+0.16	+0.18	+0.20	+0.22	+0.24	+0.26	+0.28	+0.30	+0.32	+0.40	+0.50	+0.60	+0.70	+0.80	+1.00	+1.20
0.675	+0.09	+0.14	+0.16	+0.18	+0.21	+0.23	+0.25	+0.28	+0.30	+0.33	+0.35	+0.37	+0.46	+0.58	+0.69	+0.81	+0.93	+1.16	+1.39
0.700	+0.11	+0.16	+0.19	+0.21	+0.24	+0.26	+0.29	+0.33	+0.35	+0.38	+0.40	+0.43	+0.54	+0.67	+0.80	+0.94	+1.07	+1.33	+1.61
0.725	+0.12	+0.18	+0.22	+0.24	+0.28	+0.30	+0.34	+0.37	+0.40	+0.43	+0.46	+0.50	+0.62	+0.77	+0.92	+1.08	+1.23	+1.54	+1.85
0.750	+0.14	+0.21	+0.25	+0.28	+0.32	+0.35	+0.39	+0.43	+0.46	+0.50	+0.53	+0.57	+0.71	+0.89	+1.06	+1.24	+1.42	+1.77	+2.13
0.775	+0.16	+0.24	+0.29	+0.32	+0.37	+0.40	+0.45	+0.49	+0.53	+0.58	+0.61	+0.66	+0.82	+1.03	+1.22	+1.43	+1.64	+2.04	+2.45
0.800	+0.19	+0.28	+0.33	+0.37	+0.43	+0.47	+0.52	+0.57	+0.61	+0.67	+0.71	+0.76	+0.95	+1.19	+1.41	+1.65	+1.89	+2.36	+2.84

Table 10.4. Effect size based on r and r^2

ES	r	r^2
0.10	0.05	0.00
0.15	0.07	0.01
0.20	0.10	0.01
0.25	0.12	0.02
0.30	0.15	0.02
0.35	0.17	0.03
0.40	0.20	0.04
0.45	0.22	0.05
0.50	0.24	0.06
0.55	0.27	0.07
0.60	0.29	0.08
0.65	0.31	0.10
0.70	0.33	0.11
0.75	0.35	0.12
0.80	0.37	0.14
0.85	0.39	0.15
0.90	0.41	0.17
0.95	0.43	0.18
1.0	0.45	0.20
1.1	0.48	0.23
1.2	0.51	0.26
1.3	0.54	0.30
1.4	0.57	0.33
1.5	0.60	0.36
1.6	0.62	0.39
1.7	0.65	0.42
1.8	0.67	0.45
1.9	0.69	0.47
2.0	0.71	0.50
2.25	0.75	0.56
2.50	0.78	0.61
2.75	0.81	0.65
3.00	0.83	0.69

11 Other power analytic issues and resources for addressing them*

While we have attempted to be as comprehensive in our treatment of power analytic concepts as possible in the space available to us, there are a number of issues that we have not been able to address. The purpose of this chapter, therefore, is to touch briefly on some of these issues and provide suggestions for what we consider to be the most useful resources available for investigators who (a) find themselves in need of conducting a more specialized power analysis than the tables in this book permit or (b) simply want to know more about the area.

Fortunately, there are a number of books, computer programs, and websites that can be quite helpful in these regards. While we do not aspire to be exhaustive in our treatment, we will mention those that we believe will be most likely to be of benefit or interest to the readers of this particular book.

Specialized books on power

Cohen, J. (1977). *Statistical power for the behavioral sciences* (2nd edn). New York: Academic Press. Reprinted by Laurence Erlbaum, Mahwah, NJ, 1988.

This is the best and most comprehensive text ever written on the topic. The book requires a bit more work than the present one on the part of the reader to use, but it is not exclusively geared toward experiments as ours is and covers a number of topics which we do not discuss including power/sample size tables for correlational procedures such as the Pearson *r*, multiple correlation, and χ^2.

Lipsey, M.W. (1990). *Design sensitivity: statistical power for experimental research.* Newbury Park, CA: Sage.

This is a truly excellent book which anyone contemplating the conduct of an experiment should own, although it is now rather difficult to obtain. The author makes the crucial point that power is dependent upon

* The publisher has used its best endeavors to ensure that the URLs for external websites referred to in this book are correct and active at the time of going to press. However, the publisher has no responsibility for the websites and can make no guarantee that a site will remain live or that the content is or will remain appropriate.

far more than sample size and that perhaps the most important factor of all is the care with which a trial is designed. Its treatment of power is extremely accessible and the case it makes for the importance of the topic in empirical research is truly compelling. It is an especially good resource on the statistical and conceptual underpinnings of the effect size concept.

Kraemer, H.C. & Thiemann, S. (1987). *How many subjects?: statistical power analysis in research.* Newbury Park, CA: Sage.

This brief text presents a single "master table" designed to be applicable to a wide range of designs and statistical procedures. Its unifying theory is based upon the intraclass correlation test, which allows the use of the same ES metric regardless of the design (the lack of which the authors use to criticize Cohen's text, although they, like everyone else with an interest in power, acknowledge the magnitude of his contribution to the field).

Murphy, K.R. & Myors, B. (1998). *Statistical power analysis: a simple and general model for traditional and modern hypothesis tests.* Mahwah, NJ: Lawrence Erlbaum.

This text is similar to Kraemer & Thiemann (1987) in the sense that it uses a general approach to estimating power and accordingly offers a single "one-stop *F* table" designed to fit a wide variety of designs. The book approaches power from the perspective of the general linear model. The authors make a number of excellent points regarding power (e.g., "When power is less than .50 and researchers are certain (or virtually certain) that H_0 is wrong, the test is more likely to yield a wrong answer than a right one. More to the point, the test cannot possibly produce new and useful knowledge; it can only be misleading (p. 81)"). The authors also believe that testing the null hypothesis in general is relatively old fashioned since most interventions are conceptualized, not on the basis that they have no effect at all, but that they have at least a small effect, hence they make the case that the ES should be formulated based upon the distance between this hypothesized (even if trivial) effect rather than zero.

General statistical texts of interest

In addition to these texts, many statistical textbooks provide relatively extensive treatments of power. There are so many, in fact, that we cannot name them all nor can we pretend even to be familiar with them all. Some of our favorites include the following.

Keppel, G. (1991). *Design and analysis: a researcher's handbook* (3rd edn). Englewood Cliffs, NJ: Prentice-Hall.

While not providing tables per se, this book does include clear computational approaches for most analysis of variance procedures.

Maxwell, S. & Delaney, H. (1990). *Designing experiments and analyzing data: a model comparison perspective.* Belmont, CA: Wadsworth.

This actually may be more comprehensive than Keppel with respect to certain designs.

Cohen, J. & Cohen, P. (1983). *Applied multiple regression/correlation analysis for the behavioral sciences.* Mahwah, NJ: Lawrence Erlbaum.

This classic probably provides more detail on the power of correlational procedures (e.g., power and sample size for increments to R^2) than almost any other source.

Power software

All the power analytic techniques presented in this book are available free of charge in a computer program by Bausell and Franaszczuk entitled *Power analysis for experimental research* which can be obtained by email from bbausell@compmed.umm.edu.

There are also a large number of commercial power programs and free software available, often downloadable directly from the web. In addition, the journal literature is replete with descriptions of free programs and a number of journals routinely provide software reviews.

We did not have the resources to purchase any sizable number of the former nor the time to explore the latter in any detail, although Len Thomas and Charles J. Krebs (1997) have provided a Herculean review of 29 such programs which included "13 stand-alone power and sample size programs, 11 general purpose statistics packages with built-in power capabilities, 2 programs that deal only with determining sample size, and 3 specialized power programs of interest to ecologists (the authors' substantive area of expertise)." They also provide separate, informative websites with more information (www.forestry.ubc.ca/conservation/power/ and www.bcu.ubc.ca/~krebs/power.html), the latter of which allows the downloading of the actual review article, which we recommend. There have been a number of other very excellent reviews of power analytic software in a variety of journals (e.g., Goldstein, R. (1989). Power and sample size via MS/PC-DOS computers. *The American Statistician*, **43**, 253-260; also Iwane *et al.*'s (1997) review of specialized software using survival data), but this one is relatively up-to-date and quite comprehensive.

Before discussing this review, we should mention that power programs can be quite helpful to the investigator for some of the more specialized applications for which neither we, nor any other source of which we are aware, provide simple-to-use tables. They also provide an option that tables cannot: the ability (a) to generate power and sample size estimates for *any* alpha level (which is especially helpful in experiments designed to establish bio-equivalence), (b) to allow the investigator to present his/her results graphically, and (c) to provide sample text for the reporting of the results produced.

Before discussing Thomas and Krebs' much more comprehensive review, we will mention our more idiosyncratic impressions of the two programs with which we have the most experience. The first of these we found easier to use, probably because it is largely based upon Jacob Cohen's work and he remains a co-author; the second requires more expertise on the part of the user, but also covers a wider range of statistical procedures.

Borenstein, M., Rothstein, H., Cohen, J., Schoenfeld, D., Berlin, J., & Lakatos, E. (2001). *Power and precision: a computer program for statistical power analysis and confidence intervals.* Englewood, NJ: Biostat.

As mentioned, this is a relatively easy-to-use program that covers a wide range of ANOVA/ANCOVA designs, non-parametric procedures such as the McNemar test and χ^2, as well as both linear and logistic regression. In addition there is a section on survival analysis and equivalence tests that may prove helpful.

Elashoff, J.D. (1999). *nQuery advisor.* Cork: Statistical Solutions.

This is a more biostatistically oriented package which is not quite as user friendly as *Power and precision* and, in our opinion, requires somewhat more statistical training to use effectively. We suspect that biostatisticians may prefer its approach, however, being more familiar with its input requirements. It is not quite as flexible with respect to the analysis of variance as we would prefer, however, and the parameters it requires for techniques such as logistic regression are sometimes unrealistic. The program does have a very interesting feature that enables the user actually to input sample data into a data entry screen to allow the investigator to estimate some of these parameters.

Thomas, L. & Krebs, C. (1997). A review of statistical power analysis software. *Bulletin of the Ecological Society of America*, **78**, 126-139. Interestingly, these authors picked five general purpose power programs that they found the easiest to use and both *Power and precision* and *nQuery advisor* were among these. The other three were:

(1) PASS, NCSS Statistical Software, 329 North 1000 East, Kaysville, UT 84037 (www.ncss.com). This was ranked as easiest to use by the authors' graduate students.
(2) Stat Power, Statistical Design Analysis Software, P.O. Box 12734, Portland, OR 97212 (email: QEISys@aol.com).
(3) GPOWER, which is available on the web at www.psychologie.uni-tier.de:8000/projects/gpower.html.

Other programs that were reviewed and judged to be potentially useful although not quite as user friendly (at least at the time of this review) included:

(1) PC-Size that calculates power or sample sizes for a restricted number of tests (shareware available at ftp.simtel.net/publ/simtelnet/msdos/statstcs/size102.zip).
(2) PowerPack, which is command driven and programmable (ftp.stat.uiowa.edu/pub/rlenth/Powerpack).
(3) PowerPlant, which is free and was judged to have considerable potential (CSIRO Biometrics Unit, e-mail: biometrics@ccmar.csiro.au).
(4) N, a commercial program that calculates *N* but not power (SciTech International, Inc., 2525 North Elston Avenue, Chicago, IL 60647-3011).

In addition to this 1997 review, Len Thomas routinely updates available software at the following address: www.forestry.ubc.ca/conservation/power, where he lists 58 software packages (along with addresses at which they can be obtained) under six categories: (1) general purpose statistical software that contain power analysis routines, (2) general purpose statistical software that can be used to calculate power (i.e., calculate non-central distributions), (3) specialized statistical software that contains power analysis routines, (4) standalone power analysis software, (5) stand-alone power analysis software for specialized applications, and (6) packages that calculate sample size but not power. We all owe a considerable debt to Professor Thomas for his unselfish service in making this resource available to the professional community.

The journal literature

The journal literature on power is much too voluminous to attempt to review in a single chapter. What we will attempt to do here, therefore, is to mention a *sample* of those articles that (a) discuss important topics in more detail than we have done in this book, (b) cover issues that we did not because of our focus on experiments primarily involving continuous outcome variables, and/or (c) include specialized power/sample size tables/formulas that we feel may supplement the ones presented in the present book.

It is worth repeating the fact that this list is by no means complete. We are personally aware of many other very thoughtful journal articles that make a decided contribution to the field but are not mentioned here because (a) we felt that they did not apply directly to the everyday conduct of empirical research or (b) they were too technically oriented or discussed a topic too specialized to be helpful to practicing researchers.

For those individuals who desire more technical and specialized treatments of the topic, there are a number of journals that routinely publish papers on power including *Biometrics*, *Controlled Clinical Trials*, *Journal of the American Statistical Association*, *Journal of Educational Statistics*, *The American Statistician*, and *Statistics in Medicine* among others.

To facilitate the use of this brief bibliography, we have attempted to categorize the references by topic area and indicated (via an asterisk) those papers dealing with specific statistical procedures that contain power/sample size tables *deemed to contain sufficient parameters to be helpful in the actual design of a study* (those that do not present tables usually present formulas for computing the power or sample size associated with the relevant statistical procedure). Prior to presenting this list, however, we would like to recommend a few more general treatments of the subject that we feel would be worthwhile for anyone interested in the subject to retrieve and read.

Conceptual/philosophical issues

Cascio, W.F. & Zedeck, S. (1983). Open a new window in rational research planning: adjust alpha to maximize statistical power. *Personnel Psychology*, **36**, 517–526.

After discussing ways to increase statistical power, the authors argue that a minimum useful effect size should be specified and the power analysis performed on that. They go further, however, and suggest that the alpha level be adjusted based upon the relative costs of making a false positive or false negative error. They also advise that the prior probability based upon past research should be taken into account (advice which has largely been ignored in practice) except for equivalence trials.

Cohen, J. (1990). Things I have learned (so far). *American Psychologist*, **45**, 1304–1312.

From the author's abstract: "This is an account of what I have learned (so far) about the application of statistics to psychology and the other sociobiomedical sciences. It includes the principles 'less is more' (fewer variables, more highly targeted issues, sharp rounding off), 'simple is better'. . . . I have also learned the importance of power analysis and the determination of just how big (rather than how statistically significant) are the effects we study. Finally, I have learned that there is no royal road to statistical induction, that the informed judgment of the investigator is the crucial element in the interpretation of data, and that things take time."

Cohen, J. (1992). A power primer. *Psychological Bulletin*, **112**, 150–159.

Hypothesizing that one possible reason for the continued neglect of power analysis might be the inaccessibility of resources on the topic, Professor Cohen attempts once more to remedy this problem with two very useful and succinct tables, one summarizing the different ES indices and one presenting the power of the most common statistical tests, ES values, and significance levels.

Cohen, J. (1994). The earth is round ($p < .05$). *American Psychologist*, **49**, 997–1003.

This is a lightly written paper that contains a rare commodity: wisdom. Basically Cohen says that while the null hypothesis test is not perfect, there is no real substitute for it. Instead he implies that we should use a little common sense, improve the quality of our data, and report confidence intervals in addition to the *p*-values.

Lachin, J.M. (1981). Introduction to sample size determination and power analysis for clinical trials. *Controlled Clinical Trials*, **2**, 93–113.

This is a helpful introductory article with a specific focus on biomedical clinical trials that both makes the case for the importance of performing a power analysis and provides a number of specific guidelines for accomplishing same.

Power/sample size of specific statistical procedures

ANOVA/ANCOVA

*Keselman, H.J. & Keselman, J.D. (1987). Type I error control and the power to detect factorial effects. *British Journal of Mathematical and Statistical Psychology*, **40**, 196–208.

Muller, K.E. & Burton, C.N. (1989). Approximate power for repeated-measures ANOVA lacking sphericity. *Journal of the American Statistical Association*, **84**, 549–555.

*Tiku, M.L. (1967). Tables of the power of the *F*-test. *Journal of the American Statistical Association*, **62**, 525–539.

Attrition (loss of subjects)

*Palta, M. & McHugh, R. (1970). Adjusting for losses to follow-up in sample size determination for cohort studies. *Journal of Chronic Diseases* (now *Journal of Clinical Epidemiology*), **32**, 315–316.

Bio-equivalence

Hauschke, D., Kieser, M., Diletti, E., & Burke, M. (1999). Sample size determination for proving equivalence based on the ratio of two means for normally distributed data. *Statistics in Medicine*, **18**, 93–105.

Makuch, R. & Johnson, M.F. (1986). Some issues in the design and interpretation of "negative" clinical studies. *Archives of Internal Medicine*, **146**, 986–989.
Makuch, R. & Simon, R. (1978). Sample size requirements for evaluating a conservative therapy. *Cancer Treatment Reports*, **62**, 1037–1040.
Schuirmann, D.J. (1987). A comparison of the two one-sided tests procedure and the power approach for assessing the equivalence of average bioavailability. *Journal of Pharmacokinetics and Biopharmaceutics*, **15**, 657–680.

Censored data and attrition/loss to follow-up

⋆Palta, M. & McHugh, R. (1979). Adjusting for losses to follow-up in sample size determination for cohort studies. *Journal of Chronic Diseases* (now *Journal of Clinical Epidemiology*), **12**, 315–326.
Palta, M. & McHugh, R. (1980). Planning the size of a cohort study in the presence of both losses to follow-up and non-compliance. *Journal of Chronic Diseases* (now *Journal of Clinical Epidemiology*), **13**, 501–512.
Schumacher, M. (1981). Power and sample size determination in survival time studies with special regard to the censoring mechanism. *Methods of Information in Medicine*, **20**, 110–115.
⋆Wu, M.C. (1988). Sample size for comparison of changes in the presence of right censoring caused by death, withdrawal, and staggered entry. *Controlled Clinical Trials*, **9**, 32–46.

χ^2 (also see Cohen, 1977)

Lachin, J.M. (1977). Sample size determination for rxc comparative trials. *Biometrics*, **33**, 315–324.

Correlation techniques

Tattner, M.H. & O'Leary, B.S. (1980). Sample sizes for specified statistical power in testing for differential validity. *Journal of Applied Psychology*, **65**, 127–134.

Dunnett multiple comparison procedure

Liu, W. (1997). On sample size determination of Dunnett's procedure for comparing several treatments with a control. *Journal of Statistical Planning and Inference*, **62**, 25–261.

Group means

⋆Barcikowski, R.S. (1981). Statistical power with group means as the unit of analysis. *Journal of Educational Statistics*, **6**, 267–285.

★Overall, J.E., Hollister, L.E., & Dalal, S.N. (1967). Psychiatric drug research: sample size requirements for one vs. two raters. *Archives of General Psychiatry*, **16**, 152–161.

Group sequential designs

★Case, L.D., Morgan, T.M., & Davis, C.E. (1987). Optimal restricted two-stage designs. *Controlled Clinical Trials*, **8**, 146–156.
★Lim, K. & Demets, D.L. (1992). Sample size determinations for group sequential clinical trials with immediate response. *Statistics in Medicine*, **11**, 1391–1399.
★Pasternack, P.S. (1981). Sample sizes for group sequential cohort and case–control study designs. *American Journal of Epidemiology*, **113**, 182–191.
★Wieand, S. & Therneau, T. (1987). A two-stage design for randomized trials with binary outcomes. *Controlled Clinical Trials*, **8**, 20–28.

Logistic regression

★Hsieh, F.Y. (1989). Sample size tables for logistic regression. *Statistics in Medicine*, **8**, 795–802.
★Whittemore, A.S. (1981). Sample size for logistic regression with small response probability. *Journal of the American Statistical Association*, **76**, 27–32.

Logrank test

Lan, K.K.G. (1992). A comparison of sample size methods for the logrank statistic. *Statistics in Medicine*, **11**, 179–191.
Freedman, L.S. (1982). Tables of the number of patients required in clinical trials using the logrank test. *Statistics in Medicine*, **1**, 121–129.
★Rubinstein, L.V., Gail, M.H., & Santner, T.J. (1981). Planning the duration of a comparative clinical trial with loss to follow-up and a period of continued observations. *Journal of Chronic Diseases*, **34**, 469–479.

Kappa

Block, D.A. & Kraemer, H.C. (1989). 2×2 kappa coefficients: measures of agreement or association. *Biometrics*, **45**, 269–287.

Multivariate analysis of variance

Stevens, J.P. (1980). Power of the multivariate analysis of variance tests. *Psychological Bulletin*, **88**, 728–737.

A table is presented that enables the estimation of power for the two-group MANOVA. Tables of from three to five groups are presented in the author's 1986 text:

Stevens, J.P. (1986). *Applied multivariate statistics for the social sciences.* Hillsdale, NJ: Lawrence Erlbaum.

Non-central *t* and *F*

Owens, D.B. (1965). A special case of a bivariate non-central *t*-distribution. *Biometrika*, **52**, 437–446.

Non-parametric procedures

Noether, G.E. (1987). Sample size determination for some common non-parametric tests. *Journal of the American Statistical Association*, **82**, 645–647.

Relative risks

*Blackwelder, W.C. (1993). Sample size and power for prospective analysis of relative risk. *Statistics in Medicine*, **12**, 691–698.
This gives a table for very small risks such as for vaccine trials.
*Walter, S.D. (1977). Determination of significant relative risks and optimal sampling procedures in prospective and retrospective comparative studies of various sizes. *American Journal of Epidemiology*, **105**, 387–397.

Reliability

*Feldt, L.S. & Ankenmann, R.D. (1998). Appropriate sample size for comparing alpha reliabilities. *Applied Psychological Measurement*, **22**, 170–178.

Survival endpoints

Shih, J.H. (1996). Sample size calculation for complex clinical trials with survival endpoints. *Controlled Clinical Trials*, **16**, 395–407.

Web based resources

Despite being well aware of the exponential growth of information on the web, we were still somewhat surprised by the wealth of information on power located thereon. As with the other surveys of resources presented here, this one is idiosyncratic and not particularly systematic, partly because the searcher (first author) is not an accomplished web navigator. With all of this said, one starting point is the impressive index of statistical pages in general compiled by John C. Pezzullo.

members.aol.com/johnp71/javastat.html

The site is described as a project that "represents an ongoing effort to develop and disseminate statistical analysis software in the form of web pages." It contains interactive statistical pages, links to other resources, and free software (among other things) relevant to power. Specifically, it offers access to the following power aids:

(1) DQO-PRO, which is a sample-size calculator capable of "determining the rate at which an event occurs (confidence levels versus numbers of false positive or negative conclusions), determining an estimate of an average within a tolerable error (given the standard deviation of individual measurements), and determining the sampling grid necessary to detect 'hot spots' of various assumed shapes,"
(2) logistic regression power calculation with a continuous exposure variable and an additional continuous covariate.

ebook.stat.ucla.edu

The UCLA statistics department provides this site that contains everything from a history of statistics to online statistical consulting. It also contains two power calculators:

(1) Power Calculator written by Jason Bond which provides (a) power or sample size for one- and two-sample studies, equal and unequal variances, raw scores and lognormal transformation, (b) the same information for exponential, binomial, and Poisson distributions, and (c) power and sample size calculations for the correlation coefficient,
(2) Sample Size Calculator which computes the sample size needed for confidence and maximum allowable deviation for (a) means, (b) proportions, and (c) totals.

home.clara.net/sisa

This site contains online statistical computational capability and a power/sample size program for the *t*-test.

www.stat.uiowa.edu/~rlenth/Power/index.html

Entitled "Java Applets for Power and Sample Size," this page (Russell Lenth, University of Iowa) provides an online program capable of computing the power, sample size, and confidence intervals for means, proportions, and a balanced ANOVA. It is quite easy to use and the page also contains useful links to other sites.

Additional issues

Even with the references presented in this chapter, there will occasionally be unique power analytic aspects of an experiment or an analysis that are difficult to address – and which neither we nor the authors cited in this text have specifically addressed. Examples include violations of statistical assumptions (e.g., hetereogeneity of variance in the analysis of variance), the use of random rather than fixed effects in the analysis of variance, and the use of new, specialized, or idiosyncratic statistical procedures.

Our approach to addressing these issues is relatively simple and has been stressed throughout this book. First, when power analytic approaches have not been developed for a specific statistical procedure, we suggest that the investigator either (a) use a more conventional analytic approach or (b) model his/her power based upon the closest simple analog. When assumptions are more likely to be violated than not (e.g., heterogeneous covariate–dependent variable regression slopes), then the results can be modeled using a less restrictive approach (e.g., one-way ANOVA rather than ANCOVA). When neither of these options is feasible (e.g., a within subjects design must be used but it is suspected that sphericity will be violated and Muller & Burton (1989) above does not specifically address the issue at hand), then the formulas in the Technical appendix may be used to calculate power after adjusting for the hypothesized problems (e.g., by altering the degrees of freedom based upon the Greenhouse–Geisser correction factor). This may well mean that technical statistical assistance will be required, but this is always an appropriate (and recommended) option. Regardless of the approach taken, however, an explicit description of the modeling assumptions made and the power analytic methods used to arrive at the indicated power/sample size estimates should always be provided to enable the professional consumer to understand the rationale by which the results were obtained. The power analytic process, after all, is ultimately simply an imperfect (but extremely valuable) tool for helping to ensure the ultimate success of our experimental efforts.

Conclusion

The above resources are only a sample of those available and they represent only a single snapshot in time (August 11, 2001). We expect that these will increase exponentially over the next decade due to (a) the increased use of the web (and the unstinting willingness of statisticians and research methodologists to share their expertise with the professional community) and (b) the increasing availability of non-commericial, downloadable software. Perhaps eventually the availability of these resources, more than any other factor, will make a reality of Jacob Cohen's dream of a power analysis being performed at the design stage of every experiment.

Technical appendix

The purposes of this appendix are (a) to provide the detail necessary for the evaluation and/or replication of our results and (b) to allow an investigator to tailor our tables to other values and parameters that may be encountered in the conduct of certain experiments. Different formulas for the calculation of power will produce slightly different results but, as described elsewhere in this book, a power analysis is basically an estimating modeling procedure, hence the relatively minor differences accruing from the various formulas presented in the literature are usually trivial in nature.

Chapter 1. The conceptual underpinnings of statistical power

The most basic (and perhaps the most important) formula used in our considerations of power involves the effect size (ES) concept as it applies to the difference between two means:

Formula 1.1. The effect size between two independent means

$$ES = \frac{M_E - M_C}{SD_{pooled}}$$

Most of the power calculations presented in this book are based upon adjustments and extensions related to this two-group ES. Let us therefore begin with the statistical procedure used to evaluate the statistical significance of this unadjusted ES, which is the independent samples *t*-test. To calculate its power, it is necessary to specify three parameters: (1) the alpha level, (2) the *N*/group to be employed, and (3) the hypothesized ES which is expected to accrue as a result of the proposed experiment. Once these values have been specified, it is possible to compute the power available for any two-group study (involving independent means) based upon only two statistics: the critical value of the *t* needed for statistical significance (which of course is based upon the alpha level to be employed (the first parameter), and the projected *N*/group (the second parameter)), and the hypothesized

t which will occur if the third parameter (the hypothesized ES) is appropriate.

The critical value of t is available directly from a t-distribution (we employed the COMPUTE function in SPSS) using degrees of freedom equal to $2N-2$ (where N refers here, and elsewhere, to the number of subjects per group). As indicated above, the hypothesized t can be calculated directly from the hypothesized ES:

Formula 1.2. The independent samples *t*-test

$$t_{\text{hyp}} = \frac{\text{ES}_{\text{hyp}}}{\sqrt{2/N}}$$

Once these values are computed, an intuitively attractive means of computing the power of a t-test simply involves (a) subtracting the critical value of t from the hypothesized t, (b) "pretending" that the difference is a z-statistic, and (c) ascertaining what proportion of the normal curve is to the left of the z-score (the COMPUTE function of SPSS gives exact values for this parameter).

Formally, this process is depicted by the following formula:

Formula 1.3. An intuitively attractive (and surprisingly accurate) power formula

$$\text{power} = p\ (z \leq t_{\text{hyp}} - t_{\text{cv}})$$

Since the hypothesized t is not normally distributed (it is slightly skewed because it, for example, does not have a mean of zero), Hays (1973) suggests a correction term derived by Scheffe (1959), producing Formula 1.4 which we have used to construct many of the power tables throughout this book. In actuality, the differences between the values produced by this correction and Formula 1.3 are completely trivial in all but the smallest experiments and many texts do not bother to apply it, using instead some variant of Formula 1.3 (which provides perfectly acceptable results).

Formula 1.4. Scheffe's corrected power formula

$$\text{power} = p\left(z \leq \frac{t_{\text{hyp}} - t_{\text{cv}}}{\sqrt{1 + \dfrac{t_{\text{cv}}^2}{2\text{df}_t}}} \right)$$

Chapter 2. Strategies for increasing statistical power

Strategies 1 through 4 follow directly from Formulas 1.3 and 1.4 above in the sense that the size and statistical significance of the hypothesized t is directly influenced by N and the ES. Strategy 5 (employing as few groups as possible) is based upon the fact that multiple comparison procedures are necessary when three or more group means are employed. The fact that t_{cv} is increased (and the method by which this occurs) when it is derived from the Studentized range distribution is documented in the Chapter 6 heading below. Strategy 6 is based upon the fact that the use of a covariate or blocking variable adjusts the hypothesized ES upwards and the method in which this is computed is described in the section below devoted to Chapter 7. The computational procedures demonstrating strategy 7 (employing repeated measures) are presented under Chapter 8. Techniques for assessing the power of interactions are presented under the Chapter 9 heading.

Strategy 9: Employing measures which are sensitive to change. Table 2.10 was produced by a linear adjustment to the hypothesized ES based upon proposed increases in dependent variable sensitivity. Hence if the originally hypothesized ES was 0.50, and it was possible to construct a dependent variable that was 40% more sensitive to change from baseline to end-of-treatment, then the actual hypothesized ES would be 0.50×1.4 or 0.70.

Strategy 10: Employing reliable measures. The relationship between increases in dependent variable reliability is based upon the classical measurement formula that quantifies the attenuating effect of reliability upon validity. To generate the values presented in Table 2.11, we first converted the hypothesized ES to an r via the following formula:

Formula 2.1. Relationship between r and the ES concept

$$r_{\text{hyp}} = \sqrt{\frac{\text{ES}^2}{\text{ES}^2 + 4}}$$

Once this is done, a formula presented by Nunally (1967, p. 219) and others can be used to adjust the hypothesized r (since the ES can be conceptualized as an instance of concurrent validity):

Formula 2.2. Relationship between reliability and validity

$$r_{\text{adj}} = r_{\text{hyp}} \sqrt{\frac{\text{Rel}_{\text{obt}}}{\text{Rel}_{\text{hyp}}}}$$

Where

(1) Rel_{obt} is the reliability actually obtained after collecting data, and
(2) Rel_{hyp} is the reliability the investigator expected to obtain.

Once accomplished, the adjusted *r* can be converted back to an adjusted ES:

Formula 2.3. Converting *r* to ES

$$\text{ES} = \frac{2r}{\sqrt{1 - r^2}}$$

This ES, then, becomes the ES that has been adjusted based upon *changes* or *discrepancies* in reliability between the reliability of the dependent variable upon which the hypothesized ES was based and the hypothesized reliability of the dependent variable that will be employed in the study being designed.

Strategy 11: Using direct rather than indirect dependent variables. Here again it is convenient to convert the hypothesized ES to a hypothesized *r* via Formula 2.1 and, once it is adjusted, convert it back again using Formula 2.3. The adjustment process for the hypothesized *r* is then given by the following formula:

Formula 2.4. Effects of causal distance between indirect and direct dependent variables upon *r*

$$r_{xy-\text{adj}} = r_{xy} \times r_{x,x1} \times r_{x1x2} \ldots$$

Where

(1) r_{xy} is the originally hypothesized ES (converted to *r*) between the IV and the DV of choice,
(2) $r_{x,x1}$ is the correlation between the DV of choice and the DV which is one step removed from this DV of choice,
(3) r_{x1x2} is the correlation between the DV which is one step removed with the DV which is two steps removed, and
(4) $r_{xy-\text{adj}}$ is the adjusted correlation between the IV and the DV at the final step in the causal chain.

Chapter 5. The paired *t*-test

The basic power formula for the paired *t*-test is the same as for its independent samples counterpart, except that the actual values of the following parameters change based upon differing degrees of freedom and the fact that the correlation between the paired correlations directly impacts the error term.

The hypothesized ES is adjusted based upon the projected correlation between paired observations:

Formula 5.1. ES adjustment based upon the correlation between paired observations

$$ES_{adj} = \frac{ES_{hyp}}{\sqrt{1-r}}$$

Where ES_{hyp} is the original hypothesized mean difference between the two paired groups (e.g., baseline vs. follow-up) divided by the pooled standard deviation of each group using Formula 1.1 and r is the projected correlation between pairs of observations; df now becomes the number of pairs minus 1 (i.e., $N-1$ rather than $2N-2$).

Thus, returning to our earlier example in which the ES_{hyp} was 0.50 and the N/group was 64, we obtained a difference between the t_{cv} (1.98 for 126 df) and the t_{hyp} (2.82) of 0.84, which when converted to a z yielded a power of 0.80 using Formula 1.4. Let us suppose now, however, that we had a pre–post design employing 64 subjects and the same hypothesized ES of 0.50, but that the r between our paired observations was 0.60.

As described above, both our hypothesized t and the critical value of t would change, although the hypothesized t will change the most dramatically because of the direct adjustment to the ES based Formula 5.1:

$$ES_{adj} = \frac{0.50}{\sqrt{1-0.60}} = 0.79$$

Which affects the hypothesized t as follows:

$$t_{hyp} = \frac{0.79}{\sqrt{2/64}} = 4.47$$

The t_{cv} is now computed based upon 63 df (instead of 126) and is reduced slightly to 2.00. Using Formula 1.3, the z corresponding to the difference between these two t values ($4.47-2.00$) is 2.47, which corresponds to the 99th percentile, which also means that the power has increased from 0.80 to 0.99 via the use of a paired t-test under the above conditions.

Chapter 6. One-way analysis of variance

Power of the *F*-ratio. For the power of the F-ratio, we used a formula provided by Laubscher (1960, Formula 6), which is simplified as follows:

Formula 6.1. Power of the between groups F-ratio

$$Z_{\text{power}} = \frac{\sqrt{2(\text{df}_{\text{num}} + \text{NCP}) - \dfrac{\text{df}_{\text{num}} + 2\text{NCP}}{\text{df}_{\text{num}} + \text{NCP}}} - \sqrt{(2\text{df}_{\text{num}} - 1)\dfrac{\text{df}_{\text{num}} F_{\text{cv}}}{\text{df}_{\text{den}}}}}{\sqrt{\dfrac{\text{df}_{\text{num}}\, F_{\text{cv}}}{\text{df}_{\text{den}}} + \dfrac{\text{df}_{\text{num}} + 2\text{NCP}}{\text{df}_{\text{num}} + \text{NCP}}}}$$

Where

(1) Z_{power} is the percentile value for power,
(2) df_{num} is the degrees of freedom for the numerator of the F-ratio (number of groups − 1),
(3) NCP is the nowcentraling parameter or $f^2N\,(\text{df}_{\text{num}} + 1)$,
(4) $f = \frac{1}{2}\, d$, where in this case d is the ES for the two most divergent means:

$$\frac{M_{\text{highest}} - M_{\text{lowest}}}{\text{SD}}$$

(5) df_{den} is the degree of freedom for the denominator of the F-ratio, $\text{df}_{\text{den}} = (\text{df}_{\text{num}} + 1)\,(N - 1)$, where N is the number of subjects per group, and
(6) F_{cv} is the critical value of F for df_{num} and df_{den}.

Since the significance, hence power, of F is dependent upon the pattern of means for a study of three or more groups, however, it is necessary to adjust this formula for the hypothesized pattern or spread of means. For this step, we rely upon formulas presented in Cohen (1977) that represent low, medium, and high dispersion patterns (which produce low, medium, and high power values) for F. These weights are multiplied by d (the largest pairwise ES in the experiment), thereby adjusting f in Laubscher's formula above as follows:

$$f = W_{\text{p}} d$$

where W_{p} is the weight for the hypothesized pattern of means.

The specific formulas for low, medium, and high mean dispersions are as follows:

Formula 6.2. Adjustment values for f based upon the dispersion pattern of means; K is the number of groups (3–5)

$$W\,(\text{low dispersion pattern}) = \sqrt{\frac{1}{2K}}$$

$$W \text{ (medium dispersion pattern)} = \frac{\sqrt{\dfrac{K+1}{3(K-1)}}}{2}$$

$$W \text{ (high dispersion pattern)} = \frac{\sqrt{K2-1}}{2K}$$

The weights themselves, for three-, four-, and five-group studies produced by these formulas are:

Groups	W_L	W_M	W_H
3	0.408	0.408	0.471
4	0.354	0.373	0.500
5	0.316	0.354	0.490

To illustrate the calculation of the power of F, then, let us assume the following parameters:

(1) four groups,
(2) N/group = 50,
(3) $ES_{hyp} = 0.50$ (i.e., the largest pairwise difference between means), and
(4) a high dispersion pattern for the four means.

The first step in the process is to adjust f based upon the hypothesized dispersion pattern, or $f = W_p d = 0.50 \times 0.5 = 0.25$. Next, the remaining values described above are computed as follows:

$df_{num} = \#\text{ groups} - 1 = 3$,
$N = 50$,
$f = 0.25$,
$NCP = 0.25^2 \times 50 \times 4 = 12.50$, and
$df_{den} = 4N - df_{num} - 1 = 196$.

Filling these values into Formula 6.1 produces a z of 1.044, which corresponds to a power value of 0.85 (i.e., the area of the normal curve to the left of this particular z-score).

It is important to note that the above formula is applicable to any F-ratio emanating from any design as long as (a) the degrees of freedom for the numerator and denominator (as well as the appropriate values built upon them) and (b) the ES are adjusted appropriately.

Multiple comparison procedures. We will now describe the rationale used for computing power for two multiple comparison procedures designed to protect the alpha level while permitting the statistical evaluation of all available pairwise contrasts. Both procedures employ the Studentized range statistic, for which distributions exist for experiments with different numbers of groups. The first, Tukey's HSD procedure, is designed to compensate based solely upon the number of groups involved, while the second, the Newman–Keuls procedure, is a more liberal procedure that is based upon how many group means separate a particular pairwise comparison.

Since both the Studentized range statistic (designated as q) and t, for which we have already demonstrated the computation of power, can be expressed in terms of the ES concept (see Formula 1.2 for the relationship between ES and t), it follows that the procedures used to produce the power tables for the independent samples t-test could be used for q (and thus for experiments involving different numbers of groups), if it were possible to adjust a pairwise ES based upon the appropriate Studentized range statistic rather than t.

Fortunately, a formula provided by Tukey (1953) designed to determine the mean difference between any two groups needed to achieve statistical significance (which we will designate as the critical value of the mean difference or MD_{cv}) in a multiple group study serves quite nicely for this purpose:

Formula 6.3. Critical value of mean differences based upon studentized range statistic

$$MD_{cv} = q\sqrt{\frac{MS_{error}}{N}}$$

At first glance this formula may not appear particularly helpful in an a priori consideration of power since the ANOVA term (MS_{error}) is not typically known until after the study data have been collected and analyzed. N, of course, can be specified in advance and q can be determined from the Studentized range distribution based upon N and the number of groups being employed (i.e., comparable to the degrees of freedom for the F-ratio).

As in our previous computational procedures for the overall F-ratio, however, if we assume a within group standard deviation of one (by standardizing the scores involved if nothing else), this sets the MS_{error} to 1.0. It also means that MD_{cv} automatically becomes ES_{cv} *based upon the Studentized range distribution* since the ES is the mean difference divided by the standard deviation and, when SD = 1, the ES *equals* the mean difference. Tukey's formula thus reduces to the following much more useful construction:

Formula 6.4. ES based upon the Studentized range statistic

$$\mathrm{ES}_{\mathrm{cv}} = q\sqrt{\frac{1}{N}}$$

Since we can easily calculate t_{cv} based upon $\mathrm{ES}_{\mathrm{cv}}$ using Formula 1.2:

$$t_{\mathrm{cv}} = \frac{\mathrm{ES}_{\mathrm{cv}}}{\sqrt{2/N}}$$

it follows that we now have the basic constituents to calculate the power of an MCP that employs the Studentized range distribution. In other words, what we have done in effect is to determine the critical value of t which will be necessary to achieve statistical significance for a pairwise comparison based upon any number of groups in an experiment. All we need do is calculate the hypothesized t and the corrected critical value of t between any two means that we wish to compare and then compute the available power based upon Formula 1.4.

Let us illustrate this process for the Tukey HSD procedure by employing the numerical example used to show how power is computed for the overall F-ratio in the previous section. There, it will be recalled, the relevant parameters were:

(1) N/group = 50,
(2) the number of groups is four, and
(3) $\mathrm{ES}_{\mathrm{hyp}}$ of the largest pairwise comparison is 0.50.

In order to employ either Formula 1.3 or Formula 1.4, we must calculate both the t_{hyp} and the t_{cv}. The t_{hyp} is calculated irrespective of the number of groups involved as follows:

$$t_{\mathrm{hyp}} = \frac{\mathrm{ES}_{\mathrm{hyp}}}{\sqrt{2/N}} = \frac{0.50}{\sqrt{2/50}} = 2.50$$

The t_{cv} is calculated based upon the q-distribution for four groups and the $\mathrm{df}_{\mathrm{den}}$ for the F-ratio of 196 (i.e., $\mathrm{df} = 4N - 4$), which in this case is 3.64. (This value can be obtained from many standard statistical texts or from distributions available in statistical packages such as SPSS or SAS.) Filling these values into Formulas 6.4 and 1.2, we obtain the following:

$$\mathrm{ES}_{\mathrm{cv}} = q\sqrt{\frac{1}{N}} = 3.64\sqrt{\frac{1}{50}} = 0.514$$

and,

$$t_{cv} = \frac{ES_{cv}}{\sqrt{2/N}} = \frac{0.514}{\sqrt{2/50}} = 2.57$$

Filling these values into Formula 1.3 then produces a power estimate of 0.46 and the required sample size to achieve a power level of 0.80 is 92 subjects per group. The power/sample size requirements based upon the Studentized range statistic are, of course, only relevant for the single pairwise comparison involving an ES of 0.50 and it is only relevant for Tukey's HSD. The Newman–Keuls procedure employs different q-statistics based upon how many means are hypothesized to fall between the two groups upon which the ES is based. Thus, assuming the ES remains at 0.50, if the two groups upon which this mean is based are separated only by one additional mean, the required sample size drops from 92 to 82 and if no groups intervene between the pairwise ES of 0.50, the Newman–Keuls procedure allows statistical significance to be achieved between these two groups 80% of the time using only 64 subjects per group. (Note that, not coincidentally, this is the same result that would have accrued if an independent samples t-test were to be employed for this contrast.)

The Newman–Keuls MCP can, therefore, result in significant savings with respect to required sample sizes under certain circumstances, namely when the contrast of interest does not correspond to the largest possible pairwise ES in an experiment. (Power and sample size requirements are identical for the two MCPs when this is the case, because both the Newman–Keuls and the Tukey procedures employ identical Studentized range statistics.)

Because of the number of tables involved (and the fact that many statisticians disagree with the assumptions made by the Newman–Keuls procedure, considering it too liberal), we have provided power tables only for the Tukey HSD procedure, although we have provided sample size charts for *both* the Newman–Keuls and the Tukey HSD multiple comparison procedures.

Chapter 7. One-way analysis of covariance

The computation of power for the between subject analysis of covariance (ANCOVA) is exactly comparable to the between subject ANOVA except that (a) the presence of the covariate decreases the size of the error term and (b) one degree of freedom per covariate is lost therefrom.

For power analytic purposes it is more convenient to adjust the ES than the error term, the former of which can be done via the following formula:

Formula 7.1. Adjusting the ES based upon the presence of a covariate

$$\mathrm{ES}_{\mathrm{adj}} = \frac{\mathrm{ES}}{\sqrt{1 - r^2}}$$

Thus, let us return to our four-group ANOVA example above in which the following parameters were posited:

(1) the number of groups is four,
(2) the ES for the largest mean difference among groups is 0.50,
(3) there is a maximum spread among the four means (high dispersion pattern), and
(4) N/group = 50, *and the additional parameter needed for ANCOVA.*

These parameters, it will be remembered, resulted in a power estimate of 0.85 for the overall F-ratio. If a covariate were available for this study, however, an additional parameter would need to be posited which would be the correlation between this new variable and the dependent variable. Let us assume, therefore, that this value was hypothesized to be $r = 0.60$.

In applying Laubscher's formula (6.1), the following parameters would change: (a) $\mathrm{df}_{\mathrm{den}}$ (this will now be $4N-5$ or 195, instead of $4N-4$ or 196, which would have a completely minimal effect upon the overall power calculation) and (b) f, the multiple group ES adjusted for the effects of the covariate, which it will be remembered is equal to $d/2$, or the ES for the largest difference among the four means *adjusted* for the effects of the covariate. This latter adjustment, employing Formula 7.1, would produce the following result:

$$\mathrm{ES}_{\mathrm{adj}} = \frac{0.50}{\sqrt{1 - 0.6^2}} = 0.50/0.80 = 0.625$$

From this point on, using Laubscher's formula is the same, regardless of whether an ANCOVA or ANOVA F-ratio is being computed. First, for example, it is necessary to adjust the ES (d), which is now 0.625, based upon the hypothesized pattern of means. For the high dispersion pattern, this involves multiplying by 0.50 (which actually is no adjustment at all since no power is lost for this pattern) and $f = 0.50d$:

$$f = 0.50 \times 0.625 = 0.3125$$

The other values that will be needed for Formula 6.1 are as follows:

(1) $\mathrm{df}_{\mathrm{num}} = 3$,
(2) $N = 50$,

(3) $NCP = f^2\ N \times \text{groups} = 0.31^2 \times 50 \times 4 = 19.22$
(4) $df_{den} = 195$.

Filling these values into Formula 6.1 will produce a z of 1.84, which in turn corresponds to a power of 0.97 (i.e., the area of the normal curve to the left of a z of 1.84), which constitutes a substantive increase from the previously computed power of 0.85 for the F-ratio produced without the benefit of a covariate.

Multiple comparison procedures. The rationale for calculating the power of individual MCPs is exactly the same as the development presented earlier for the one-way analysis of variance except that here, again, the presence of the covariate changes a number of the relevant parameters: namely (a) each of the individual pairwise ES values must be adjusted for the covariate–dependent variable r via Formula 7.1 and (b) the df_{den} of the Studentized range statistic (q) must be adjusted for the one degree of freedom consumed by the covariate (which again is minimal except for the very smallest N values).

Power and sample size tables for ANCOVA. For convenience, the power and sample size tables for both the overall ANCOVA F-ratio and the multiple comparison procedures are constructed based upon the unadjusted ES (d). Thus, while the tables themselves employ the unadjusted ES, the power and sample size values were computed based upon the adjusted ES to avoid the necessity of the user to employ Formula 7.1 each time he/she wished to perform a power/sample size analysis for a design employing a covariate.

Chapter 8. One-way repeated measures analysis of variance

The power of the repeated measures (within groups) F-ratio is, not surprisingly, also computed by the use of the Laubscher formula (6.1) that was employed for both the one-way between subject ANOVA and the one-way between subject ANCOVA, although of course the df_{den} and the ES (f) were adjusted based upon the design and the correlation between the dependent observations. As in the independent samples ANOVA, f is based upon the largest pairwise difference among groups and, in the present case, is adjusted in the same manner as described for the paired t-test:

Formula 8.1. Within subject adjustment of the ES (d)

$$ES_{adj} = \frac{ES_{hyp}}{\sqrt{1 - r}}$$

where r is now the correlation between the repeated observations.

In way of illustration, let us use the same example and parameters employed in the ANCOVA discussion above, except now r will be used as the correlation among the repeated measures instead of the correlation between each subject's covariate and dependent variable score. These parameters, then, are:

(1) the number of groups is four,
(2) the ES for the largest mean difference among groupsis 0.50,
(3) there is a maximum spread among the four means (high dispersion pattern),
(4) N/group = 50 (which, of course, is the *same* 50 subjects represented in each of the four groups), *and the additional parameter needed for RM ANOVA*, and
(5) $r = 0.60$ between the matched group scores.

Our first step is to adjust d (which is the ES between the largest pairwise group difference) based upon the r of 0.60, which we have already done for an ES of 0.50 in computing the power of the paired t-test and found to be:

$$\mathrm{ES}_{\mathrm{adj}} = \frac{0.50}{\sqrt{1 - 0.60}} = 0.50/0.63 = 0.79$$

(Note that this is a considerably more dramatic ES adjustment than was observed using Formula 7.1.)

The next step is to convert this value to f by multiplying it by the appropriate weight presented in Chapter 6 for a four-group study with a high dispersion pattern (or 0.500) or: $0.79 \times 0.5 = 0.40$.

Once these two steps are completed, the appropriate values are filled into Formula 6.1, although it should be noted that $\mathrm{df}_{\mathrm{den}}$ employs the following formula:

$$\mathrm{df}_{\mathrm{den}} = (\#\ \text{groups} - 1) \times (N - 1) = (4 - 1) \times (50 - 1) = 3 \times 49 = 147$$

(which is considerably less than the 195 present for the between subject ANCOVA). These values, then, produce an estimated power in excess of 0.99 which is considerably higher than either the value obtained for the one-way between subject ANOVA (0.85) or the one-way ANCOVA (0.97), even though substantively fewer subjects were employed (i.e., 50 as opposed to 200).

Multiple comparison procedures. The rationale for calculating the power of individual MCPs is exactly the same as the development presented earlier for the one-way analysis of variance, except that the ES is adjusted for (a) the effects of the correlation among the matched observations and

(b) the df_{den} for the q-statistic is adjusted based upon the value used for the overall F-ratio discussed above. For our RM ANOVA example, the power for detecting a pairwise comparison involving 0.50 between two most extreme groups (employing the Tukey HSD procedure) would thus be 0.91 as compared to 0.46 for a between subject ANOVA and 0.70 for an ANCOVA (assuming a covariate–dependent variable r of 0.60).

The power and sample size tables. As with the ANCOVA, it is the adjusted ES values based upon Formula 8.1 that are actually represented in the columns presented in Chapter 8, although it is the unadjusted ES values that are actually presented for the reader's convenience. (In other words, while the user specifies an ES of 0.50 for the above analysis and locates the power at the intersection of this column value and the N/group row, the table itself has been constructed based upon an ES of 0.79 instead of 0.50.)

Chapter 9. Interaction effects for two-factor between and mixed analysis of variance

Since interaction effects deal with patterns among *cell* means that occur *independently* of the main effects that comprise them, their ES values must be computed independently thereof as well. The general formula for an interaction ES, however, is conceptually the same as for a t-test, although it is helpful to remember that an ANOVA ES (often designated as f) is interpreted on a different scale than is the case for the two-group ES that we use throughout this book. (Specifically, for one degree of freedom effects, $d = 2f$).

Conceptually, then, it is convenient to view the interaction ES in terms of the standard deviation of each cell's *interaction effect*, since we can (and always do) set the within cell standard deviation to 1.0:

Formula 9.1. Conceptual formula for an interaction ES

$$ES_{interaction} = \frac{\text{standard deviation of each cell's interaction effect}}{\text{within cell standard deviations}}$$

Prior to calculating the interaction effect's standard deviation, however, it is important to note that any given cell's contribution to the overall interaction term must first have the contribution due to each main effect factored out, which can be accomplished computationally as follows:

Formula 9.2. Individual cell interaction effect

$$\text{Cell interaction effect} = M_{cell} - M_{1st\ factor} - M_{2nd\ factor} + M_{total}$$

A very useful formula provided by Cohen (1977) can then be adapted to provide the standard deviation of the interaction effects (or the interaction ES):

Formula 9.3. ES for a two-factor interaction

$$ES_{interaction} = \sqrt{\frac{\Sigma(\text{interaction effects for each cell})^2}{(\#\text{rows})(\#\text{columns})}}$$

Which can be generalized to higher levels (e.g., a three-way interaction) as follows:

$$ES_{interaction} = \sqrt{\frac{\Sigma(\text{interaction effects for each cell})^2}{(\text{total } \# \text{ cells})}}$$

Remembering, that both of these formulas assume that the standard deviation within each cell is 1.0. That is:

$$ES_{interaction} = \frac{SD_{interaction}}{SD \text{ cells}} = SD_{interaction}$$

To illustrate numerically, let us consider a 2×2 example in which the following standardized means were hypothesized. (Standardized means from an interaction perspective are calculated by subtracting the smallest hypothesized unstandardized mean from each cell and then dividing by the pooled standard deviation.)

	Males (B1)	Females (B2)	Gender main effect
Experimental (A1)	1.00 (A1B1)	0.00 (A1B2)	0.50
Control (A2)	0.00 (A2B1)	0.00 (A2B2)	0.00
Treatment main effect	0.50	0.00	Total mean = 0.25

Filling these values into Formula 9.3, we obtain:

$$\text{Effect}_{A1B1}{}^2 = (1.0 - 0.5 - 0.5 + 0.25)^2 = +0.25^2 = 0.0625$$

$$\text{Effect}_{A1B2}{}^2 = (0.0 - 0.5 - 0.0 + 0.25)^2 = -0.25^2 = 0.0625$$

$$\text{Effect}_{A2B1}{}^2 = (0.0 - 0.0 - 0.5 + 0.25)^2 = -0.25^2 = 0.0625$$

$$\text{Effect}_{A2B2}{}^2 = (0.0 - 0.0 - 0.0 + 0.25)^2 = +0.25^2 = 0.0625$$

Total = 0.25

or,

$$ES_{interaction} = \frac{\sqrt{0.25/4}}{1} = 0.25$$

Which when converted to *d* (since we are dealing with only one degree of freedom and we employ *d* throughout this book), becomes $0.25 \times 2 = 0.50$.

This, in turn, encouraged us to suggest that all interactions be visualized in terms of collapsed 2×2 shells, thereby providing a more appealing conceptual basis for interpreting the interactions and the far simpler computational procedure for 2×2 interactions, or simply subtracting the ES values for each level of B at each level of A (or vice versa) and dividing the difference by 2 to express the ES in terms of *d*:

$$[(A1B1 - A1B2) - (A2B1 - A2B2)]/2 =$$
$$[(1.0 - 0.0) - (0.0 - 0.0)]/2 = 0.5$$

The *N*/group for a between subject interaction, then, is given in Formula 9.4:

Formula 9.4. Generalized formula for calculating *N*/group in ANOVA designs

$$N/\text{group} = \frac{df_{error}}{[df_{numerator} + 1]} + 1$$

Where the df_{error} is calculated via the following formula:

$$df_{error} = [(\#\ \text{cells}) \times (N/\text{cell})] - [\#\ \text{cells}]$$

The power tables for the designs presented in Chapter 9 are based upon these values of *N*/group, although for convenience sake the tabled values are presented as *N*/cell. The same basic ANOVA power formula (6.1) is employed here as in Chapter 6, except that the parameters employed (e.g., df_{error}, F_{cv}) are based upon the specific design employed (hence for the present 2×2 design with an *N*/cell of 25, the df_{error} would be 96, the critical value of F would be 3.94, the df_{num} would be one, and so forth). The *N*/group, here, would be $(4 \times 25) - 4$ or $96/2 = 48$.

The interpolated power could be estimated using either Table 4.1 (using the appropriate adjustments) or Table 9.1, although for convenience sake the interaction tables presented in Chapter 9 are based upon the *N*/cell, hence the power (0.70) is found at the intersection of the *N*/cell row of 25 and the ES = 0.50 column.

Between subject ANCOVA interactions. The same computational procedures were employed for the ANCOVA interaction tables as just discussed, although the df_{error} was reduced by one based upon the existence of a covariate and the interaction ES was reduced based upon the covariate–dependent variable correlations of 0.40 and 0.60 via Formula 7.1.

Mixed design two-factor interactions. Since the interaction in a mixed design is a within subject (or repeated measures) effect, the degrees of freedom for the error term (and hence the critical value of *F*) and the interaction ES (*f*) must be adjusted accordingly.

For convenience, all of the interaction tables, including mixed designs, are constructed based upon *N*/cell (instead of *N*/group) and the unadjusted ES expressed in terms of *d* (rather than the adjusted ES based upon *f* which is inputted into Formula 6.1).

Chapter 10. Power analysis for complex designs

The extensions to other designs discussed in this chapter are based upon previous formulas presented earlier in this Technical appendix. For example, the values of *N*/group for calculating the power of main effects in multifactor designs (and hence the values presented in Table 10.1) are based upon Formula 9.4. The adjustments to ES values other than 0.40 and 0.60 presented in Tables 10.2 and 10.3 are based upon Formulas 7.1 and 8.1 respectively. The formula for converting ES to *r* (and vice versa) is provided by Formula 2.1 and so forth. In general, then, while a number of shortcuts have been suggested in Chapter 10 to facilitate a quick estimate of power for the more complex designs discussed, more precise estimates may be obtained by actually computing the power based upon the formulas already presented in this appendix.

Bibliography

Bausell, R.B. (1986). *A practical guide to conducting empirical research.* New York: Harper & Row.

Bausell, R.B. (1991). *Advanced research methodology: an annotated guide to sources.* Metuchen, NJ: Scarecrow Press.

Bausell, R.B. (1994). *Designing meaningful experiments: 40 steps to becoming a scientist.* Thousand Oaks, CA: Sage.

Bergstralh, E.J., Kosanke, J.L., & Jacobsen, S.J. (1996). Software for optimal matching in observational studies. *Epidemiology,* **7**, 331–332.

Betensky, R.A. & Tierney, C. (1997). An examination of methods for sample size recalculation during an experiment. *Statistics in Medicine,* **16**, 2587–2598.

Birkett, M.A. & Day, S.J. (1994). Internal pilot studies for estimating sample size. *Statistics in Medicine,* **13**, 2455–2463.

Borenstein, M., Rothstein, H., & Cohen, J. (1997). *SamplePower.* Chicago, IL: SPSS, Inc.

Bradley, D.R. & Russell, R.L. (1998). Some cautions regarding statistical power in split-plot designs. *Behavioral Research Methods, Instruments, and Computers,* **30**, 319–326.

Bradley, D.R., Russell, R.L., & Reeve (1996). Statistical power in complex experimental designs. *Behavior Research Methods, Instruments, and Computers,* **28**, 319–326.

Brewer, J.K. (1972). On the power of statistical tests in the American Research Journal. *American Educational Research Journal,* **9**, 391–401.

Browne, R.H. (1995). On the use of a pilot sample for a sample size determination. *Statistics in Medicine,* 14, 1933–1940.

Chase, L.J. & Baran, S.J. (1976). An assessment of quantitative research in mass communication. *Journalism Quarterly,* **53**, 308–311.

Chase, L.J. & Chase, R.B. (1976). A statistical power analysis of applied psychological research. *Journal of Applied Psychology,* **61**, 234–237.

Chase, L.J. & Tucker, R.K. (1975). A power-analytic examination of contemporary communication research. *Speech Monograms,* 42, 29–42.

Clark-Carter, D. (1997). The account taken of statistical power in research published in the British Journal of Psychology. *British Journal of Psychology,* **88**, 71–83.

Cleary, T.A., Linn, R.L., & Walster, G.W. (1970). Effect of reliability and validity on power of statistical tests. In B.F. Borgatta & G.W. Bohrnstedt (Eds.), *Sociological methodology.* San Francisco, CA: Jossey-Bass.

Cohen, J. (1962). The statistical power of abnormal-social psychological research: a review. *Journal of Abnormal and Social Psychology*, **65**, 145–153.

Cohen, J. (1977). *Statistical power for the behavioral sciences* (2nd edn). New York: Academic Press.

Cohen, J. (1988). *Statistical power analysis for the behavioral sciences* (2nd edn). Hillsdale, NJ: Lawrence Erlbaum.

Cohen, J. (1992). A power primer. *Psychological Bulletin*, **112**, 155–159.

Cohen, J. (1992). Fuzzy methodology. *Psychological Bulletin*, **112**, 409–410.

Davis, S.C. & Gaito, J. (1984). Multiple comparison procedures within experimental research, *Canadian Psychology*, **25**, 1–13.

Dayton, C.M. & Schafer, W.W. (1973). Extended tables of *t* and chi square for Bonferroni tests with unequal error allocation. *Journal of the American Statistical Association*, **68** (34), 78–83.

Dunn, O.J. (1961). Multiple comparison among means. *Journal of the American Statistical Association*, **56**, 52–64.

Elashoff, J.D. (1999). *nQuery advisor.* Cork: Statistical Solutions.

Feldt, L.S. (1961). The use of extreme groups to test for the presence of a relationship. *Psychometrika*, **26**, 307–316.

Fisher, R.A. (1935). *The design of experiments*. Edinburgh: Oliver & Boyd.

Freiman, J.A., Chalmers, T.C., Smith, H.J., & Kuebler, R.R. (1978). The importance of beta, the type II error and sample size in the design and interpretation of the randomized control trial: survey of 71 "negative" trials. *New England Journal of Medicine*, **299**, 690–694.

Gail, M., Williams, R., Byar, D.P., & Brown, C. (1976). How many controls? *Journal of Chronic Diseases*, **29**, 723–731.

Gaito, J. & Norberga, J.N. (1981). A note on multiple comparisons as an ANOVA problem. *Bulletin of the Psychometric Society*, **17** (3), 169–170.

Games, P.A. & Howell, J.F. (1976). Pairwise multiple comparison procedures with unequal *n*'s and/or variances: a Monte Carlo study. *Journal of Educational Statistics*, **1**, 113–125.

Glass, G.V., McGaw, B., & Smith, M.L. (1981). *Meta-analysis in social research.* Beverly Hills, CA: Sage.

Goldstein, R. (1989). Power and sample size via MS/PC-DOS computers. *The American Statistician*, **43**, 253–260.

Haase, R.F. (1974). Power analysis of research in counselor education. *Counselor Education and Supervision*, **14**, 124–132.

Haase, R.F., Waechter, D.M., & Solomon, G.S. (1982). How significant is a significant difference? Average effect size of research in counseling psychology. *Journal of Counseling Psychology*, **29** (1), 58–65.

Hager, W. & Moller, H. (1986). Tables and procedures for the determination of power and sample sizes in univariate and multivariate analyses of variance and regression. *Biometrical Journal*, **28**, 647–663.

Hays, W.B. (1973). *Statistics for the social sciences.* New York: Holt, Rinehart, and Winston.

Hopkins, K.D. & Hopkins, B.R. (1979). The effect of reliability of the dependent variable on power. *Journal of Special Education*, **13**, 463–466.

Iwane, M., Palensky, J., & Plante, K. (1997). A user's review of commercial sample size software for design of biomedical studies using survival data. *Controlled Clinical Trials*, **18**, 65–83.

Katzer, J. & Sodt, J. (1973). An analysis of the use of statistical testing in communication research. *Journal of Communication*, **23**, 251–265.

Keppel, G. (1991). *Design and analysis: a researcher's handbook.* Englewood Cliffs, NJ: Prentice-Hall.

Keuls, M. (1952). The use of the Studentized range in connection with an analysis of variance. *Euphytica*, **1**, 112–122.

Keselman, H.J. & Keselman, J.D. (1987). Type I error control and the power to detect factorial effects. *British Journal of Mathematical and Statistical Psychology*, **40**, 196–208.

Knoll, R.M. & Chase, L.J. (1975). Communication disorders: a power analytic assessment of recent research. *Journal of Communication Disorders*, **8**, 237–247.

Kopans, D.B., Halpern, E., & Hulka, C.A. (1994). Statistical power in breast cancer screening trials and mortality reduction among women 40–49 years of age with particular emphasis and the national breast screening of Canada. *Cancer Diagnosis, Treatment, Research*, **74** (4), 1196–1203.

Kosciulek, J.F. & Szymanski, E.M. (1993). Statistical power analysis of rehabilitation counseling research. *Rehabilitation Counseling Bulletin*, **36** (4), 212–219.

Kraemer, H.C. (1991). To increase power in randomized clinical trials without increasing sample size. *Psychopharmacology Bulletin*, **27** (3), 217–224.

Kraemer, H.C. & Thiemann, S. (1987). *How many subjects: statistical power analysis in research.* Newbury Park, CA: Sage.

Laubscher, N.F. (1960). Normalizing the noncentral *t* and *F* distributions. *Annals of Mathematical Statistics*, **31**, 1105–1112.

Li, Y.-F. (1997). The development of power estimates and sample size requirements for seven multiple comparison procedures. Unpublished dissertation, University of Maryland, Baltimore, MD,

Lipsey, L.W. (1990). *Design sensitivity: statistical power for experimental research.* Newbury Park, CA: Sage.

Lipsey, M.W. & Wilson, D.B. (1993). Educational and behavioral treatment: Confirmation from meta-analysis. *American Psychologist*, **48**, 1181–1209.

Maxwell, S.E., Cole, D.A., Arvey, R.D., & Salas, E. (1991). A comparison of methods for increasing power in randomized between subjects designs. *Psychological Bulletin*, **110** (10), 328–337.

Mazen, A.M., Graf, L.E., Kellogg, C.E., & Hemmasi, M. (1974). Statistical power in contemporary management research. *Academy of Management Journal*, **30**, 369–380.

McNemar, Q. (1947). Note on the sampling error of the difference between correlated proportions or percentages. *Psychometrika*, **12**, 153–157.

Miller, R.B., Jr. (1966). *Simultaneous statistical inference.* New York: Springer-Verlag.

Milton, S. (1986). A sample size formula for multiple regression studies. *Public Opinion Quarterly*, **50**, 112–118.

Moyer, D., Dulberg, C.S., & Wells, G.A. (1994). Statistical power, sample size, and their reporting in randomized controlled trials. *Journal of the American Medical Association*, **272**, 122–124.

Murphy, K.R. & Myors, B. (1998). *Statistical power analysis: a simple and general model for traditional and modern hypothesis tests*. Mahwah, NJ: Lawrence Erlbaum.

Newman, D. (1939). The distribution of the range in samples from the normal population expressed in terms of an independent estimate of the standard deviation. *Biometrika*, **31**, 120–30.

Neyman, J. & Pearson, E.S. (1933). On the problem of the most efficient tests of statistical hypotheses. *Transactions of the Royal Statistical Society, Series B*, **33**, 218–250.

Nunally, J. (1967). *Psychometric theory*. New York: McGraw-Hill.

Odeh, R.E. & Fox, M. (1975). *Sample size choice: charts for experiments with linear models*. New York: Marcel Dekker.

Petrinovich, L.F. & Hardyck, C.D. (1969). Error rates for multiple comparison models: some evidence concerning the frequency of erroneous conclusions. *Psychological Bulletin*, **71**, 43–54.

Polit, D. & Sherman, R.E. (1990). Statistical power in nursing research. *Nursing Research*, **39** (6), 365–369.

Reed, J.F. & Slaichert, W. (1981). Statistical proof in inconclusive "negative" trials. *Archives of Internal Medicine*, **141**, 1307–1310.

Rosenbaum, P.R. (1989) Optimal matching for observational studies. *Journal of the American Statistical Association*, **84**, 1024–1032.

Sawyer, A.G. & Ball, D. (1981). Statistical power and effect size in marketing research. *Journal of Marketing Research*, **18**, 275–290.

Scheffe, H. (1959). *The analysis of variance*. New York: Wiley.

Sedlmeier, P. & Gigerenzer, G. (1989). Do studies of statistical power have an effect on the power of studies. *Psychological Bulletin*, **105**, 309–316.

Sidak, Z. (1967). Rectangular confidence regions for the means of multivariate normal distributions. *Journal of the American Statistical Association*, **62**, 626–633.

Stevens, J.P. (1980). Power of the multivariate analysis of variance tests. *Psychological Bulletin*, **88** (3), 728-737.

Stevens, J.P. (1992). *Applied multivariate statistics for the social sciences*. Hillsdale, NJ: Lawrence Erlbaum.

Thomas, L. & Krebs, C.J. (1997). A review of statistical power analysis software. *Bulletin of the Ecological Society of America*, **78**, 126–139.

Tukey, J.W. (1953). The problem of multiple comparisons. Unpublished paper, Princeton University, Princeton, NJ.

Tukey, J.W. (1984). The problem of multiple comparisons. In H.I. Braun (Ed.), *The Collected Works of John W. Tukey*. New York: McGraw-Hill.

Winer, B.J. (1962). *Statistical principles in experimental design*. New York: McGraw-Hill.

Index

ANCOVA, one-way between subjects
  correlations other than 0.40 and 0.60 315–316
  effect size 118
  guidelines for use 113
  multiple comparison procedures 120–122
  power tables 131–148
  sample size tables 149–160
  templates 126–130
ANOVA, one-way between subjects
  effect size 74
  guidelines for use 71–72
  multiple comparison procedures 72, 78–79
  power tables 89–96
  sample size tables 97–101
  templates 84–88

Borenstein, M. 332

Cascio, W.F. 334
Cohen, J. 9, 329, 331, 334, 335, 355
Cohen, P. 331
correlation coefficients
  ANCOVA, guidelines 114–115
  paired *t*-test, guidelines 58
  repeated measures ANOVA, guidelines 181–182
cross-over designs 26

effect size
  adjustments to 316–318
  binomial effect size display 4–6
  definition 6
  estimation 8–9
  extreme groups effect upon 21
  influence on power 19–22
  proportion of subjects helped 6
  shared variance 7
  sensitivity of dependent variable effect upon 29–30
Elashoff, J.D. 332

factorial ANOVA
  interactions, ANCOVA
    power tables 267–276
    sample size tables 290–295
    templates 260
  interactions, between subjects
    collapsing 251–257
    conceptual basis 240–244
    effect sizes 244–248
    power tables 262–266
    sample size tables 287–289
    templates 258–260
    three-way and higher 308–312
  interactions, repeated measures ANOVA
    power tables 277–286
    sample size tables 296–301
    templates 261
  main effects 302–308
Fisher, R.A. 1

Goldstein, R. 331

Hays, W.B. 342

Keppel, G. 330
Keuls, M. 349–350
Kraemer, H.C. 330
Krebs, C.J. 331

Lachin, J.M. 335
Laubscher, N.F. 345–346, 351
Lipsey, M.W. 21, 329

matched subjects 54
Maxwell, S. 331
multiple comparison procedures (see Newman–Keuls and Tukey HSD)
Murphy, K.R. 330
Myors, B. 330

Newman–Keuls sample size tables
- between subject ANCOVA 167–178
- between subject ANOVA 105–110
- repeated measures ANOVA 227–238

Newman, D. 349–350
Neyman, J. 1

one- vs. two-tailed tests 49

Pearson, E.S. 1

repeated measures ANOVA (one-way)
- effect size 184
- guidelines for use 179
- multiple comparison procedures 185–187
- power tables 193–208
- sample size tables 209–220
- templates 188–192

sample size
- relationship to power 17
- unequal groups (harmonic mean *N*) 18

Scheffe, H. 342
software
- GPOWER 332
- N 333
- nQuery 332
- PASS 332
- PC-Size 333
- Power analysis for experimental research xi, 331
- Power and precision 332
- PowerPack 333
- PowerPlant 333
- Stat Power 332

statistical power
- alpha level, effect upon 18–19
- definition 14
- factors affecting
  - alpha level 40–41
  - blocking variables 25–26
  - covariates 25
  - extreme groups 21–22
  - number of groups 22–24
  - pairwise contrasts 41–44
  - proximity to true outcome 32
  - reliability of dependent variable 30–31
- interpolation guidelines 38–40
- modeling guidelines 44
- reporting results 47–48

statistical tests, choice of 46–47

Thomas, L. 331
Thiemann, S. 330
t-test, independent samples
- guidelines for use 50
- power table 55
- sample size table 56
- templates 51

t-test, paired
- correlation coefficients, guidelines 57
- power tables 63–67
- sample size tables 68–70
- templates 62

Tukey, J.W. 348
Tukey HSD power tables
- between subject ANCOVA 161–166
- between subject ANOVA 102–104
- repeated measures ANOVA 221–226

Tukey HSD sample size tables
- between subject ANCOVA 167–178
- between subject ANOVA 105–110
- repeated measures ANOVA 227–238

Wilson, D.B. 21

Zedeck, S. 334